About Island Press

Since 1984, the nonprofit Island Press has been stimulating, shaping, and communicating the ideas that are essential for solving environmental problems worldwide. With more than 800 titles in print and some 40 new releases each year, we are the nation's leading publisher on environmental issues. We identify innovative thinkers and emerging trends in the environmental field. We work with world-renowned experts and authors to develop cross-disciplinary solutions to environmental challenges.

Island Press designs and implements coordinated book publication campaigns in order to communicate our critical messages in print, in person, and online using the latest technologies, programs, and the media. Our goal: to reach targeted audiences—scientists, policymakers, environmental advocates, the media, and concerned citizens—who can and will take action to protect the plants and animals that enrich our world, the ecosystems we need to survive, the water we drink, and the air we breathe.

Island Press gratefully acknowledges the support of its work by the Agua Fund, Inc., Annenberg Foundation, The Christensen Fund, The Nathan Cummings Foundation, The Geraldine R. Dodge Foundation, Doris Duke Charitable Foundation, The Educational Foundation of America, Betsy and Jesse Fink Foundation, The William and Flora Hewlett Foundation, The Kendeda Fund, The Forrest and Frances Lattner Foundation, The Andrew W. Mellon Foundation, The Curtis and Edith Munson Foundation, Oak Foundation, The Overbrook Foundation, the David and Lucile Packard Foundation, The Summit Fund of Washington, Trust for Architectural Easements, Wallace Global Fund, The Winslow Foundation, and other generous donors.

The opinions expressed in this book are those of the author(s) and do not necessarily reflect the views of our donors.

The **National Council for Science and the Environment** (NCSE) improves the scientific basis of environmental decisionmaking through collaborative programs with diverse communities, institutions and individuals. While an advocate for science and its use, the Council does not take positions on environmental policy issues and is dedicated to maintaining and enhancing its reputation for objectivity, nonpartisanship, and achievement. The Council has programs in five strategic areas:

Strengthening Education and Career Development

NCSE brings members of the academic community together to improve their environmental programs and increase their value to society. Programs in this area include:

- The **University Affiliate Program** provides services to advance programs at 150+ member schools ranging from large private and public research institutions to smaller liberal arts institutions.

- The **Council of Environmental Deans and Directors** brings academic leaders together to improve the quality and effectiveness of environmental programs on the nation's campuses.

- The **EnvironMentors Program** prepares high school students in underserved communities for college programs and careers in science and environmental professions. There are six chapters and more planned.

- The **Campus to Careers** program partners with government agencies, businesses, and foundations to advance young people in environmental careers through fellowships, internships, and other means.

- The **Council of Energy Research and Education Leaders** fosters interdisciplinary collaboration among leaders of university-based energy programs to advance the role of higher education in the energy field.

Science Solutions for Environmental Challenges

The Council brings stakeholders together to develop and implement science-based solutions to specific environmental challenges. Programs in this area include:

- The **National Commission on Science for Sustainable Forestry** provides practical information and tools to serve the needs of forest managers and policymakers to improve sustainable forestry.

- The **Wildlife Habitat Policy Research Program** produces information and tools to accelerate the conservation of wildlife habitat in the United States through State Wildlife Habitat Plans.
- The **Outlook Forest Research Dialogue** enhances research coordination, collaboration, and partnership within the forestry community.

National Conference on Science, Policy and the Environment

Each year over 1,100 leaders from science, government, corporate and civil societies develop strategies to improve decisionmaking on a major environmental theme. Strategies are disseminated and catalyze new initiatives with key communities.

The Encyclopedia Of Earth (www.eoearth.org)

Through the Environmental Information Coalition, the Council engages partnering organizations, scientists, professionals and others from around the world to educate the public with free access to high quality, reviewed, and attributed information on every environmental issue.

Science Policy

NCSE builds understanding of, and support for, environmental science and its applications, and the programs that make it possible. The Council educates the Congress and other key decisionmakers, and works to promote funding for environmental programs at numerous federal agencies.

The Climate Solutions Consensus

The Climate Solutions Consensus

David E. Blockstein
and
Leo A.W. Wiegman

THE NATIONAL COUNCIL FOR SCIENCE
AND THE ENVIRONMENT

ISLANDPRESS

Washington • Covelo • London

Library of Congress Cataloging-in-Publication Data
  The climate solutions consensus / by the National
  Council for Science and the Environment ;
  David E. Blockstein, Leo Wiegman, editors.
     p.   cm.
  Includes bibliographical references and index.
  ISBN-13: 978-1-59726-636-9 (cloth : alk. paper)
  ISBN-10: 1-59726-636-1 (cloth : alk. paper)
  ISBN-13: 978-1-59726-674-1 (pbk. : alk. paper)
  ISBN-10: 1-59726-674-4 (pbk. : alk. paper)
     1. Climatic changes—International coopera-
  tion. 2. Carbon dioxide mitigation—International
  cooperation.   3. Intergovernmental Panel on
  Climate Change.   I. Blockstein, David E., 1956–
  II. Wiegman, Leo.   III. National Council for
  Science and the Environment (U.S.)
  QC903.C58   2010
  551.6—dc22   2009026252

♻ Printed on recycled, acid-free paper

Manufactured in the United States of America

10 9 8 7 6 5 4 3 2 1

Keywords: Greenhouse gases, global warming,
renewable energy, carbon pricing / markets,
green building, geoengineering, sequestration,
sea level rise, ecosystem services, climate change
mitigation / adaptation, climate science.

CONTENTS

PART II How to Think About Climate Solutions

PART IV   **Thirty-Five Immediate Climate Actions**

FIGURES, TABLES, AND INSIGHT BOXES

Figures

Tables

Insight Boxes

PREFACE

The Climate Solutions Consensus presents the consensus among the environmental science community that the scientific evidence for human caused global warming and disruption of the global climatic system is unequivocal, the consequences are dangerous and potentially catastrophic to life on Earth, that humanity's only choices are mitigation, adaptation and suffering, and that therefore the time for massive action is immediate.

It further presents the consensus of the more than 1,350 scientists, educators, students, environmentalists, policymakers, business people and other citizens who attended the Eigth National Conference on Science, Policy and the Environment in January 2008 with a theme Climate Change: Science and Solutions that much of what we need to do now is known, many actions are technically feasible, although massive investment is needed to develop the next set of solutions, and that these actions are economically and socially beneficial and necessary.

This book is drawn from the presentations and discussions at that national conference convened by the National Council on Science and the Environment (NCSE). It elaborates on the conclusions drawn by these leading scientists and decision makers, presenting the underlying science in a way that is both understandable by the layperson and well-documented.

The book further presents 35 areas of action necessary for the U.S. and other nations to reverse our contamination of the atmosphere and devastation of the planet. Each action is divided into 6 to 15 tasks with recommendations on who should take on these tasks. This set of recommendations was developed by the participants at the 2008 national conference. They include controversial topics such as nuclear energy, geoengineering and ocean fertilization as well as green building, transportation, education, economics, research and other more conventional topics. As such they constitute a comprehensive agenda for action.

This text, which presents many websites for further information, is accompanied by a website www.nsceonline.org/climatesolutions that includes video and power point presentations from the conference, the list of speakers, sponsors, and other participants, and considerable background material, including material for classroom discussion that has been designed to accompany this book. The website is part of the Encyclopedia of the Earth www.eoearth.org, organized by NCSE, which includes more than 5000 hyperlinked articles, written and reviewed by experts about climate change, climate solutions and many other environmental topics. It also contains a set of educational resources for those teaching about climate science and solutions prepared by NCSE's Council of Environmental Deans and Directors (CEDD). The website will continually be updated. We invite readers of this book to contribute their own ideas and report on their own actions.

It is the fervent hope of the authors, producers and the thousands of individuals whose work and insights are represented in the Climate Solutions Consensus that we can help not only to avoid a climate catastrophe, but to redirect the course of human actions to a sustainable pathway that provides improved quality of life for all inhabitants of planet Earth.

THIRTY-NINE REASONS
WHY WE HAVE TO ACT NOW

1. Global climate change is not a future or hypothetical situation; it is occurring now, with many of its effects happening more rapidly than the scientific models have predicted. The main causes are economic and population growth, which are powerful drivers not easily reversed.

2. From 2000 to 2006, the global carbon dioxide (CO_2) emission growth rate has accelerated to 3.3% per year, the fastest growth rate in recent history. This makes it particularly important to implement an aggressive mitigation program immediately to avoid the introduction of huge quantities of long-lived CO_2 and other greenhouse gases (GHGs).

3. Because of the GHGs already released into the atmosphere by human activity, a substantial amount of global warming—at least 2 degrees Celsius (2°C) plus or minus 0.5°C—appears to be inevitable.

4. At or beyond this level, major damage to the world's ecosystems, biodiversity, and humans is a near certainty.

5. The acidification of the ocean is particularly alarming, with potential effects spanning the entire marine food web.

6. We are now experiencing the beginning of dangerous and potentially catastrophic climate disruption, with people experiencing fatal heat stress, destructive storms, drought, and floods, as well as indirect effects such as increased disease and decreased harvests. As with so many environmental problems, the first victims are those who contribute the least to the problem, the impoverished, the young, the elderly, and the indigenous populations, and the ultimate victims will be future generations—thus climate change is a moral issue of justice.

7. The destabilizing nature of climate change makes it a serious issue of national and global security by causing increased strife over increasingly scarce natural resources and because of the effects of increased natural disasters.

8. As characterized by Dr. John Holdren, past president of the American Association for the Advancement of Science and now Science Advisor to President Obama, the only choices available to humanity are "mitigation, adaptation, and suffering." The more mitigation and adaptation we do, the less suffering will occur.

9. The seriousness of the problem is under-recognized, and the costs of inaction or insufficient action are undervalued.

10. US consumer behavior alone accounts for more GHG emissions than the total emissions of almost every other country in the world.

11. Although China recently passed the United States as the major CO_2 emitter, the exponential growth in China's emissions is driven in part by consumer behavior in the United States and the European Union (EU).

12. Rapid reduction of GHG emissions to near-zero levels over the next four decades is needed to prevent a "dangerous" situation from becoming "catastrophic." Many scientists and policymakers are calling for an 80% reduction by 2050, which may not be sufficient albeit very difficult to achieve.

13. With carbon remaining unpriced and China and India building carbon-based economies mostly fueled by coal, it is unlikely that fossil fuel use will change significantly in the next two decades, making it impossible to assure that target.

14. With each passing month and year of inaction, it will be harder to prevent the catastrophic effects. The inertia of political inaction only compounds the inertia of human-induced climate change.

15. The key question is Will the political tipping point to implement climate change solutions occur before the climatic tipping point, where irreversible changes create catastrophe?

16. These two tipping points have time lags in opposite directions—once policy decisions are made, it will take years to implement them, whereas results of past actions are already "loaded" into the climate system but not yet expressed.

18. The magnitude of the problem means we all must do something, individually and collectively, and immediately.

19. Many of the solutions to help prevent radical climate change are known; many provide win-win-win opportunities to improve health and provide economic opportunities while simultaneously battling climate change. Ironically, the urgency of the climate issue may push the United States and other societies onto the pathways toward sustainability that is necessary for long-term prosperity and security.

20. These "no regrets" solutions offer increased conservation and efficiency, including in the home energy, building, transportation, agriculture, and consumer sectors. Cost curves for energy efficiency show that more than half of the actions that could be taken would directly lead to cost reduction. These "low-hanging fruit" can provide between 25% and 50% of the required GHG mitigation.

21. There are many opportunities in job creation and enhancement of our nation's global competitiveness that come with actions to minimize dangerous environmental change.

22. Emission-reduction scenarios that use a more diversified approach including methane and other GHGs in addition to CO_2 show us meeting climate targets at substantially lower costs and with greater flexibility compared with strategies to control CO_2 alone.

23. Reduction of tropical deforestation, which accounts for more than 20% of global emissions, provides many additional benefits, including biodiversity conservation and provision of ecological goods and services such as climate moderation, and sustainable development of tropical nations.

24. There are considerable unstudied opportunities to reduce GHG emissions in many sectors, such as agriculture and transportation and in many fields, from information technology to the social sciences.

25. Factors often not considered in GHG reduction strategies, including population, consumption, land use and planning, and forestry, all need to be reexamined and can provide substantial co-benefits in the context of climate change.

26. To prevent climatic catastrophe, there must be significant and rapid transformation in most economic sectors, especially the power-generation and mobile-source sectors.

27. The climate problem is serious enough to warrant an objective analysis of geoengineering options, that is, those technologies that temporarily add a cooling component to deliberately modify Earth's heat balance, which could potentially buy us time as we make dramatic reductions in our GHG emissions for more-permanent climate moderation. Particular focus should be on efficacy, economic, environmental, and ethical issues.

28. Given the inevitability of a certain degree of disruptive climate change, much more attention should be given to adaptation—particularly of vulnerable populations and ecosystems, such as polar and coastal areas and urban areas. It is likely that all species and all areas will be affected, so analysis of vulnerability and resilience should guide adaptation strategies.

29. Although available technology and energy conservation offer important near-term opportunities, the wide-scale development and deployment of new technology will be needed to avoid potentially catastrophic impacts in the longer term.

30. The current research and development (R&D) effort is woefully inadequate and not always directed at the most critical needs.

31. Significantly greater investment in science of all types and in technology is necessary to develop the technologies and approaches needed for transformation to a sustainable, low-carbon society. At least a doubling of current investment is necessary in the very near term. The US Global Change Research Program and the federal Climate Change Technology Program should be

expanded, and new institutional arrangements such as the proposed national climate service should be considered.

32. Major investments in public education, formal and informal education at all levels, and expanded communication pathways between scientists and decision makers are all necessary to provide both the information and the motivation for needed social changes. The investments should be based on high-quality educational and social science research.

33. Current information management systems are also insufficient, and mechanisms such as a national climate effects network are needed to better manage, analyze, and distribute information.

34. Much of the leadership on climate change in the United States to date has come from local and state government. For example, Salt Lake City, Utah, has already reduced its GHG emissions in its municipal operations by 31% from 2001 levels. Since the US Mayors Climate Protection Agreement was launched in mid-2005, nearly 1,000 mayors have signed on, representing nearly 84 million Americans.*

35. The university community has recently become extremely active in reducing its own carbon footprint. More than 640 presidents have signed the American College and University Presidents Climate Commitment. Students are responsible for much of the energy and impetus toward climate-neutral campuses.†

36. Additional leadership has come from the business and financial communities, but there are limits to what they will do without strong policy and price signals from the federal government. Lacking a price on carbon, capital markets cannot help, and investors remain in the dark on the potential for returns.

37. Cross-sectoral partnerships such as the US Climate Action Partnership (USCAP) between business and nongovernmental communities are promoting policies to tackle climate change.

38. The United States has a moral and a security obligation to assist developing nations and to work in partnership internationally.

39. As the premier economic and technological world power with the highest GHG emissions, the US must lead at the federal level because of the scale and urgency of the problem and the urgent need to encourage action by other nations such as China.

---

*Is your municipality on the list? To find out, visit http://usmayors.org/climateprotection/map.asp.

†Is your campus on the list? To find out, visit http://www.presidentsclimatecommitment.org/html/signatories.php.

Thirty-Nine Reasons Why We Have to Act Now

This Is Not Global Warming!

The planet has a fever. If your baby has a fever, you go to the
doctor. If the doctor says you need to intervene here, you don't
say, "Well, I read a science fiction novel that told me it's not a
problem." If the crib's on fire, you don't speculate that the baby
is flame retardant. You take action.

AL GORE, 2007

In 1998 the most severe flooding the world
had seen in centuries swallowed most of
low-lying Bangladesh, the Earth's seventh
most populous nation. The Brahmaputra, Ganges, and Meghna rivers overflowed their banks
when unusually heavy monsoon rains made
landfall from the Bay of Bengal. The swollen rivers swiftly swallowed 300,000 homes. More than
1,000 Bangladeshi men, women, and children
were killed in the initial flooding and another
1,000 in the days that followed. An additional 30
million people became homeless as two-thirds
of the country was under water at some point.
The flooding blocked or destroyed some 9,700
kilometers (km), or 6,000 miles, of roadway
and 2,600 km (1,600 miles) of embankment,
delaying evacuation and the arrival of help. The
impact on livestock was profound, as 135,000
cattle died in their fields.

A People in Peril

Two natural factors caused the flooding: unusually high monsoon rainfalls and increased

spring snow melt in the Himalayas. Human
activities made the impact of flooding much
worse. The deforested hillsides along the delta's
tributaries failed to hold back the rain and meltwater. Degraded soils and erosion further sped
up the flooding.

In the future, if meltwater flowing from the
mountains and monsoon rain falling from the
sky combine with sea level rise along the coast,
the devastating 1998 floods in Bangladesh will
look tame in comparison. Its vast Ganges delta
offers some of the most fertile farmland in the
world. The alluvial plain contains the largest
mangrove forest in the world, known as the
Sunderbans, which is home to the royal Bengal
tiger and diverse flora and fauna. It is also home
to one of the world's most densely populated
countries. In the past 100 years, Bangladesh
has become 0.5 degree Celsius (°C), or 1 degree
Fahrenheit (°F), warmer and has suffered a
0.5 meter (m), or 1.5 foot, rise in sea level. [2]
A rise of just an additional meter (3 feet) in
sea level would submerge half of the nation's
delta, where most of its people now live and

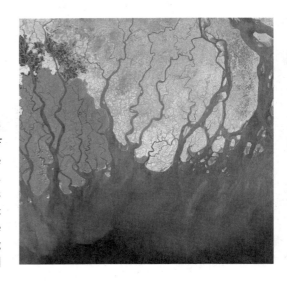

FIGURE 0.1 **Bangladesh's Ganges River delta**
A 1 m rise in sea level may flood huge sections of Bangladesh. The Ganges River forms an extensive delta where it empties into the Bay of Bengal. Roughly 120 million people live on the Ganges delta under threat of repeated catastrophic floods due to heavy runoff of meltwater from the Himalayas and due to the intense rainfall during the monsoon season. Source: [3]

work.* Ironically, despite Bangladesh's "water, water everywhere," the nation does not have clean drinking water for its population. Shifting rain patterns will only make the existing water shortage worse. South Asia's monsoon seasons are already wetter and its dry seasons are already drier, as climate records of the past 40 years clearly show. So, through no fault of their own, millions of Bangladeshis already live in areas that are and will remain increasingly vulnerable to climate extremes, as we can glimpse in Figure 0.1. [1,4]

This Is Not Global Warming!

Global warming has caught the public podium as the label we most often use for the accelerated disruptions in climate unfolding around us. But "global warming" is the wrong label. The term is both too benign and too narrow. Not every place is becoming equally warm, and temperature is not the only factor that is changing. Some places, like the northeastern United States and northern Europe, are becoming wetter. Some

places, like the American Southwest and sub-Saharan Africa, will be become drier.

The stark evidence shows dramatic warming at the Earth's formerly ice-bound polar regions: disappearing glaciers in Greenland, dramatically shrinking ice in the Arctic, melting permafrost, disintegrating ice sheets in the Antarctic, shrinking habitat for penguins. The climate disruption signs are everywhere from the tropics to the middle latitudes. Coral reefs are dying. Rain forests are drying. Heat waves are more common. And rising sea levels threaten island nations and shorelines everywhere.

When we average the surface temperatures over the whole planet, we see a warming trend. But the impacts are different in different locations. A rise in the average annual temperatures of 2°C (3°F) would make much more difference to the 55,000 inhabitants of Greenland than to the residents of Washington, DC, as an example of a city that still sees some wintertime snow and is closer to the equator than Greenland. When the air temperature is just below freezing (e.g., at −1°C), water falls as snow crystals, and glacier surfaces stay frozen. But when the air temperature rises a tiny bit to just above freezing (e.g., at +1°C), snowfall turns to rain, and glacier surfaces begin to melt.

Melting at the poles is both a symptom of warming and a trigger that begets more melting,

---

*A 1 meter rise means the United States loses large swathes of southern Florida, such as the Everglades, to the ocean. See for yourself with these interactive maps: http://www.treehugger.com/files/2008/04/map-sea-level-rise-global-warming-climate.php.

which in turn disrupts the climate in other areas of the world. These trigger points show how small changes can set in motion huge effects.

Global Weirding

The better and more accurate term is *global climate disruption*. The whole truth is simple. We humans have let slip the dogs of war—a war that attacks the very cycles in nature upon which human civilization depends for health, safety, and welfare. This book's opening chapters explain some of these cycles in nature and how humans have interfered. After that, we turn to solutions for both slowing down climate disruption and softening its blows.

Planet Earth is in the early stages of "global weirding."[†] For example, poison ivy is becoming more abundant and more allergenic. Rising carbon dioxide levels are increasing the potency of urushiol, the oil that puts the "poison" in poison ivy and the robustness in the growth of the poison ivy plant. This is one tiny symptom of global weirding—there are many more. The habitat ranges of invasive insects, such as fire ants, are increasing as the planet warms. Beetles that are voracious eaters of turf and trees live longer with higher levels of carbon dioxide. Disease-bearing mosquitoes are spreading farther north because of milder winters. New England's sugar maple trees are disappearing from the region. Asthma rates are skyrocketing.

Global climate patterns in place for thousands of years are being disrupted more rapidly now than at any prior time in the Earth's history. About 10,000 years ago, while *Homo sapiens* were perfecting the art of cultivating and

cooking starchy plants as high-energy food, the last Ice Age ended. Cooking our food helped us fuel our brains. [7] Climate patterns fixed since the last Ice Age have allowed humans to invent agriculture, domesticate animals, build cities, develop trading networks, and populate the entire globe.

Yet, the climate is being thrown off balance and will continue to shift in ways that will alter the location and volume of the one commodity upon which all life depends: water. The amount of water that arrives as rain, snow, or cloud cover in any given region is changing before our very eyes. No place on Earth will escape climate disruption. Every place will get wetter, drier, warmer, windier, or stormier, and just plain weirder weather. [5] In short, growing our food and recharging our reservoirs will become much less predictable as the Earth's moisture distribution radically changes.

America's 19th century humorist Mark Twain said climate was what we expect, weather is what we get. As this 21st century unfolds, we are moving into decades where we will not know what precise climate to expect, let alone what weather to predict. Recent human industriousness has already released so many warming gases into the atmosphere that, even if we were to stop driving cars entirely today, the planet would continue to warm up over the next decades. Based on today's advanced science, we can predict with a high degree of certainty that a rise of 2°C (3°F) is the minimal amount of atmospheric warming that human activity has already locked in place. This 2-degree rise in average annual air temperature will cause tropical air masses to capture more heat energy and deliver that energy to land more often.

The "one hundred year" storm will occur much more frequently. Hurricane Katrina was one of the five deadliest hurricanes in the history of the United States. Think of Hurricane Katrina, which devastated New Orleans and the American Gulf Coast in 2005, happening once every 10 years. A severe heat wave hit central Europe in 2003, killing 35,000 people. (We will

[†]First used in the *New York Times* in 2002, the term *global weirding* implies supernatural or unpredictable events. But the pattern we face has been predicted for decades by scientists and is not a matter of superstition. It is real and happening in nature all around us and will affect our health, safety, and welfare. The term *global weirding* is often attributed to Hunter Lovins (www.hunterlovins.com). No one claims coining the term *climate disruption* yet!

discuss more about this disaster in chapter 2.) What if that kind of extreme heat became common?

In other words, we will see, and are already seeing, events that we just simply have not seen before, such as hail-battering crops where it had never hailed like that before. And we will see undesirable "natural" disasters occurring much more frequently, such as forest fires in the American West.

What's in a Name?

In this book, we will use *global warming* when we mean the overall rise in the average surface temperature of the Earth—the symptom of the underlying pathology. We will use the more complete term *climate disruption* when we mean the global package of cooler, warmer, wetter, drier, windier, cloudier, and stormier weather that our human activities have exacerbated. We will use *climate change* in those cases where the researchers whose work we are examining use the term. But the change in climate is one fraught with disruption of long-standing climate patterns in which snowfall, rain volume, cloudiness, wind persistence, and a host of other natural events will become far less predictable, but where extreme events will become more common.

Disruptions and Solutions

Geologists will tell us that the Earth as a physical system has always changed. So "global change" is nothing new. The Earth's crust slowly and inexorably over millions of years is continually recycling itself. But it is the speed of the change now underway for which we have no prior example.

There is no reason to sugarcoat the future. But there is also no reason to languish in doom and gloom. We have options, some better than others, as we explore in the second half of this book.

This book is less about global climate disruptions than about what to do about their causes and effects. Many excellent works on the details of climate change itself already exist. Instead, we focus on potential solutions, the science behind these options, and the policies that could enact such solutions.

For this book and—more importantly—society itself, two critical questions clamor for action and discussion: How can we reduce the expected climate disruptions? How can we prepare to survive the climate changes by adapting in advance? The answer to the first question is that we have many good options for reducing the negative, but each requires immediate action on a society-wide scale. The answer to the second question is more nuanced. More-affluent nations will be able to cope better with the spread of new diseases, the shortage of potable water, the disruption of local growing seasons, the flooding of coastal plains. Less-affluent nations will be more vulnerable to human suffering caused by climate change. However, the globalized shipment of food and consumer goods alone and the international financial networks that trade these goods in a wired world mean that a big problem anywhere ripples out to everywhere. If we maintain business as usual, every nation will face steep consequences.

Therefore, we hope some of the ideas in this book help you begin an earnest discussion of what we can do now to reduce climate disruption and prepare ourselves for the climate patterns that our children and grandchildren will experience.

Let's start today!

The Encylopedia of Earth

As you begin reading, we hope you'll refer often to our companion web site, the Encyclopedia of Earth (http://www.eoearth.org). This site is constantly updated and offers more depth than can be contained within the covers of one book. It presents news, images, a forum, and links to a growing encyclopedia of authoritative information. The Encyclopedia also includes many helpful summaries, such as time lines for specific related events. See Appendix 1, Climate Change Time Line, at the end of this book as one example of the helpful learning tools waiting for you.

The researchers and teachers collaborating through the National Council for Science and the Environment (NCSE), a not-for-profit organization dedicated to improving the scientific basis for environmental decision making, bring this book and the Encyclopedia of Earth to you.

Online Resources

Climate Solutions (book site), http://ncseonline.org/climatesolutions
Encyclopedia of Earth, www.eoearth.org
Intergovernmental Panel on Climate Change, www.ipcc.ch
National Council for Science and the Environment, http://ncseonline.org

Works Cited and Consulted

[1] Chadwick MT (2004) The 1998 flood: coping with flood events in Bangladesh, PhD dissertation, University of Leeds, Department of Geography, Leeds, UK. http://www.leeds.ac.uk/cwpd/index.htm

[2] Khan TMA, Singh OP, Rahman S (2000) *Recent sea level and sea surface temperature trends along the Bangladesh coast in relation to the frequency of intense cyclones.* Marine Geodesy 23(2). www.informaworld.com/smpp/content˜content=a713833139˜db=all

[3] NASA (2008) Visible Earth. http://visibleearth.nasa.gov

[4] Ramamasy S, Bass S (2007) *Climate Variability and Change: Adaptation to Drought in Bangladesh.* 66 pp. (FAO, Rome). www.fao.org/documents/index.asp?lang=en

[5] Rosenzweig C, Casassa G, Karoly DJ, Imeson A, Liu C, Menzel A, Rawlins S, Root TL, Seguin B, Tryjanowski P (2007) Assessment of Observed Changes and Responses in Natural and Managed Systems (in *Climate Change 2007: Impacts, Adaptation and Vulnerability. Contribution of Working Group II to the Fourth Assessment Report of the Intergovernmental Panel on Climate Change,* 79–131, eds Parry ML, Canziani OF, Palutikof JP, van der Linden PJ, Hanson CE) ar4-wg2-chapter1.pdf: www.ipcc.ch

[6] Schwarzenegger A (2008) Gov. Schwarzenegger Tours Damage Caused by Humboldt Fire. *Press Release.* Office of the Governor, State of California 07.07.08. Sacramento (read October 15, 2008). http://gov.ca.gov/press-release/10111/

[7] Wrangham R (2008) *Cooking up bigger brains.* Scientific American. www.scientificamerican.com/article.cfm?id=evolving-bigger-brains-th

PART I

What We Know About Climate

CHAPTER 1

The Dance of the Mice and Elephants

We must not waste time and energy disputing the IPCC's report or debating the right machinery for making progress. The International Panel's work should be taken as our signpost, and the United Nations Environment Programme and the World Meteorological Organization as the principal vehicles for reaching our destination.

MARGARET THATCHER, Prime Minster, United Kingdom, Second World Climate Conference, 1990

By the late 1970s, both the scientific and diplomatic communities had become alarmed at patterns emerging in the natural world that seemed hazardous to humans and unexplained by natural causes alone. From the spread of diseases to out-of-control forest fires, the changes in climate patterns had no central clearinghouse for information on what was happening. The World Meteorological Organization (WMO) held the first ever World Climate Conference in 1979 to explore concerns that human activities were interfering with regional and global climate patterns. In 1985, the United Nations (UN) established the Advisory Group on Greenhouse Gases. By the time NASA scientist James Hansen testified to the US Senate's Energy Committee in June 1988 that global warming was occurring unequivocally, the United Nations Environment Programme (UNEP) and WMO needed better data on climate in order to advise citizens and governments on what to expect. The two organizations were sufficiently concerned to form the Intergovernmental Panel on Climate Change (IPCC) in November 1988. In the UN's words,

UNEP and WMO established the Intergovernmental Panel on Climate Change to provide independent scientific advice on the complex and important issue of climate change. The Panel was asked to prepare, based on available scientific information, a report on all aspects relevant to climate change and its impacts and to formulate realistic response strategies. [4]

Who could have guessed then that less than 20 years later these scientists and diplomats would share the Nobel Peace Prize simply for providing "an objective source of information about the causes of climate change, its potential environmental and socio-economic consequences and the adaptation and mitigation options to respond to it"? [4] So who are these 4,000 Nobel laureates, and how do they work?

How an Obscure Panel Organized Itself for Action

Just 2 years after its founding, the Intergovernmental Panel on Climate Change (alternately called the IPCC or the Panel) issued its First Assessment Report on the last day of August 1990 in Sundsvall, Sweden. Though the IPCC is headquartered in Geneva, Switzerland, it convened meetings all over the world. Given the enormity of assessing climate on a global scale within a short 2-year time frame, the Panel divided the chores among three working groups, each of which would employ a broad international base of scientists with specialized knowledge in its delegated arena. Working Group I would assess a broad range of scientific topics including "greenhouse gases and aerosols, radiative forcing, processes and modeling, observed climate variations and change, and detection of the greenhouse effect in the observations." Working Group II would summarize "the scientific understanding of climate change impacts on agriculture and forestry, natural terrestrial ecosystems, hydrology and water resources, human settlements, oceans and coastal zones and seasonal snow cover, ice and permafrost." Working Group III would study possible response strategies and establish subgroups to "define mitigative and adaptive response options in the areas of energy and industry; agriculture, forestry and other human activities; and coastal zone management." [4]

The Panel's scientific staff can be pictured as an international jury of top scientists, borrowed from leading universities and research institutions from all over the world. They weigh the best available information from all the ongoing scientific research streams and collectively assess which evidence is the most reliable and most relevant—and how that evidence fits in with other evidence on related topics. This is why the Panel's major reports—four in 17 years

(1990, 1995, 2001, 2007)—are called Assessment Reports.*

The Panel itself does not conduct any original research. Individual members are researchers at their home institutions, but when they are on loan to the Panel and huddled in the conference rooms, they participate as peer reviewers of research results. There is plenty of excellent research already being generated by researchers all over the globe every day. The service that the Panel and its members provide is the critical collection and synthesis of information.

The Panel is constantly asking, "What does all this information mean?". No single scientist, university, or national science academy could possibly read and evaluate the technical merits and likely relevance of the thousands of research reports, refereed journal articles, data collections, and theory-building proposals that pour forth each month that deal with some aspect of climate change. Only a vast international coordinated effort could do that. In the Panel's words,

> The role of the IPCC is to assess on a comprehensive, objective, open and transparent basis the scientific, technical and socio-economic information relevant to understanding the scientific basis of risk of human-induced climate change, its potential impacts and options for adaptation and mitigation. Review by experts and governments is an essential part of the IPCC process. The Panel does not conduct new research, monitor climate-related data or recommend policies. It is open to all member countries of WMO and UNEP. [4]

A key aspect of the scientific community greases the skids of this international effort: the need for collaboration. Modern science often involves highly specialized knowledge, expensive methods, and difficult-to-access data that require parties to pool resources. A composer

---

*See the online appendix to this chapter for a short summary of the prior reports.

may sit at a piano and create a masterpiece. While she will need others to accept and play it, composition is largely a solo act. A scientist who wants to study the atmosphere may need to acquire access to high-altitude research balloons, radio equipment to communicate with the balloons, and atmosphere-measuring instruments that hang from the balloons, and all that costs much more than a Steinway piano and requires more than just a good piano tuner to maintain.

Mice and Elephants

In addition to scientists, the Panel includes diplomats representing all the member nations. Every IPCC report calls for a strict and multipart protocol involving both these contingents. Three rules govern the creation of the reports: Only the best possible scientific, technical advice should be included; the circulation of draft chapters must include experts not involved in the preparation of that chapter from both developed countries and those in development transition; and the whole process must be open and transparent.‡ This last goal explains why so much of the Panel content is available for all to read online.

Reports go through three formal drafts and reviews: first an expert review, second a government/expert review, and finally a government review of the plain-English Summary for Policymakers. In the early stages of designing and collecting data for a working group's Assessment Report, the scientists are the proverbial

elephants, as the major force doing the heavy lifting while sifting through mountains of research to prepare a first-order draft report. Experts—from national science academies, industry, and government research—review the draft to comment on whether it accurately and adequately represents the state-of-the-art knowledge on the subtopic. Again the Panel scientists revise the draft report, based on all the peer reviews, and issue a second-order draft. This second draft is then reviewed both by experts and by the participating governments, whose diplomats we can consider the proverbial mice, scurrying along the edges of the report, observant but staying out of the way.

Once the experts of the Panel's three working groups prepare a final draft of the complete Assessment Report, the diplomats meet to extract the Summary for Policymakers from these findings. At this point, the mice and elephants switch roles, according to Stanford scientist and 2007 IPCC lead author Stephen Schneider. Now the diplomats take the lead on writing the summary about the implications for government policy, and the originating scientists act as observers. The Summary for Policymakers is inevitably watered down as diplomats from over 150 nations reach compromise wording.

When the Panel announces the final Summary to the press, the roles reverse once more. When Schneider described this process to the Eighth National Conference on Science, Policy, and the Environment in 2008, he noted that journalists are not interested in listening to government functionaries, and they "want to know if the report is fair. And we had to remind certain governments that we scientists would be reporting their behavior."† Schneider's chief point, echoed by many scientists, is that the transparency of the entire Panel's process allows the strongest scientific data to become broadly known and forces a consensus among nations

---

‡"Three principles governing the review should be borne in mind. First, the best possible scientific and technical advice should be included so that the IPCC Reports represent the latest scientific, technical and socio-economic findings and are as comprehensive as possible. Secondly, a wide circulation process, ensuring representation of independent experts (i.e., experts not involved in the preparation of that particular chapter) from developing and developed countries and countries with economies in transition should aim to involve as many experts as possible in the IPCC process. Thirdly, the review process should be objective, open and transparent." [3]

---

†For more from the NCSE conference participants, see their complete talks at http://ncseonline.org/climatesolutions/.

The Intergovernmental Panel on Climate Change (IPCC) is a scientific intergovernmental body set up by the World Meteorological Organization (WMO) and the United Nations Environment Programme (UNEP) in 1988. It has three components:

(1) The governments: The IPCC is open to all member countries of WMO and UNEP. Governments participate in plenary sessions of the IPCC, in which main decisions about the IPCC work program are made and reports are accepted, adopted, and approved. The governments also participate in the review of IPCC reports.

(2) The scientists: Thousands of scientists all over the world contribute to the work of the IPCC as authors, contributors, and reviewers.

(3) The people: As a UN body, the IPCC aims to promote the UN's human development goals. [4]

The IPCC is further organized into working groups and a task force (see Table 1.1). The charter of the IPCC does not empower it to recommend specific courses of action to lower climate risks to the world's governments. It can only lay out scientifically grounded paths to lower emissions of greenhouse gases or lessen the impacts. But governments are not required to follow these paths. Only we—the people as citizens—can make that happen.

TABLE 1.1 The IPCC's Three Working Groups and a Task Force

These groups recruit and assign lead authors, contributing authors, and reviewers for specific chapters of reports:
Working Group I assesses the scientific aspects of the climate system and climate change. Cochairs: Dahe Qin and Susan Solomon (Technical support unit in the United Kingdom)
Working Group II assesses the vulnerability of socioeconomic and natural systems to climate change, negative and positive consequences of climate change, and options for adapting to it. Cochairs: Osvaldo Canziani and Martin L. Parry (Technical support unit in the United States)
Working Group III assesses options for limiting greenhouse gas emissions and otherwise mitigating climate change. Cochairs: Ogunlade Davidson and Bert Metz (Technical support unit in the Netherlands)
The Task Force on National Greenhouse Gas Inventories is responsible for the IPCC National Greenhouse Gas Inventories Programme. Cochairs: Thelma Krug and Taka Hiraishi (Technical support unit in Japan)

Source: [3]

that might have stonewalled if their objections had remained secret. After the scientists and diplomats complete the work, both the Assessment Report and the Summary are distributed to all the governments for review.

The Scale of the Science: 4,000 Scientists Summarizing

The scale of the science that the Panel uses is stunning in its breadth, depth, and collaborative nature. The IPCC reports must reflect a consensus among all the Panel's scientific and diplomatic participants. Therefore, the Panel's methods tend to be very rigorous, and its findings tend to be quite conservative. For example, for *Climate Change 2007*, Working Group I examined the research results of about 80,000 different sets of data compiled in 577 different studies that show significant change in many physical and biological systems. Of the more than 29,000 observational data series that passed the stringent quality controls that Working Group I used, more than 89% were consistent with the direction of change expected as a response to warming. [1]

From the IPCC to International Law

When, in 1990, an obscure, newly formed panel of international experts with a long, cumbersome name, the Intergovernmental Panel on Climate Change, issued its first Assessment Report on

climate change, the Panel's statements shook up the policymakers at UNEP who had commissioned the work. The Panel's first assessment on climate change was so persuasive that it served as the basis for a completely new international treaty, the United Nations Framework Convention on Climate Change (UNFCCC). Member states and UN diplomats negotiated this agreement, often called the Framework treaty, the first ever on global change due to climate, between 1990 and 1994.

As a ratified international treaty, the UNFCCC entered into force on March 21, 1994. It was eventually signed by 192 nations. Therefore, it is the law of the land, to which all the 192 nations that signed it are bound, including the United States. That does not mean all 192 nations follow the letter or even the spirit of the law they have signed. Specifically, UNFCCC requires all signing nations to achieve "stabilization of greenhouse gas concentrations in the atmosphere at a level that would prevent dangerous anthropogenic interference with the climate system." (UNFCCC, Article 2)

How well nations have been doing in working toward this goal is a different story. The 1994 Framework treaty was intended as a beginning step "to consider what can be done to reduce global warming and to cope with whatever temperature increases are inevitable." [8] Any principle to change behavior needs specific, quantifiable targets and commitments to reach those targets.

A second important step occurred in 1997, with the adoption in Kyoto, Japan, of the text of the Protocol to the Framework. This addition to the original treaty—known as the Kyoto Protocol—contains more powerful and legally binding measures. Whereas the 1994 Framework treaty encouraged industrialized nations to reduce greenhouse gas emissions, the 1997 Kyoto Protocol required them to do so, with specific emission targets and compliance dates. By 1999, only 84 nations had ratified the Kyoto Protocol, which entered into force in 2005 for those early adopters. Nonetheless, the UN left the accep-

---

FIGURE 1.1 Online at http://ncseonline.org/climatesolutions

(A) Online edition of the Panel's Summary for Policymakers, part of the IPCC report *Climate Change 2007*.

(B) After the final draft of the Assessment Report is completed, scientists (seen in the back rows) carefully track the language being approved for the Summary. View at http://ncsconline.org/climatesolutions

---

FIGURE 1.2 Online at ncse.org/climate solutions

Global map shows 29,000 collection sites for data used by the IPCC in preparing *Climate Change 2007*.

FIGURE 1.3 Online at ncse.org/climate solutions

Graph from *Climate Change 2007* tracks the rising average global temperature, the rising average sea level, and the falling average area of winter snow cover in the northern hemisphere.

TABLE 1.2 The Progression of Confidence by the IPCC in Its Findings

Over the course of 17 years and four assessment reports, the IPCC has made the following statements about whether the data show the Earth is warming and whether human activities are part of the warming:

1990: "Earth has been warming, and continued warming is likely." (First Assessment Report)

1995: Balance of evidence suggests discernible human influence." (Second Assessment Report)

2001: "Most of warming of past 50 years [is] likely (odds 2 out of 3) due to human activities." (Third Assessment Report)

2007: "Most of warming [is] very likely (odds 9 out of 10) due to greenhouse gases." (Fourth Assessment Report)

Source: The IPCC's First, Second, Third, and Fourth Assesment Reports, respectively. See www.ipcc.ch/ipccreports/assessments-reports.htm.

tance book open at its New York headquarters. By mid-2008, a total of 181 nations and one regional economic regime, the European Economic Community (EEC), had ratified or accepted the lowered-emission targets of the Kyoto Protocol. [9]

Only one significant emitter of greenhouse gases, the United States, refused for years to ratify, accept, adopt, or even acknowledge the Kyoto Protocol. This lack of action by the United States has been particularly glaring. Many of the mechanisms by which nations could meet their Kyoto targets specifically included market-based approaches that the United States favored during the drafting of the protocol. For example, under the Kyoto Protocol treaty, each country must meet its targets primarily through national measures, which reduce emissions within that nation. Nations are also allowed to meet their targets by way of three market-based mechanisms: (1) emissions trading that allows any country to establish a carbon market for emitters within the country, (2) the Clean Development Mechanism that allows any country to get credit for implementing reductions in a developing nation, and (3) joint implementation that allows any country to get credit for a joint project in a different nation. [9]

Nonetheless, large international efforts did get underway after UNFCCC went into effect. In 2005, the largest emission-trading market in the world, the European Union Emission Trading Scheme (EU ETS), opened for business. After some early bumps, EU ETS developed into an effective tool to track and reduce all greenhouse

gas emissions. The EU ETS is mandatory for 10,000 European installations that spew greenhouse gas emissions, from factories to power plants. The EU ETS benefitted by learning from experiences of the earlier voluntary United Kingdom ETS that operated from 2002 to 2006. [2]

Even though the US federal government has avoided ratifying the Kyoto Protocol, there is strong movement within the United States that may provide models for a national effort. For example, in the fall of 2008 the Regional Greenhouse Gas Initiative (RGGI) of the northeastern states kicked off a carbon emission–trading market based in New York City. RGGI is a cooperative effort to reduce atmospheric carbon dioxide, a gas that will be discussed further in Chapter 3. This interstate effort may in turn serve as a model for a larger American federal greenhouse gas market, much as the initial UK emission market served as a precursor to the larger EU scheme. [6]

A Growing Consensus

As the United States considers action on global climate change, we should be guided by the sobering realization that international consensus on this topic has been steadily mounting. After its First Assessment Report in 1990, the IPCC issued a Second Assessment Report in 1995, which strongly confirmed the initial report. The Panel issued a Third Assessment Report in 2001, which stated there is newer and stronger evidence that most of the warming of the past 50 years is attributable to human activities. The Panel's Fourth Assessment Report in 2007, *Climate Change 2007*, found unequivocal evidence for human causes in climate disruption. In table 1.2 we can see the steady strengthening of conviction in the scientific community about the causes of warming activity.

Public opinion has also shifted. The release of the Fourth Assessment Report happened to coincide with the release of a documentary film, *An Inconvenient Truth*, narrated by former US vice president Al Gore. Mr. Gore's film used much of the same science that the Panel's report did to explain the global climate disruption that humans are causing. In an unprecedented decision by year's end, the Nobel Peace Prize committee in Oslo, Norway, awarded the 2007 Peace Prize to the entire IPCC organization and Al Gore jointly.

It is the first time that 4,000 scientists and one politician have shared the prize, and it will probably be the only time. For science to be awarded a humanitarian prize normally bestowed on diplomats and peacemakers truly indicates the profound link that the Nobel committee saw between avoiding future climate disruptions and ensuring human well-being and security.

The bad news is that some of the indicators cited in *Climate Change 2007* have worsened even in the short time since that Assessment Report. Specifically, shrinking sea ice and expanding coral damage are two indicators that are already beyond what the Panel had projected. We are running out of time to make a meaningful difference in curbing future climate disruption.

CONNECT THE DOTS

- In 1988, UNEP and WMO organized the IPCC and commissioned it to report on the best

In this first chapter the Connect the Dots focuses on a chronology of key events leading up to today. In future chapters, Connect the Dots will contain facts that capture conclusions reached in the chapter and lead to new additional activities, some of which await you online.

available scientific and technical data on climate change.

- In 1990, IPCC issued its First Assessment Report, finding global warming trends and likely human causes.

- In 1994, the 191 nations adopted the UNFCCC treaty that encourages nations to reduce greenhouses gas emissions, based on the warning in the first IPCC report to avoid "dangerous anthropogenic interference" in climate.

- In 1997, the IPCC issued the Second Assessment Report, which includes a new area of analysis, the socioeconomic aspects of climate change, and finds "the balance of evidence suggests a discernible human influence on global climate."

- In 1997, the Kyoto Protocol required signatory nations to meet specific emission targets and compliance dates. By 2008, there were 181 nations that adopted the Kyoto Protocol. Among major emitters, only the United States did not.

- In 2001, the IPCC issued the Third Assessment Report, which states that "the atmospheric climate change will persist for many centuries" and which found "new and stronger evidence that most of the warming over the last 50 years is attributable to human activities."

- In 2005, the Kyoto Protocol requirements entered into force, with many signatory nations struggling to meet their targets. In addition, the European Union launched the largest emission-trading market in the world.

- In 2007, the IPCC issued its Fourth Assessment Report, finding that "warming of the climate system is unequivocal" and "very high confidence" that human activities intensify the warming. The 2007 Nobel Peace Prize was awarded to the IPCC (and its 4,000 science experts) and Al Gore jointly.

- Since 2007, key climate symptoms have gotten worse than even the most recent IPCC projections, for example, shrinking sea ice and coral damage.

Online Resources

www.eoearth.org/article/Kyoto_Protocol

www.eoearth.org/article/Kyoto_Protocol_and_the_
United_States

www.eoearth.org/article/Global_Climate_Change%
3A_Major_Scientific_and_Policy_Issues

www.eoearth.org/article/Intergovernmental_Panel_
on_Climate_Change_%28IPCC%29

The Intergovernmental Panel on Climate Change,
www.ipcc.ch

The Kyoto Protocol, http://unfccc.int/kyoto_protocol

See also extra content for Chapter 1 online at http://
ncseonline.org/climate solutions.

Works Cited and Consulted

[1]   Bernstein L, Bosch P, Canziani O, Chen Z,
Christ R, Davidson O, Hare W, Huq S, Karoly
D, Kattsov V, et al. (2007) Synthesis Report. (in
*Climate Change 2007: Fourth Assessment Report of
the Intergovernmental Panel on Climate Change*,
74 pp, eds Allali A, Bojariu R, Diaz S, Elgizouli I,
Griggs D, Hawkins D, Hohmeyer O, Pateh Jallow
BP, Kajfež-Bogataj L, Leary N, Lee H, Wratt D)
ar4_syr.pdf: www.ipcc.ch

[2]   European Commission (EC) (2008) Emission
Trading Scheme. *Europa*. European Commission
(read August 24, 2008). http://ec.europa.eu/
environment/climat/emission/index_en.htm

[3]   IPCC (2007) Assesment Report Archive. UNFCCC
IPCC (read August 24, 2008). www.ipcc.ch/ipcc
reports/index.htm

[4]   IPCC (2004) About IPCC. UNFCCC, Geneva (read
August 24, 2008). www.ipcc.ch/about/index.htm

[5]   Parry ML, Canziani O, Palutikof JP, van der
Linden PJ, Hanson CE (2007) Technical Summary: Working Group II (in *Climate Change 2007:
Impacts, Adaptation and Vulnerability. Contribution
of Working Group II to the Fourth Assessment Report
of the Intergovernmental Panel on Climate Change*
23–78). ar4-wg2-ts.pdf: www.ipcc.ch

[6]   Regional Greenhouse Gas Initiative (RGGI) (2008)
Home Page (read September 23, 2008). www.rggi
.org

[7]   Revkin A (2008) "1988–2008: Climate Then
and Now." *New York Times*, June 23, 2008 (read
October 15, 2008). http://dotearth.blogs.nytimes
.com/2008/06/23/1988-2008-climate-then-and-now

[8]   UNFCCC (2008) Essential Background (read
August 24, 2008). http://unfccc.int/essential_
background/items/2877.php

[9]   UNFCCC (2008) Kyoto Protocol (read August
24, 2008). http://unfccc.int/kyoto_protocol/
items/2830.php

CHAPTER 2

Three Questions Every Citizen Should Ask

The three questions that lay persons need to ask experts to be more literate in the environmental policy debates are (1) what can happen? (2) what are the odds? and (3) how do we know? [15]

STEPHEN SCHNEIDER, Stanford University

W hile the big story of global climate disruption is stark, the technical details all boil down to some basic laws of nature taught in introductory science courses.* But we need not become scientists to be literate on environmental topics. In this book, we will offer short refreshers—as we go—on the most essential facts needed to boost our environmental literacy. All kinds of decisions that affect our communities will require our informed input—from the size of agricultural subsidies and fishing quotas to the expansion of electric power plants. How can we help participate in the myriad public policy decisions that climate change necessitates? How can we gain enough environmental literacy to be well-informed enough about the facts? Let's find out how.

Actors and the Stage

The stage on which very real action will take place to address climate disruption is as big as the world and as intimate as a grandparent peel-

---

*Some of these basic science principles include the following: In the absence of another force, a body at rest remains at rest and a body in motion continues in motion. If several forces act on a body, the result is the sum of these individual forces. Whenever a body exerts a force on another body, the latter exerts an equal and opposite force on the former. Disorder in a closed system always increases. When humans make a watch from metals, we may seem to create order, but the mining of the ores and the machining of the metals demand energy that is converted into waste heat and therefore disorder. In other words, we continually convert useful ordered energy (such as sunlight or chemical energy) into useless, disordered energy (such as waste heat). Hotter substances expand and become less dense. Denser substances that mix with less-dense substances will be separated into layers by the Earth's gravity, with the denser matter at the bottom. That is why air floats above water and why warmer water at the ocean surface floats above the layer of cooler water below. A light or heat source twice as far away supplies one-fourth as much heat. Heat or any energy is always conserved, and it may be converted to a different kind of energy (or to matter), but it is never lost.

ing an apple with a grandchild. Three key groups of actors play roles in deciding which courses of action to take: scientists, policymakers, and citizens. Scientists gather the facts and follow them wherever they may lead. Policymakers weigh the merits of specific policy choices, from doing nothing (which we will call "business as usual" in this book) to trying a bit of everything. Citizens at large will ultimately have to pay for any course of action and enjoy its benefits or suffer its consequences. Regardless of how much we know, this last group includes all of us.

We all tend to organize ourselves into groups. That is what humans do. Members of interest groups voluntarily cooperate with each other out of mutual self-interest toward a common goal. In a daily avalanche of press releases and reports, hundreds of groups representing organized interests within the business, science, and environmental communities offer analyses and recommendations. Some of the most successful interest groups mobilize collaborations across the sectors of the businesses, scientists, governments, nonprofits, and citizenry.

Any given environmental topic may draw the attention of many of these interest groups. For example, the public discussion on how to reduce powerful greenhouse gases from municipal waste landfills may see input from several sectors. An interest group may comprise a set of scientists organized to do joint research or compile research findings. Or an interest group may represent businesses that are impacted by proposed legislation, for example, the National Solid Waste Management Association, or like the Natural Resources Defense Council, it may represent citizens. Or it may represent an alliance of policymakers, who represent a particular constituency of decision makers, for example, the Municipal Waste Management Association.

A fierce battle for the ear of the public—and the climate-related policy made on our behalf—has been taking place. Longtime environmental advocates, such as Al Gore, the World Wildlife Fund, and the Sierra Club, raise awareness about climate disruption and the need for big changes in our addiction to fossil fuels. Other groups with vested interests in keeping the status quo in place sometimes create front organizations to spread misinformation and doubt about climate change. These front groups fund "researchers" who do not disclose who paid for the research. And in the middle, scientific organizations—such as the National Council for Science and the Environment, which prepared this book—present scientific, verifiable information on climate to inform environmental decision making. But interest groups from both sides often drown out the voice of science. So that leaves us all wondering who to believe and why?

The Three Questions of Environmental Literacy

A leading climate scientist from Stanford University, Stephen Schneider argues that the key for citizens is to understand the interaction between science and policymakers—two sectors with different roles to play. In other words, we need to become familiar with the context in which scientists produce facts and explanations for the facts and then propose potential solutions. Likewise, as citizens we need to become familiar with the context in which policymakers use science to figure out what to do. We do not need to be able to judge the technical merits of the opposing positions on a given policy, for example, where to set the fishing limits for a given species. We do need to be able to rate the credibility of processes in which the scientific claims or expertise sources are assessed. How do we do that?

Schneider, whom we met in Chapter 1 as contributor to the Intergovernmental Panel on Climate Change (IPCC) reports, poses three questions that lay audiences need to ask experts in order to become better informed in environmental policy debates:

1. What can happen?

2. What are the odds that it will happen?

3. How are such estimates made?

INSIGHT 2: WHERE THE WASTE GOES, SO GO THE EMISSIONS

Americans produce a lot of waste. We lead the world in the volume of garbage we send to be buried. Americans generate about 1 ton (2,000 pounds, or 0.9 metric tons) of municipal solid waste per resident per year. According to the Office of Resource Conservation and Recovery of the US Environmental Protection Agency (EPA), Americans generated 4.62 pounds (lb), or 2 kilograms (kg), per person per day in 2007, slightly less than the prior year. That is almost twice as much waste per person as in 1960. The good news is the recycling rate has improved. In 2007 Americans recycled 1.54 lb per person (0.7 kg) per day—an increase of 2.7% over the year before. The bad news is that over half of all our solid waste is still being discarded in landfills, rather than being recovered for energy or recycling uses. [19]

The most common waste management practice, landfilling, results in the release of methane from the anaerobic (without oxygen) decomposition of organic materials. Methane from landfills can be a source of energy. Some landfills capture and use methane for energy. But most do not. Landfills remain the single largest source of methane emissions in the United States, accounting for over one-third of all methane emitted and producing even more emissions than the nation's natural gas distribution system. The EPA's Landfill Methane Outreach Program reports that every 1 million tons (0.9 million metric tons) of municipal solid waste is the energy equivalent of 0.8 megawatts (MW) of electricity or 432,000 cubic feet (12,232 cubic meters) per day of landfill gas. Over 450 projects in 43 states now capture landfill gas either for generating electricity or for direct use as gas to fuel pumps, heaters, or other equipment. [19]

Source reduction, that is, alteration of the design, manufacture, or use of products and materials to reduce the amount and toxicity of what gets thrown away, is the best way to reduce landfill use. Recycling as much as we can from the remaining waste is another. Nationwide today, 6 out of 10 American households have curbside recycling pickup, one of the most effective ways to raise recycling volume. (In the Northeast states, 84% of households have curbside recycling. But only 3 of out 10 households in the South have curbside recycling pickup.) Finally, composting the organic waste, such as food scraps and yard trimmings, can further reduce the waste stream that reaches landfills, while producing rich soil for local gardens. [19]

Another alternative to landfilling waste is combusting it to recover energy from the resulting heat. The waste-to-energy (or energy-from-waste) potential is quite large. This alternative involves burning the waste after all the recoverable and recyclable materials have been removed. About 90 municipal waste-to-energy facilities operate in the United States today. Next-generation energy-from-waste combustion technology in Germany increases the usable by-products while reducing the resulting emissions and remaining ash. [20]

For more on municipal landfills, gas emission, and energy from waste, see the following sources:

US EPA, http://www.epa.gov/osw/nonhaz/municipal/index.htm

NDRC, www.nrdc.org/air/energy/lfg/execsum.asp

NSWMA, http://www.environmentalistseveryday.org/about-nswma-solid-waste-management/index.php

USCOM MWMA, http://www.usmayors.org/mwma/

These questions will help us assess the validity of various consequences and their probabilities, without needing a PhD in statistics or science. Schneider sums up, "Such literacy does require the ability to discern what components of the debate deal with factual and theoretical issues and which are political value judgments." [15]

For instance, going back to the "dump" example: The vast majority of municipal solid waste in North America is interred in landfills, sometimes hundreds of miles from the city that collected that waste from its citizens. Out of sight, out of mind. Except that garbage decomposing underground without air produces a gas called methane, also known as marsh gas. Methane has commercial value, burns well, and also is produced when the world's millions of cows belch or when marshes decompose or layers of permafrost begin to thaw.

What can happen? The scientific answer is that methane escapes to the atmosphere. Due to its molecular structure and durability, a methane molecule traps heat in the atmosphere 21 times more effectively than a carbon dioxide molecule. What are the odds that the methane will contribute to global warming? The answer is that it is virtually certain. When the odds that something will happen are high enough that it is virtually assured, scientists refer to the likelihood as having a "very high confidence" degree. How are estimates of methane emissions from landfills made? Scientists and engineers would explain that we know roughly how much material is buried in landfills, how quickly methane gas develops, and how many landfills have methane capture systems (to fuel local equipment or sell as bottled gas). We also know that the United States has 3,091 active landfills and over 10,000 old municipal landfills. Before modern landfill techniques were developed in the 1930s, every town and many businesses and factories had their own dumps. US landfills consist of 40% to 50% paper waste, 20% to 30% construction debris, and 1% disposable diapers. So the question of how much methane from landfills is escaping to the atmosphere, where it accelerates global warm-

ing, is one we can answer relying on the facts, without using any opinions. The engineers can also describe various options for capturing the methane. And from there, it is up to policymakers and citizens to sift through the options and choose a methane capture solution that best fits the local situation and availability of funds.

Question 1: What Can Happen?

What can happen flows from what is already happening. Temperatures are rising. Coral reefs are dying. Deserts are expanding. Tropical cyclones are intensifying. Oceans are acidifying.

In the summer of 2003, an extreme heat wave and drought hit Europe. While the cultural habits and homes of southern Europeans have long ago adapted to hot weather, this heat wave hit northern and central Europe with special severity. In the modern, wealthy nation of France, 14,800 deaths attributed to heat occurred in just 14 days in mid-August. Temperatures elsewhere? In that same period they were extreme, 10 degrees Celsius (°C) higher than in the prior 3 years, approaching or exceeding 32°C, or 90 degrees Fahrenheit (°F), in separate waves. In nine other well-to-do nations, another 20,000 deaths were attributed to the sudden, prolonged heat. Belgium, the Czech Republic, Germany, Italy, Portugal, Spain, Switzerland, the Netherlands, and the United Kingdom all reported excess mortality during the heat-wave period, with total heat-related deaths in Europe in the range of 35,000 persons. In France, 60% of the heat wave deaths between August 1 and 15 were those aged 75 years and older. Beyond the tragedy of these deaths, outdoor air pollution spiked, as did smoke from forest fires exacerbated by drought. The economic impact of this heat wave

---

FIGURE 2.1 Online at ncse.org/climate solutions

Temperature map for the summer of 2003 shows the heat wave was hottest in large areas of central and southern France.

---

was severe as well. The European Union estimated that 10% of its total grain harvest (10 million metric tons) was lost. [2]

If humans continue emitting greenhouse gases in business as usual, what can happen? Over 60 scientists from over 25 countries helped write the section of the IPCC's 2007 Fourth Assessment Report that focused on the potential consequences of climate change. Specifically, these experts—and the diplomats who reviewed every line of the final summary—agreed on the "sensitivity, adaptive capacity, and vulnerability of national and human systems to climate change." [4]

The conclusions they came to are very sobering. For example, sudden severe heat waves are becoming more frequent and less predictable. Indeed, independent research has uncovered the fact that extreme heat waves in Europe are already two times more frequent today than in the past century and will be "normal" for the continent as the mid-range of heat extremes by 2050. If. . . . [18]

Here is the kicker: In searching high and low for what could have caused the 2003 weather anomaly, the Panel's authors agreed that "the excess deaths of the 2003 heat-wave in Europe are likely to be linked to climate change." [2] Specifically, "the observed higher frequency of heat-waves is likely to have occurred due to human influence on the climate system." In other words, in disrupting our planet's temperature-control system, we are waking an unpredictable giant so much larger than our human systems that we can only provoke it, but not control it, once unleashed.

A two-week summer heat wave may be the least of the problems in store for humans if climate change is unchecked. Other potential key vulnerabilities lurk in the near future for us if the average global temperature rises by just 2°C above its average from 1990 to 2000. We will consider 2°C (which is 4°F) for the simple reason that 2 degrees is the amount of warming we have already locked in place due to our profligate burning of fossil fuels in the past.

Intensity of Tropical Cyclones[†]

Tropical cyclones form when warm air rises over the tropical oceans, releasing heat as water vapor condenses. That is why they are called "warm core" storm systems. Regardless of where they form, cyclones produce very high winds and torrential rain and can push storm surges of high water toward land, magnifying the coastal damage they wreak. Tropical cyclone storms occupy such a place in our collective conscience that we give them different family names depending on where they form. We call them hurricanes if they form in the Atlantic or northeastern Pacific, typhoons if they form in the northwestern Pacific, or cyclones if they form in the Indian Ocean or southern hemisphere. As each storm forms, we give it a personalized, alphabetized first name, such as Charley or Katrina.

To qualify as a category 4 storm, a cyclone must have sustained wind speeds of 210 to 249 kilometers per hour (km/h), or 131–155 mph, and cause storm-surge rises in sea level of 4.0–5.5 meters (13–18 feet). Category 4 storms are exceedingly dangerous and cause extensive building-wall failure and coastal erosion. In August 2004, Hurricane Charley, which made landfall twice in Florida and once in South Carolina, causing 10 deaths and $15.4 billion in damage, was a category 4 storm. Category 5 storms are even more dangerous. They exhibit wind speeds of up to 250 km/h (156 mph) and can cause storm surges of 5.5 meters (18 feet) or greater. Storms of this strength cause very heavy damage to all structures except those built with steel and concrete reinforcement. The present decade has seen more category 5 hurricanes than any before, with eight such storms: Hurricanes Isabel (2003), Ivan (2004), Emily (2005), Katrina (2005), Rita (2005), Wilma (2005), Dean (2007), and Felix (2007). [11]

[†]The source in this section of the chapter for each of the confidence level impacts is Chapter 19 in IPCC's *Climate Change 2007*, for which Stephen Schneider was—coincidentally—one of the lead authors.

An increase of as little as 2 degrees of warming will produce an increase in category 4 and category 5 tropical cyclone storms. The warmer the ocean water, the more heat energy the storm gathers up, thus increasing severity, wind speed, and moisture content. In addition, the storms' impacts on coastal regions will be exacerbated by sea level rise. [The likelihood has a medium to high confidence degree according to IPCC.] Such tropical cyclone intensity levels will exceed what most infrastructure has been designed to withstand. [Medium to high confidence] These storms will bring large economic costs and threaten many lives. [High confidence] [16]

Flooding

A warming of as little as 2 degrees will increase the frequency and magnitude of both flash flooding in many regions and large-scale floods in the mid and high latitudes. Why? Because rainstorms will be more intense, dropping more water in less time. Increased winter rainfall and loss of winter snow storage will exacerbate flooding in North American and Europe. [High confidence] Unlike with tropical cyclones, floods do not have a uniform classification system for intensity. The risk of increasing dam bursts in mountain glacial lakes, critical to drinking-water supply in many regions such as the Andes, will rise. Floods often create long-lasting problems as they erode arable soils critical to farming and destabilize banks of rivers critical to transportation. In addition, we continue to shrink or destroy natural wetlands through overdevelopment of our suburbs, reducing their capacity to absorb excess runoff, as happened in the American Midwest flooding in 2008.

Extreme Heat

As we learned with the 2003 European drought, extreme heat events can be dangerous and lethal over large swaths of land. A warming of as little as 2 degrees will lead to "increasing heat stress and heat waves, especially in continental areas." [16] [Very high confidence] While inland areas will be safer from the effect of tropical cyclones,

they will be more vulnerable to extreme heat. Extreme heat causes human mortality to rise, crops to fail, and forests to die back and succumb to fires, along with damage to other ecosystems. Ironically, heat can lead to more fossil fuel consumption as people use air conditioners in their cars, offices, and homes to combat excessive heat—a vicious cycle, like so many we face in our changing climate.

Drought

The absence of adequate rain and surface water runoff leads to drought. Drought conditions are already increasing in frequency. [Medium confidence] "The early spring shift in runoff leads to a shift in peak river runoff away from summer and autumn, which are normally the seasons with the highest water demand, resulting in consequences for water availability." [13] A warming of as little as 2 degrees will increase droughts in the mid-latitude continental areas as inland summer drying becomes more prevalent. Such droughts can lead to vegetation die-offs as soil dries out. Drier soil also exacerbates extreme heat waves. [High confidence]

Fire

Where drought and heat waves occur, fire often follows. A warming of as little as 2 degrees will increase the frequency and intensity of forest or grassland fires, especially in inland areas that suffer from drought. [High confidence] In addition to direct loss of habitat, such fires cause airborne smoke plumes that affect much larger areas than the fire zone itself and lead to human respiratory problems.

The convergence of "natural" disasters may sound like the four horsemen of the apocalypse arriving in unison. We'll talk more about the horsemen of climate change in Chapter 5. The consequences for how we grow our food or live in areas "safe" from natural disasters like drought or flood are dire. How is a 2-degree rise in temperature likely to affect our global social systems that we rely upon for our health, safety, and welfare?

Food Supply Changes

With a warming of as little as 2 degrees, some areas in the tropics and subtropics will experience a decrease in productivity for some cereal grains, which may be offset by an increase in productivity in the middle to high latitudes.‡ In plain English, the breadbaskets of the world will shift, with less grain being harvested closer to the equator and more being produced in areas that have four seasons. This shift may further weaken the ability of tropical nations, which have younger populations and less overall wealth, to feed themselves. [Medium confidence]

Infrastructure Damage

A warming of as little as 2 degrees will bring more frequent and more severe extreme weather events that will likely inflict exponentially increasing damage on housing, transportation, and agricultural infrastructure. [High confidence] While the severity of Hurricane Katrina cannot be linked with certainty to human influence on climate patterns, storms of such magnitude are likely to arrive onshore more frequently.

Health Risks

The World Health Organization estimates that climate change caused the loss of 150,000 lives in the year 2000 alone and that weather-related natural disasters killed approximately 600,000 people worldwide during the 1990s. Disproportionate numbers of these deaths are among the poor, the sick, the young, and the elderly, especially in the developing world. [21] With a warming of as little as 2 degrees, malnutrition

‡A region's climate varies very roughly with its distance from the equator. Geographers and meteorologists refer to the region around the equator, between the Tropic of Cancer and the Tropic of Capricorn (23.5 degrees), as the "low latitudes." They call the areas between the low latitudes and the polar regions (23.5 to 66. 5 degrees) the "middle latitudes" and the region between the middle latitudes and the poles (66.5 to 90 degrees) the "high latitudes."

(defined as the nonavailability of recommended daily calorie intake), infectious diseases, episodes of diarrheal diseases, cases of malaria, and direct fatal accidental injuries in coastal floods, inland floods, landslides, droughts, and extreme heat events would expand the risks to human health. [Medium to high confidence] These risks are directly related and sensitive to the status of the public health systems. [Very high confidence] See Figure 2.2 for a summary of the health-environment connections.

Water Resource Scarcity

With as little as a 1°C temperature increase, some mid-latitude regions and semiarid low-latitude regions will experience decreased water availability and increased drought. At 2 or more degrees warmer, a series of water-related calamities will unfold. Floods, drought, and erosion will increase, and water quality will decrease. [Very high confidence] Sea level rise will expand the salination of groundwater, decreasing freshwater in coastal regions. [Very high confidence] The reduction in water supplies will affect hundreds of millions of people. Also, as glaciers and year-round snow in high elevations melt at a faster rate, they will reduce the water available to societies depending on them as natural dams. [High confidence]

Human Migration and Conflict

With a warming of up to 2 degrees, coastal and river flooding, drought, and water and food shortages will cause suffering in many regional populations, most strongly among those already living at the margins of economic viability. [High confidence] This suffering will lead, as it does now in less-frequent cases, to people's seeking to relocate, which is likely to exacerbate regional conflicts over water resources and migration pressures. [Medium confidence]

Impact on Market Economies

At less than 2 degrees of warming, many higher-latitude areas—such as North America's prairie states and provinces—may see short-term eco-

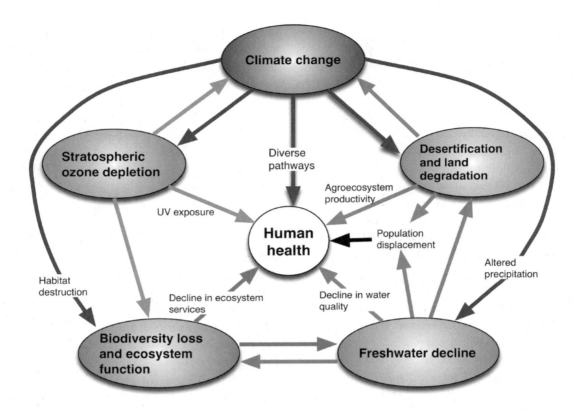

FIGURE 2.2 Global climate change and human health connections

As regional weather changes, the impact on human health comes from many sources, including alterations in microbial pathways, disease transmission patterns, water cycle for agriculture, and ecosystems
that supply food and clean water. The health effects of these changes, many of which are already underway, are enormous. Source: [18]

nomic benefits as they avoid the worst of the early negative impacts and see growing seasons extended. However, changes in the water cycle may cause problems as precipitation becomes less predictable. And many low-latitude areas and the highest-latitude areas approaching the arctic will see net economic losses. [Medium

FIGURE 2.3 online at ncse.org/climate solutions

Graph of temperature distribution shows that as the mean global temperature increases, the probability of more record hot weather increases. Conversely, there will probably be less cold weather.

confidence] A warming of 2 or more degrees creates net negative impacts in market econo-mies in all latitudes, with most people being negatively affected. [Medium confidence]

If we experience an increase in our body tem-perature of 1°C (2°F), we run a fever and stay in bed. If we experience a fever with a 2°C rise in temperature (4°F), our doctors will send us straight to the hospital. Similarly, with nature, small shifts in temperature can have very big effects.

We can think of our climate as a bell-shaped curve, with most days in the middle at average temperature and precipitation, and fewer days at either extreme, with very hot or very cold temperatures. The insight shown in Figure 2.3

is that when the whole bell-shaped distribution of events shifts to the hotter, drier side, the probability of hot and very hot weather is much larger than before, as shown in the shaded areas on the right side of this figure.

The reason our social systems will undergo such stress is that the natural systems upon which we depend for clean air, clean water, and food will themselves be subjected to rapid and debilitating stresses. As we examine the Earth's biological systems, we have to introduce a new concept, irreversibility. The extreme weather events discussed above will vary in timing and location and strength. They will come and go. The disruptions to our social systems will be severe, but our social systems will likely recover if the disruptions stop or reach a new "normal." For example, we can rebuild roads above the floodplain. But many of nature's plants and animals may no longer find new homes if their current homes are destroyed or made too inhospitable. For the plants and animals, the global change may be irreversible. They may simply die out forever—something we will talk more about in Chapter 5.

For example, the 2003 European heat wave stressed vegetation and ecosystems through heat and drought, and wildfires. Forest trees experienced crown damage in their tops as their growth faltered. Crown damage inhibits future growth, so such effects are long lasting and cumulative. Overall, plants grew less robustly, reducing the amount of carbon they took up. Freshwater lakes suffered from prolonged oxygen depletion as their deeper colder layers warmed. Rivers saw a decline in the number of mollusk species. Forest fires burned roughly 650,000 hectares (1.6 million acres) across the continent. In Portugal alone, over 5% of the nation's total forest area burned, representing an economic impact of over 1 billion euros ($US 1.5 billion). Those heat-wave conditions will become the norm with a 2°C temperature rise. [18] [5]

If a few weeks of peak heat on one continent provoke such damage in the natural world, what are the global ecosystem impacts projected with 2 degrees of warming?

Land-Based Ecosystems and Biodiversity

Many ecosystems are already affected. [Very high confidence] With a warming of as little as 2 degrees, about 20% to 30% of species will be at an increasingly high risk of extinction. [Medium confidence] At a warming of 2 or more degrees, the land-based plants will tend toward becoming a new carbon source. [High confidence] Translation: With a warmer planet, more plants that convert large amounts of atmospheric carbon into trunks and leaves will reach a point of maximum growth and may be replaced by smaller plants that absorb less carbon, as happens when shrubs and grassland replace forests.§

Marine Ecosystems and Biodiversity

Increased coral bleaching, in which coral dies and only its calcium skeleton remains, is already underway. This bleaching is caused by a convergence of disruptions from elevated temperatures in coastal waters, changes in salinity, an increase in the acidity of the water, and a decline in plankton. [High confidence] Coral reefs are zones of high biodiversity and rich habitat for many fish species historically important in human diets. Reef bleaching is happening even faster and more extensively than the IPCC predicted just 2 years ago. This elevates the extinction rates for species dependent on coral.

Question 2: What Are the Odds that It Will Happen?

When scientists estimate the likelihood or probability that something will happen, they consider two different aspects of such a prediction.

---

§This effect is an example a "positive feedback" climate loop in which the factors that cause a change (more plant growth) create the conditions (plants beginning to absorb less carbon) that exacerbate the underlying cause of the effect (higher levels of carbon in the air).

TABLE 2.1 Some Likelihood Definitions and Examples

Terminology	Likelihood of the occurrence/outcome	Example of events IPCC projects with this probability
Virtually certain	>99% probability	Cold days and nights will be less frequent and warmer over most land areas.
		Hot days and nights will be more frequent and warmer over most land areas.
		Insect outbreaks will increase.
Very likely	90% to 99% probability	If the atmospheric CO_2 level stabilizes at double today's level, average global temperature will rise by 1.5°C.
		The frequency of heavy precipitation events will increase.
		The frequency of warm spells or heat waves will increase over most land areas.
Likely	66% to 90% probability	If the atmospheric CO_2 level stabilizes at double today's level, average global temperature will rise by between 2°C and 4.5°C.
		Areas affected by drought will increase.
		Intense tropical cyclone activity will increase.
		Extreme high-sea-level events will increase.
About as likely as not	33% to 66% probability	(none given)
Unlikely	10% to 33% probability	(none given)
Very unlikely	1% to 10% probability	(none given)
Exceptionally unlikely	>1% probability	(none given)

Source: [12: table SPM 1]

The first is the likelihood that the prediction is correct. And the second is their confidence in the information that formed the basis for the prediction.

If we watch a bird return to its nest to feed its young Monday through Friday, on Saturday morning we can estimate it is *likely* the bird will feed its young that day. In other words, we think it is more likely than not that the bird will fly back to the nest with food. Because we observed this behavior 5 out of the 5 prior days, we have *high confidence* in being correct that the bird will return on Saturday. But say we were out on an errand for most of Friday, so we do not know whether the chicks may have fledged or been eaten by a predator while we were away. Our upstairs neighbor, who can see into the nest, makes the stronger projection that the adult bird will *very likely* return to the nest to feed the young. He has even better information. He can see whether or not there are still young unfledged birds in the nest, whereas we cannot. Therefore, he has a *very high confidence* level,

higher than ours, that he is correct. Data that are more complete and more reliable lead to both more-accurate probabilities and higher confidence levels that those projections are correct.

What does *likely* mean precisely? Likelihood refers to an assessment of the probability "of some well-defined outcome having occurred or occurring in the future, and may be based on quantitative analysis or an elicitation of expert views." [12] Throughout the IPCC's *Climate Change 2007* report, the authors assign a likelihood to events that are occurring now or are projected to occur. The phrases on the left in Table 2.1 reflect the probabilities listed in the middle column. Sample events with some of these corresponding likelihoods are listed on the right.

The authors of the IPCC reports assign confidence levels to the major statements, based on their assessment of the current knowledge on that topic. Each impact in the text above is accompanied by the authors' confidence level, in square brackets. The levels of confidence are as

TABLE 2.2 Some Confidence Degree Definitions and Examples

Terminology	Degree of confidence in being correct	Example of events predicted with this level of confidence
Very high confidence	At least 9 out of 10 chance	At or above 2°C warmer, floods, drought, and erosion will increase and water quality will decrease.
		Sea level rise will increase the salinization of groundwater, decreasing freshwater in coastal regions.
High confidence	About 8 out of 10 chance	A warming of up to 2°C will increase the frequency of forest or grassland fires and their intensity, especially in areas that suffer from drought.
		Most coral reefs will bleach and die off.
Medium to high confidence	N/A	A warming of up to 2°C will produce an increase in category 4 to category 5 tropical cyclone storms. In addition, their impact on coastal regions will be exacerbated by sea level rise.
		Malnutrition, infectious and diarrheal diseases, malaria, and direct fatal accidental injuries from flood, heat, and drought will increase.
Medium confidence	About 5 out of 10 chance	For a warming between 1°C and 3°C, some areas in low latitudes will experience a decrease in productivity for some cereals, which may be offset by an increase in productivity in the middle to high latitudes.
		Current effects on human health are small but discernible.
Low confidence	About 2 out of 10 chance	(none given)
Very low confidence	Less than 1 out of 10 chance	(none given)

Source: [16: table 19.1]

follows: very low, low, medium, high, and very high, as we see defined in Table 2.2.

In short, there is a much-greater-than-chance probability that we collectively are going to experience serious bad consequences, from heat waves and drought to flood and disease. Scientists have a high degree of confidence in that prediction, and we are already experiencing some of such predicted consequences.

Question 3: How Do We Know?

Schneider cautions us to ask how the experts make such estimates. This question gets to the heart of the scientific process. How is scientific research conducted?

Science does not start with an opinion about whether the globe is warming or cooling. A scientist starts with the facts and a rigorous method of testing whether those facts are reliable and accurate. A scientist only uses facts that any other scientist could also use to replicate the same tests with the same method. A conclusion

becomes more widely accepted only after many others have examined the same or similar facts with the same or different methods and come to the same or similar results. To date, no credible research has yet shown that planet Earth has cooled down since 1850. To date, all the credible research shows, instead, that the Earth is warming with unprecedented speed.

The basic test of objectivity in science is whether another researcher comes to the same results with the same method of study. In 2007, IPCC Working Group I, which is responsible for reporting the basic science of climate change, compiled a staggering 29,000 data series about temperature from about 75 studies representing regions all over the globe.

Working Group I started with about 80,000 data series from 577 different studies of temperature records and used only those in which the data (1) ended in 1990 or later, (2) spanned a period of at least 20 years, and (3) showed a significant change in either direction, as assessed in individual studies. And 70 of the 75 studies analyzed had been undertaken and published since 2001. On the matter of whether global temperatures are changing significantly, all 29,000 data sets pointed to similar conclusions: The Earth is getting significantly warmer on average. The Earth is now at an average surface temperature of 14.4°C. The last time the Earth was this warm, 120,000 years ago, alligators swam in London's River Thames and palm trees grew on Greenland, and hundreds of thousands of years were required for the temperature to reach that level. We can see the location of these data series in Figure 1.2.

Thermometer data to provide a global temperature record are available only back to the middle of the 19th century. Temperature data of a different—but nonetheless accurate—kind are available from other, natural sources for a much longer time period than that. For example, sediments settle each year on the bottoms of lakes and seas, continuously recording the conditions in the environment around them. Warmer temperature means more microorganisms live, die, and fall to the bottom as sediment. Ocean mud accumulates every year and can provide key information about past climate. Marine sediment cores can offer reliable temperature proxies going back 6,000 to 6 million years, depending on the location. On land, snowfall causes ice to build up each winter on glaciers in our mountain and in the polar regions. Colder temperatures create more ice and a different balance of oxygen isotopes that become trapped in minute air bubbles within the ice. Ice cores from Greenland are reliable proxies going back 100,000 years. Similarly, a tree's rings accurately record the amount of its annual growth, which is dependent on both proper moisture and temperature for its species. This annual banding is also apparent in coral growth as each year's new hard skeleton of coral is laid down over that of the prior year. The exact chemical makeup of the mollusk shells depends on the water temperature in which they form. This makeup is captured when the shells fossilize. Reading how much calcium and magnesium the fossilized shells contain allows scientists to determine the likely water temperatures present in ancient seas.

Ice cores, lake sediments, tree rings, fossilized plants, and marine shells all offer us glimpses into ancient climate conditions. By comparing these multiple sources, we can reconstruct the likely average annual temperatures for large areas of the globe. These natural records are proxies for direct measurements. For example, we know from tree ring and other natural records that Europe was relatively warm during the Middle Ages, allowing the Vikings to colonize Greenland, but not as warm as the trend unfolding today. Natural records from other continents show this Medieval Warm Period was confined to Europe. Indeed, other areas such as the Pacific were colder than average at this same time, as the coral reef banding there shows.

In the eyes of scientists, each potential proxy must leap over a stiff set of hurdles before being considered worth using. First, a proxy must be able to show a temperature-based change on an annual to decadal basis, what researchers call "temporal resolution." Second, a proxy must allow itself to be exactly dated so that the proxy record may be calibrated against instrumental data or other credible sources. If we have a tree stump, we have to know when it started or stopped growing in order to use its ring data. Third, the proxy needs to be reliable, especially if it is relatively uncommon, as we will have few other points of corroboration. Fourth, a set of proxies must be gathered that collectively represent different regions of the planet. This spatial diversity helps us take into account any variation across regions in past climate variability. Fifth, proxies need to be factored in a way that keeps in mind the seasonal variation of their indicators within a single year. In sum, climate researchers need data sources with sufficient spatial and seasonal sampling, sufficient temporal resolution, and sufficient retention of millennial scale variation, or the data get thrown out.

Two other aspects of the progress in climate science show powerful advances. First, the resolution of the models in time and space is much finer than ever before. And, second, scientists are learning how to feed more aspects of the complex physical and biological world itself into the models.

In Figure 2.4 we can see both the increase in the number of environmental components used in modeling and the decrease in the size of the "box" for which projections can be made with some degree of confidence. An analogy from improvements in weather forecasting may be helpful. The National Hurricane Center in Miami, Florida, makes predictions about the likely path and expected wind speeds of a major tropical storm, once the storm cell forms in the Atlantic. The Hurricane Center's ability to predict the storm path has steadily improved each passing year. More accurate numerical models, more observations over the open ocean, and a better understanding of the physics of hurricane movement have lead to 3-day forecasts today that are as accurate as 2-day forecasts were 15 years ago.** Of course, forecasting daily weather occurs on a much shorter scale of time and in a smaller physical place than projecting annual climate, which must be done for large regions over many years.

One dramatic illustration of how the science keeps improving is the reduction in the area of the average "square" for an air or water sample for which climate scientists are able to build computer models. Since the IPCC began its reports, the climate models for Europe have become much better at estimating ever-smaller surface area interactions, starting from around 500 km (310 miles) on a side in 1990 and improving to around 110 km (70 miles) on a side now.

Is the science on climate change settled? Only one reasonable answer is possible: yes! Opinions may vary on how to respond to climate change, but the scientific fact that planet Earth is warming at a very fast rate is irrefutable.

(See online Figures 2.5 and 2.6 for more on the warming trend of the past 2,000 years.) So is the fact that the amount of warming during the present, industrial era is unprecedented for that short a time span in all of the history of life on Earth. So is the fact that the past 150 years of human fossil fuel–based industry coincides with a massive transfer of carbon from its underground, sequestered, and fossilized form to the unleashed, atmospheric, and gaseous form. So is the fact that carbon dioxide and other gases in the Earth's atmosphere act as a powerful insulator around the planet, trapping heat that would otherwise escape into space. As a matter of science, our planet's warming trend is settled and accepted as fact. What is not at all settled is what we are going to do about this self-inflicted circumstance.

CONNECT THE DOTS

- Three groups of actors play a role in the decision making about which courses of action to take: scientists, policymakers, and citizens.
- Lay audiences need to ask experts (1) What can happen? (2) What are the odds that it will happen? and (3) How are such estimates made?

FIGURE 2.4 online at ncse.org/climate solutions

Resolution of climate models has improved almost fourfold in 17 years.

FIGURE 2.5 online at ncse.org/climate solutions

Most temperature information for the past 2,000 years is about the northern hemisphere.

FIGURE 2.6 online at ncse.org/climate solutions

Temperature data for the past 2,000 years show a definite warming trend since 1850.

**See Frequently Asked Questions at http://www.aoml.noaa.gov/hrd/ for more on hurricane forecasting improvements.

- The Intergovernmental Panel on Climate Change assessment describes recent warming as "unequivocal."
- Nineteen of the warmest years on record have occurred within the last 25 years.
- Science involves open access to data from repeatable experiments that produce verifiable results.

Online Resources

www.eoearth.org/article/Climate_change_FAQs
www.eoearth.org/article/
    Daily_and_annual_cycles_of_temperature
www.eoearth.org/article/Ecosystems_and_Human_
    Well-being_Synthesis~Preface
www.eoearth.org/article/Global_Climate_
    Change%3A_Major_Scientific_and_Policy_Issues
www.eoearth.org/article/Human_variability_to_
    global_environmental_change
www.eoearth.org/article/Monitoring
World Health Organization Global Environmental
    Change, www.who.int/globalchange/en/
See also extra content for Chapter 2 online at http://
    ncseonline.org/climatesolutions

Climate Solution Actions

Action 20: Climate Change, Wildlife Populations, and
    Disease Dynamics
Action 27: Looking into the Past to Understand Future
    Climate Change
Action 34: Building People's Capacities for Implement-
    ing Mitigation and Adaptation Actions

Works Cited and Consulted

[1] Carpenter KE, et al. (2008) *One-third of reef-building corals face elevated extinction risk from climate change and local impacts.* Science 321(5888):560–563. www.sciencemag.org

[2] Confalonieri U, Menne B, Akhtar R, Ebi KL, Hauengue M, Kovats RS, Revich B, Woodward A (2007) Human Health (in *Climate Change 2007: Impacts, Adaptation and Vulnerability. Contribution of Working Group II to the Fourth Assessment Report of the Intergovernmental Panel on Climate Change,* 391–431, eds Parry ML, Canziani OF, Palutikof JP, van der Linden PJ, Hanson CE) www.ipcc.ch/

[3] Havenith G (2002) *Interaction of clothing and thermoregulation.* Exogenous Dermatology 1(5):221–230. www-staff.lboro.ac.uk/~hugh/

[4] IPCC (2007) Summary for Policymakers (WG2) (in *Climate Change 2007: Impacts, Adaptation and Vulnerability. Contribution of Working Group II to the Fourth Assessment Report of the Intergovern-*

mental Panel on Climate Change, 7–22, eds Parry ML, Canziani OF, Palutikof JP, van der Linden PJ, Hanson CE) ar4-wg2-spm.pdf: www.ipcc.ch

[5] Kirilenko AP, Sedjo RA (2007) *Climate change impacts on forestry.* Proceedings of the National Academy of Sciences 104(50):19697–19702. www.pnas.org

[6] Koppe C, Jendritzky G, Kovats S, Menne B (2004) *Heat-waves: impacts and responses.* Health and Global Environmental Change Series 2, 123 pp. www.euro.who.int/publications

[7] Le Treut H, Somerville R, Cubasch U, Ding Y, Mauritzen C, Mokssit A, Peterson T, Prather M (2007) Historical Overview of Climate Change (in *Climate Change 2007: The Physical Science Basis. Contribution of Working Group I to the Fourth Assessment Report of the Intergovernmental Panel on Climate Change,* 92–127, eds Solomon S, Qin D, Manning M, Chen Z, Marquis M, Averyt KB, Tignor M, Miller HL) www.ipcc.ch

[8] Mann ME, Jones PD (2003) *Global surface temperatures over the past two millennia.* Geophysical Research Letters 30(15). www.agu.org/journals/gl/

[9] Mann ME, Zhang Z, Hughes MK, Bradley RS, Miller SK, Rutherford S, Ni F (2008) *Proxy-based reconstructions of hemispheric and global surface temperature variations over the past two millennia.* Proceedings of the National Academy of Sciences 105(36):13252–13257. www.pnas.org

[10] NASA (2003) *Land surface temperature difference during the heat wave in western Europe for July 20 through August 20, 2003.* The Earth Observer 2003(08-20). http://eospso.gsfc.nasa.gov/eos_homepage/for_scientists/earth_observer.php

[11] NOAA NHC (2008) National Hurricane Center. *National Weather Service.* National Oceanic and Atmospheric Administration (read October 1, 2008). www.nhc.noaa.gov

[12] Parry ML, Canziani O, Palutikof JP, van der Linden PJ, Hanson CE (2007) Technical Summary: Working Group II (in *Climate Change 2007: Impacts, Adaptation and Vulnerability. Contribution of Working Group II to the Fourth Assessment Report of the Intergovernmental Panel on Climate Change,* 23–78, eds Parry ML, Canziani OF, Palutikof JP, van der Linden PJ, Hanson CE) ar4-wg2-ts.pdf: www.ipcc.ch/

[13] Rosenzweig C, Casassa G, Karoly DJ, Imeson A, Liu C, Menzel A, Rawlins S, Root TL, Seguin B, Tryjanowski P (2007) Assessment of Observed Changes and Responses in Natural and Managed Systems (in *Climate Change 2007: Impacts, Adaptation and Vulnerability. Contribution of Working Group II to the Fourth Assessment Report of the Intergovernmental Panel on Climate Change,* 79–131, eds Parry ML, Canziani OF, Palutikof JP,

van der Linden PJ, Hanson CE) ar4-wg2-chapter1 .pdf: www.ipcc.ch

[14] Schmidt G, Wolfe J (2009) *Climate Change: Picturing the Science.* (W. W. Norton & Company, New York). www.wwnorton.com

[15] Schneider SH (1997) *Defining and teaching environmental literacy.* TREE 475. http://stephen schneider.stanford.edu

[16] Schneider SH, Semenov S, Patwardhan A, Burton I, Magadza CHD, Oppenheimer M, Pittock AB, Rahman A, Smith B, Suarez A, Yamin F (2007) Assessing Key Vulnerabilities and the Risk from Climate Change (in *Climate Change 2007: Impacts, Adaptation and Vulnerability. Contribution of Working Group II to the Fourth Assessment Report of the Intergovernmental Panel on Climate Change,* eds Parry ML, Canziani OF, Palutikof JP, van der Linden PJ, Hanson CE) (Cambridge University Press, Cambridge, UK) www.ipcc.ch

[17] Solomon S, Qin D, Manning M, Alley RB, Berntsen T, Bindoff NL, Chen Z, Chidthaisong A, Gregory JM, Hegerl GC (2007) Technical Summary: Working Group I (in *Climate Change 2007: The Physical Science Basis. Contribution of Working Group I to the Fourth Assessment Report of the Intergovernmental Panel on Climate Change,* 19–91) www.ipcc.ch

[18] Stott PA, Stone DA, Allen MR (2004) *Human contribution to the European heatwave of 2003.* Nature. www.nature.com/nature/journal/v432/ n7017/abs/nature03089.html

[19] US EPA (2008) Municipal Solid Waste in the United States 2007 Facts and Figures. EPA530-R-08-010. *Wastes.* www.epa.gov/osw/nonhaz/ municipal/msw99.htm

[20] USCOM (2008) Fall Summit: Waste to Energy Goes Green. *Municipal Waste Management Association.* www.usmayors.org/mwma/

[21] WHO (2008) Climate Change and Human Health. *WHO,* Geneva. www.who.int/globalchange/en/

CHAPTER 3

Human Carbon as the Smoking Gun

350 is the red line for human beings, the most important number on the planet. The most recent science tells us that unless we can reduce the amount of carbon dioxide in the atmosphere to 350 parts per million, we will cause huge and irreversible damage to the earth. [18]

BILL MCKIBBEN, 2008

Carbon is a marvelous element. It is the fourth most abundant element in the universe—after hydrogen, helium, and oxygen. Carbon is a simple, stable molecule of six protons and six neutrons surrounded by six electrons. Because the outer four of its electrons are easily shared with other molecules, carbon forms more compounds than any other element. The abundance of life on Earth is almost entirely due to biological photosynthesis, which depends on light energy. In plants, photosynthesis converts carbon as gas into solid form. For example, the wood of the massive redwood tree is built from carbon and other gases taken from the air. Even in the light-starved thermal vents on the dark ocean floor, bacteria convert the heat, methane, and sulfur compounds provided by "black smoker" vents into energy through a process called chemosynthesis. But either way—with or without direct sunlight—life on Earth depends on using carbon for both structure and nourishment. Carbon is the second most common element in the human body (18.5%)

after oxygen (65%). In living organisms, carbon forms the basis for organic molecules, the building blocks of life itself. The carbon dioxide gas in our atmosphere is a major reason why Earth is neither too hot nor too cold to sustain life. Yet, carbon gas in the atmosphere is a principal culprit in driving the temperature up at the Earth's surface. We will explore the carbon-climate link in this chapter.

The Airborne Carbon

The magic of planet Earth's atmosphere is that water is present in our atmosphere in vapor form and, hence, on our surface in liquid form. Earth is located a bit too far from the Sun for our planet's surface to be the right temperature for liquid water. If Earth had no atmosphere at all, it would be a dry ball like our nearest neighbor, the Moon, with a surface temperature that would fluctuate from daytime highs above the boiling point of water to nighttime lows well below the freezing point of water. On the Moon, without an atmosphere, surface temperatures during a lunar day average 107 degrees Celsius (°C), or 224 degrees Fahrenheit (°F), and during a lunar night, −153°C (−243°F). But unlike the Moon, Earth has a molten core that led to outgassing and volcanic eruptions that produced an atmosphere, initially of water vapor and carbon dioxide.*

The composition of the Earth's atmosphere has changed considerably since the planet's formation 4.6 billion years ago. In its first billion years or so, before photosynthesis and life began, carbon dioxide concentrations in the Earth's atmosphere used to be much higher than they are now. Heavy volcanic activity spewed ash, carbon dioxide, and water vapor into the atmosphere. This layer served to wrap the Earth in a cloudy blanket that both let heat from the Sun pass through the clouds to reach Earth and radiated its own heat back down to the Earth. As the Earth slowly warmed during its first billion years, water vapor condensed and fell as rain, formed oceans with the help of liquid water from ice delivered by asteroids and comets, and lowered the concentration of water vapor in the atmosphere. The oceans lowered the concentration of carbon dioxide in the atmosphere as the airborne carbon dioxide dissolved in the ocean water. Once dissolved, carbon dioxide combined with calcium to form carbonate minerals. Once plant life emerged in the ocean, photosynthesis used the carbon dioxide as well and added oxygen to the atmosphere. It was not until the last half billion years that shelled organisms emerged in the ocean as the precursors to all animal life. These organisms used the carbonate minerals to form their shells and eventually fell to the seafloor when they died. As oxygen increased in the atmosphere and carbon dioxide fell, the stage was set for life on land to emerge. As a result of these processes over more than 4 billion years, concentrations of carbon dioxide fell to about 0.033%, or 330 parts per million by volume (ppm), of the Earth's atmosphere. Or that is where it was until we started burning so much ancient carbon as fossil fuels.[†] In so doing, we are returning to the atmosphere carbon dioxide that the Earth's natural systems had long ago removed.

Without any greenhouse gases in its atmosphere, planet Earth would have an average surface temperature of −16°C (3 degrees above 0°F). That is a very chilly place! Today our mean surface temperature, 14.4°C (58°F), is over 30°C higher than that. In the last ice age, when the Earth's average temperature was 8.4°C (47°F)—just 6°C colder than now—glaciers 3 kilometers (km), or 2 miles, thick covered much of Europe and North America. Some amount of greenhouses gases is a good thing, as they allow water to exist in a liquid state and thus make life on Earth possible, as we know it.

Greenhouse gases come in many varieties. The most abundant molecules in our atmosphere, nitrogen (N_2) and oxygen (O_2), are not big heat blockers because of their simple atomic structure. Gases with larger, more complicated molecules are necessary to block the most infrared wavelengths of heat from escaping the atmosphere. The most abundant of these larger molecules are the ones composed of three atoms, for example, water (H_2O) and carbon dioxide (CO_2).

---

*Venus is the planet next closest to the Sun compared with Earth. The atmosphere of Venus is 96.5% carbon dioxide. The high concentration of carbon dioxide makes Venus so hot that soft metals can melt on its surface. But Venus, unlike Earth or Mars, has no seasons. On Mars, the planet next farthest from the Sun, the thin atmosphere contains about 95% carbon dioxide, similar to that of Venus. Surface temperatures on Mars vary from lows of about −140°C (−220°F) during the polar winters to highs of up to 20°C (68°F) in summers.

[†]Throughout this book we will often refer to gas concentrations as parts per million by volume (e.g., for carbon dioxide) or parts per billion by volume (e.g., for methane). We will use the shorthand of "ppm" for the former and "ppb" for the latter. If you see "ppmv" or "ppbv" with "v" for "by volume" as the unit cited in other research reports, these mean the same as what we mean by "ppm" or "ppb."

Both occur naturally, but both are also sent aloft in vast quantities by human activities. Other common greenhouse gases are methane (CH_4), ozone (O_3), and nitrous oxide (N_2O).

The three reasons for deep concern now about climate disruption are (1) the size of the carbon dioxide concentration in the lower atmosphere, currently at over 380 parts per million (ppm), (2) the rapid increase of this concentration, and (3) the direct causal correlation between atmospheric carbon and global temperature. Over a period of 800,000 years and eight glacial cycles, Earth's atmospheric carbon dioxide levels have ranged from about 170 ppm, during colder periods when glaciers expanded, to about 300 ppm in warmer interglacial periods. [16] In addition, the planet's atmospheric methane levels are now at 1,770 parts per billion by volume (ppb). [4] The planet's past methane levels have ranged between 350 and 800 ppb during these same 800,000 years. [15]

Such past shifts always took place over thousands of years while radically altering plant and animal life. For example, the most recent natural rise of 80 ppm of atmospheric carbon dioxide (to roughly 270 ppm) unfolded over a period of 5,000 years and ended the last ice age (see online Figure 3.1). In the preindustrial era before 1750, atmospheric carbon levels were roughly 280 ppm. Today they are 380 ppm with most of that spike coming since 1950. So, in a matter of decades, humans have achieved—if that is the right word—a rise in atmospheric carbon dioxide that is greater than what nature needed 5,000 years to produce. This very rapid, very recent change in the composition of our atmosphere—and the rapid atmospheric warming it brings—has no precedent in the Earth's geological record. We can find no natural causes to explain this change in atmospheric conditions. We can use a few simple graphs to capture all these numbers—one set showing what is in the atmosphere and a second picture of how many humans might be putting that stuff into the air.

The present atmospheric concentrations of carbon dioxide, methane, and nitrous oxide are higher than ever measured in the Antarctic ice core record of the past 800,000 years. These records show the large and increasing growth in human greenhouse gas emissions during the industrial era. The ice core records show that during the industrial era, the average rate of increase in carbon dioxide, methane, and nitrous oxide is greater than at any time during the past 10,000 years. [14]

So let's explore the human era a bit more by looking at the past 10,000 years representing the period during which human population has grown so much. Online Figure 3.1 depicts levels of atmospheric carbon dioxide (as well as methane and nitrous oxide) over the past 10,000 years based on ice-core data. These years encompass the entire span during which humans shifted from hunting and gathering to farming and city building. The human population on Earth grew 277% from 1900, when it stood at 1.8 billion, to 2000, when it reached 6 billion. Most of this growth occurred in the developing world. Online Figure 3.2 shows the human population growth from 10,000 years ago to the present.

The steep rise of carbon dioxide, methane, and nitrous oxide on the righthand side of the figures at the present time is hard to miss (Figure 3.7). Carbon dioxide concentrations rose from 280 ppm in the preindustrial era to 380 ppm in 2005, exceeding the prior natural range of variability by at least 25%. About 80% of the carbon dioxide increase is tied to fossil fuel use and the rest to changes in land use that reduce

---

FIGURE 3.1 online at ncse.org/climate solutions

Levels of atmospheric carbon dioxide for the past 20,000 years show a steep increase, and its warming effect, beginning with the industrial age.

FIGURE 3.2 online at ncse.org/climate solutions

The number of people on Earth has grown sixfold in the last two centuries.

vegetation volume. Meanwhile, methane rose from 750 to 1,750 ppb in 2005, exceeding the prior natural range of variability by at least 125%. This increase is predominantly due to agriculture and fossil fuel combustion. Nitrous oxide concentration increased from a preindustrial value of about 270 to 319 ppb in 2005, with a relatively constant growth rate since 1980. Over a third of all nitrous oxide emissions are primarily due to agriculture.‡

The Intergovernmental Panel on Climate Change (IPCC) sums up the increase in atmospheric carbon dioxide as follows:

The average rate of increase in atmospheric CO_2 was at least five times larger over the period from 1960 to 1999 than over any other 40-year period during the two millennia before the industrial era. The average rate of increase in atmospheric CH_4 was at least six times larger, and that for N_2O at least two times larger over the past four decades, than at any time during the two millennia before the industrial era. Correspondingly, the recent average rate of increase in the combined radiative forcing by all three greenhouse gases was at least six times larger than at any time during the period AD 1 to AD 1800. [14, p. 447]

How Are Atmosphere, Greenhouse Gases, and Temperature Linked?

Did you ever leave a sealed bottle of water in the sun too long? It became quite hot to the touch, didn't it? Thermal energy from the Sun's rays

---

‡Carbon dioxide is present in much larger volumes than methane. But both are present in very small quantities relative to nitrogen and oxygen. Therefore we measure carbon dioxide in parts per million (ppm) and methane in parts per billion (ppb). One part per billion is one one thousandths part per million. One part per million is 1,000 ppb. One part per million is 0.0001 percent. Graphs of these other two gases over the past 20,000 years are available at http://www.ipcc .ch/graphics/graphics/ar4-wg1/jpg/fig-6-4.jpg.

that passed through the container's walls heated that liquid inside. It could not escape, so it in turn heated the bottle, which in turn heated more liquid. The same bottle without the seal in the sun might become warmer too. But some of the water inside would evaporate as it heated. As this evaporated water escaped through the top of the bottle, it would have the effect of cooling the bottle.

Planet Earth is like the bottle sealed with a nearly airtight cork. Some meteorites arrive and some satellites become space junk, but these cancel each other out. Some lighter gases leak out into space, but they are replaced by gases from volcanic eruptions and by plant respiration. What closes the Earth system is the atmosphere surrounding the planet. The atmosphere is a mixture of gases we call air. Completely dry air is a mixture of roughly 78% nitrogen gas and 21% oxygen gas. That leaves about 1% for trace gases of all other kinds, including the greenhouse gases, such as water vapor, carbon dioxide, methane, nitrous oxide, and ozone. Air also contains aerosols, tiny solid or liquid drops that remain suspended for a long time, such as soot or dust.

Gaseous water (H_2O), carbon dioxide (CO_2), methane (CH_4), nitrous oxide (N_2O), and ozone (O_3), among others, all have molecular structures large enough to act as targets to encounter radiated heat and thereby heat up. When radiated heat from the Earth's surface is absorbed in the lower atmosphere by molecules, aerosols, or cloudiness (i.e., condensed water vapor) and radiated back to the surface in excess of the planet's normal heating and cooling balance, the harmful warming of the greenhouse effect occurs.

Among the trace gases, water vapor is the most common and most important greenhouse

---

FIGURE 3.3 online at ncse.org/climate solutions

Energy radiated from the Earth's surface is largely absorbed in the atmosphere and reradiated back to Earth. This is the greenhouse effect.

gas. Carbon dioxide is the second most common and the most important greenhouse gas. Others include methane, nitrous oxide, ozone, and other less common human-made halocarbon gases (such as chlorofluorocarbons). In humid tropical areas, the greenhouse effect is largely due to water, as the air is already so laden with water vapor. At the colder, drier polar areas, a small added amount of carbon dioxide or water vapor in the atmosphere has a much greater greenhouse effect than that same amount would have near the equator.

Similarly, adding more water vapor or carbon dioxide, even in small amounts, to the higher, colder, drier layers of the atmosphere has a greater greenhouse effect than adding them to the lower, denser, more humid layers of the atmosphere. While there is always some water vapor in the atmosphere, its concentration level fluctuates greatly: More heat means more humidity. But water vapor precipitates out of the atmosphere very quickly as rain, snow, hail, dew, and frost, whereas trace gases such as carbon dioxide reside in the atmosphere for 50 years or more. Put another way, water vapor has a significant influence in warming the air around it by absorbing infrared (solar) radiation. But its concentration in the atmosphere mainly depends on air temperature. Unlike with carbon and some other greenhouse gases, we have no way to directly influence atmospheric water vapor concentration.

If greenhouse gases are such small compo-

FIGURE 3.4 online at ncse.org/climate solutions

Human activities and natural processes are components of the climate change system.

FIGURE 3.6 online at ncse.org/climate solutions

Deep water is a huge reservoir for heat and carbon. The ocean holds enormous quantities of water in its deepest parts, such as the Puerto Rico Trench.

Atmosphere

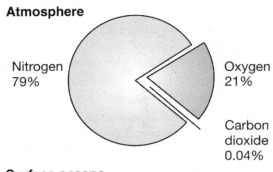

Surface oceans

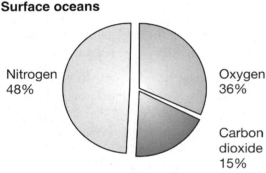

Total oceans including deep water

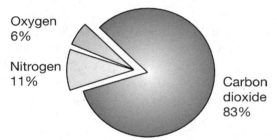

FIGURE 3.5 Distribution of gases in the atmosphere and dissolved in seawater

Carbon dioxide (CO_2) comprises only a very small overall portion of our atmosphere but 15% of the surface water of our oceans and an astounding 83% of the deep, cold, salty water in our oceans. More than 90% of all the water in our oceans is deep water. The oceans are deeper than the land is tall. For example, the average height of land above sea level is 840 meters (2,800 feet), but the average depth of the ocean floor is 3,800 meters (12,000 feet). Source: Adapted from [23: p. 143]

nents of the atmosphere, why is there a connection between such gases and global surface temperature? The short answer is that adding greenhouses gases to the atmosphere has a self-intensifying effect on temperature. This effect is a positive feedback loop. But *positive* here means "additive" rather than "good."§ For example, when we add more carbon dioxide to the atmosphere from burning fossil fuels or clearing forests, these molecules enter clouds, heat up, and radiate (add) more heat from the lower atmosphere back to Earth. Warmer air is able to hold more water vapor than colder air. More water vapor in the atmosphere means more heat radiated back to Earth, which causes more warming on the surface, which causes a higher concentration of water vapor in the air. And so on. Even long after we may have stopped adding any more carbon dioxide to the atmosphere, these internal feedback loops between temperature and increased water vapor will keep intensifying the greenhouse effect.

We just mentioned the lower atmosphere as most active in the greenhouse effect. The Earth's atmosphere is not a uniform shell of air. It has layers, like an onion. Most of the greenhouses gases that have an impact on climate are found in the 50 km just above sea level in our atmosphere.

Both Colder and Warmer, but Never So Quickly

The Earth has been both warmer and cooler than it is now. But it has never changed temperature so quickly before. Ice-core data extend back almost 1 million years and show recurring patterns of cooler, longer periods of glacier buildup (ice ages) followed by shorter, warmer interglacial periods. Ocean sediment cores allow us to extend the temperature record back millions more years.** In the period of about 10,000 years ago to the present, temperatures fluctuated mildly with swings of less than 0.5°C over a 3,000-year period. During this warmer period, the glaciers retreated to their current locations. A Medieval Warm Period lasted from AD 800 to 1400, coinciding with Viking voyages to Greenland, but temperatures were still not as warm as they are today. This warm-up was followed by a relatively cool period from about AD 1400 to 1800 called the Little Ice Age. [17] Landscape paintings by the famous Flemish masters of winter scenes with frozen harbors and winter hunting parties depict this era.

If we line up all the temperature records available, we can see that whenever the Earth was warmer than it is today, it took a very long time to reach that temperature. Over the thousands of years it took to reach that new average temperature, whole continents developed vastly different coastlines, plant and animal species disappeared, and others arose to take their place. Finally, none of the prior temperature peaks happened while our species, *Homo sapiens sapiens*, was present. We can also usually find a set of likely natural causes for the warming or cooling. In any case, we can say with certainty that humans did not cause any global warming prior to the onset of the modern industrial age. We simply were not around in sufficient numbers and had not yet begun to burn fossil fuels in such enormous quantities.

---

§We say a factor causes a positive feedback in climate change when the effects of it magnify the action that causes it in the first place. The opposite is negative feedback, in which a factor tends to reduce the conditions that cause it, for example, soot laden clouds from fires tend to block sunlight and cool the atmosphere below them, which may reduce heat and fire-prone conditions, temporarily.

**In deep ice cores, more oxygen-18 isotopes relative to oxygen-16 isotopes means lower temperatures, the opposite relationship to that found in seafloor cores. In seabed cores, the opposite holds, as more oxygen-18 relative to oxygen-16 means a higher temperature.

How Do We Know Humans Cause Warming?

Until the onset of the industrial age, all the human activities of farming, burning wood, and decomposing organic matter combined did not measurably cause any increase in carbon dioxide levels. Why? Burning wood and farming activities cause the release of carbon that is already in circulation on the surface of the Earth. Agriculture does contribute methane and nitrogen, but the scale of agriculture was too small until recently for these to make much difference. With the invention of the steam engine and the massive increase in the mining and burning of coal to stoke the engines, atmospheric carbon dioxide levels began to rise steadily from 1850 onward. Coal is, after all, carbon that nature had removed from the atmosphere millions of years ago and buried deep underground.

Scientists are able to determine the amount of naturally occurring atmospheric content, due largely to volcanic activity and solar variability (or solar irradiance, as it is called). They are also able to determine how much temperature gain is due to natural causes. No natural factors are enough to explain the global warming that has occurred over the past century.

While fire occurs naturally as a result of lightning strikes and volcanic eruptions, humans have used combustion as the basis for the entire process of civilization. The simple equation for what happens when we burn something can be summed up as follows:

$$\text{Fuel} + \text{Air} \rightarrow \text{Heat} + \text{Carbon Dioxide (CO}_2\text{)} + \text{Water (H}_2\text{O)} + \text{Nitrogen}$$

Air has lots of hydrogen, oxygen, and nitrogen. Fuel, whether wood, coal, or oil, has lots of carbon. Combustion reshuffles the chemicals so that fire converts the fuel and air into heat, water, carbon dioxide, and nitrogen compounds. The smoke of a fire is water vapor heated to the point of rising plus some soot particles that it carries along for the ride. It is all the carbon

dioxide our fires have released that is coming back to haunt us.

Let's take a look at the past century. We know the actual average global temperature for the 20th century. We can add up the annual cumulative effect of all the natural forcings on temperature. And we can add up the annual heating or cooling caused by human activity. Is nature's work enough to explain the rise of temperature in the past 100 years? No! If we add up all the natural warming or cooling effects, we discover that the Earth's temperature in AD 2000 should have been virtually the same as in AD 1900.

The Atmospheric Overshoot

But we know from our weather stations that the temperature climbed by three-fourths of a degree Celsius in the past 100 years. As James Hansen and his colleagues from around the world write, "If the present overshoot of this target carbon dioxide is not brief, there is a possibility of seeding irreversible catastrophic effect." [8]

What overshoot? Researchers have feverishly been examining our planet's climate history to learn how much atmospheric carbon yields how much climate disruption and warming. The short answer is that a cooling trend began about 50 million years ago when the atmospheric carbon level dropped down to about 450 ppmv (plus or minus 100 ppmv). The Earth was warmer then, as fossilized palms in Greenland attest. But at the present rate that atmospheric carbon is climbing (10 times faster now than ever before), only prompt policy changes and massive actions will help us avoid going above 385 ppmv. Hansen and his colleagues put it bluntly:

If humanity wishes to preserve a planet similar to that on which civilization developed and to which life on Earth is adapted, paleoclimate evidence and ongoing climate change suggest that CO_2 will need to be reduced from its current 385 ppm to at most 350 ppm. The largest uncertainty in

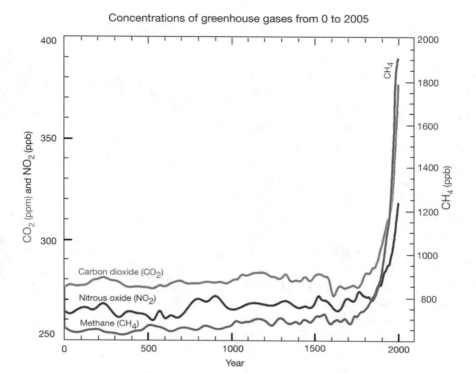

Concentrations of greenhouse gases from 0 to 2005

FIGURE 3.7 Two thousand years of atmospheric concentrations of key greenhouse gases

Since about 1750, increases in atmospheric concentrations of important long-lived greenhouse gases are attributed to human activities in the industrial era. Concentration units are parts per million (ppm) or parts per billion (ppb), indicating the number of molecules of the greenhouse gas per million or billion air molecules, respectively, in an atmospheric sample. Source: [13]

the target arises from possible changes of non-CO_2 forcings. An initial 350 ppm CO_2 target may be achievable by phasing out coal use except where CO_2 is captured and adopting agricultural and forestry practices that sequester carbon. [8]

James Hansen, et alia, 2008

The "non-CO_2 forcings" Hansen refers to include the other greenhouse effects whose cycles in nature we will not be able to control once triggered. Methane (whose chemical structure makes its warming impact per molecule released much more powerful than carbon dioxide's) will increase as tundra thaws. Ocean warming will accelerate once polar ice shrinks too much to stop polar seas from heating up. The decreasing reflectivity of the Earth's surface as the winter world becomes less white will accelerate the Earth's absorbing of heat. The interaction of these massive Earth systems with each other means climate disruption could unfold very quickly in a cascading effect. If we adopt a business-as-usual approach and make no significant changes, the carbon emission concentrations could rise unchecked to 550 or 600 ppmv—double their preindustrial level of 280 ppmv.

In Figure 3.8a we see the results for the 20th century of the likely amount of temperature change forced by natural causes, with the lighter shading showing the range. The thick black line is the actual observed temperature

FIGURE 3.8 Natural versus human forcings versus actual temperature in the 20th century

Both charts here show the observed actual temperature changes (with AD 1900 as 0 degrees of anomaly) in heavy black.

(a) In the top chart, the likely amount of temperature forced up or down by all natural causes is shown in blue, with the lighter shading showing the range for each year. Clearly, the natural causes do not match the actual temperature changes.

(b) The bottom chart shows the predicted impact on temperature of natural forcings combined with all the human-caused forcings in red, with the lighter shading showing the range for each year. Adding the human impact to the underlying natural variations produces a predicted temperature fluctuation that mirrors the actual rise almost perfectly.

Source: [10: fig. 9.5b]

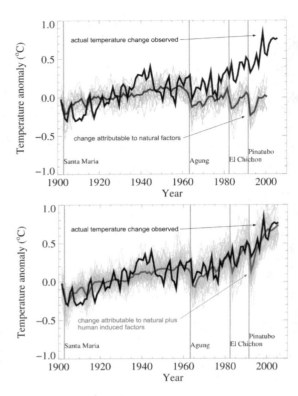

variation (anomaly) as recorded by instruments. The thin vertical bars are major volcanic eruptions, which add aerosol particles to the atmosphere and cool the Earth slightly by reflecting sunlight. Notice the slight dip in temperatures after each volcano. As we can see, all the natural forcings combined produce a likely temperature in AD 2000 that is identical to what is was in AD 1900. The natural forcing of solar variability, volcanic activity, and other factors were all taken into the calculation. But we know from our weather stations that the temperature climbed by an unprecedented three-fourths of a degree Celsius in the same period.

In Figure 3.8b we see a plot that mirrors the temperature changes nearly perfectly. In this figure, the human forcings on temperature are added to the natural forcings. So this line now collects the rapid rise in greenhouse gas emission and deforestation, among other human impacts, as well as solar variability and other natural causes. The result is a model that produces an accumulation of the same three-

fourths of a degree Celsius as the weather instruments collected. This closeness of fit between the hypothesis—that human activity explains a significant amount of global warming—and the results of the model are extremely persuasive.

Only when we calculate the amount of human causes of warming does the sum of the small natural forcing and the much larger human forcing add up to the actual temperature curve we have observed as fact. Climatologist Gabriele Hegerl of Duke University and her IPCC colleagues, along with Britain's Nicholas Stern, sum this up as follows:

> The fact that climate models are only able to reproduce observed global mean temperature changes over the 20th century when they include anthropogenic forcings, and that they fail to do so when they exclude anthropogenic forcings, is evidence for the influence of humans on global climate. [10]
>
> *IPCC, 2007*

I: What We Know About Climate

It is going to be very difficult to keep temperature increases down to between 2 and 3 degrees centigrade [3.6–5.4°F]. We should work very hard to do that. [24]

Nicholas Stern, 2006

The mean global surface temperature has warmed by about 0.74°C (about 1.3°F) between AD 1900 and 2000. That may not sound like much. But global average temperatures are rising now at more than 10 times the rate that they have risen since humans began forming societies around cities and agriculture and trade about 10,000 years ago. No other natural phenomena that might drive climate make sense. Sunspots happen every decade or so and are not more frequent now than before. The wobble of the Earth's rotation that places us slightly closer to the Sun happens very slowly and over thousands of years, so that is not the cause. Volcanic eruptions cause short-term cooling. Hence, the only remaining explanation is that humans have contributed to the pace of this temperature increase.[††]

In addition to the forcings model discussed above, there is a smoking gun that proves beyond reasonable doubt that human activity is forcing temperature disruption: the chemical fingerprint of the carbon isotopes we find in the atmosphere. The character of airborne carbon dioxide has changed in a way than can only be attributed to the burning of fossil fuel. A heavy form of carbon with one extra electron, the carbon-13 isotope, is less common in vegetation or fossil fuels formed from long-dead vegetation and is more abundant in carbon found in the oceans and in volcanic or geothermal emissions. The amount of carbon-13 in the atmosphere has been declining relative to other forms of carbon. This decrease in carbon-13 would happen if the sources with less carbon-13 were the cause of more of the carbon emissions. Such sources are fossil fuel combustion or vegetation

losses. Humans are clearly responsible for a very significant increase in fossil fuel combustion. Humans are also responsible for deforestation and other changes in land use that release carbon dioxide into the atmosphere from vegetation. [13, FAQ 7.1]

Fossil fuel combustion is the single largest cause of higher CO_2 concentrations in the atmosphere. Curbing the amount of fossil fuel we burn is the most effective way to limit the growth of CO_2 concentrations in the atmosphere. Fossil fuels are essentially ancient sunlight, captured by plants, buried in swamps and wetlands, sealed long ago by tectonic movements of the Earth's crustal plates, and cooked under tremendous pressure and heat over millions of years. Coal, oil, and natural gas are called fossil fuels because they are derived from fossilized plant and animal material. The decaying organic matter under the most intense heat and pressure became solid lumps of almost pure carbon that we know as coal. Above the coal, we often find organic matter, cooked under lower heat and pressure, that turned into liquid petroleum oil. Above the oil, processed under even less heat and pressure, there is often natural gas (methane), caught when the decaying organic matter got trapped under the impermeable salt lid of an ancient seabed.

Each of these fossil fuels is a less concentrated form of carbon than the next. Burning any of these fossil fuels produces carbon dioxide and water vapor. The burning process bonds the carbon molecule in the coal, oil, or gas to the oxygen abundant in the air, which produces CO_2. The tiny particles of carbon and other impurities that do not burn become soot. But soot emission is easily captured. The emissions of the colorless, tasteless carbon dioxide and nitrous oxide gases are the climate culprits.

For example, a coal-fired power plant that produces the equivalent of 1 gigawatt of electricity emits 1,000 tons of carbon dioxide greenhouse gas per hour and about 75 tons of air pollutants and ash per hour. A house with a 200-ampere line at 110 volts may use a maximum of 22,000

[††]For more details on what can force climate, see the Chapter 3 online addendum.

INSIGHT 3: EMISSION SCENARIOS FOR THIS CENTURY

First published in 2000, the *IPCC Special Report on Emission Scenarios* (SRES) examines 40 distinct future emission scenarios by dividing the analysis into two different families, each of which is examined under two different conditions. Each scenario makes different assumptions for future greenhouse gas emissions, land use, and other forces that drive climate change. In the first family of scenarios, researchers assume different degrees of maximizing conventional economic growth (business as usual) versus maximizing environmental sustainable practices. Each of these sets of scenarios is further examined in terms of how integrated they would be worldwide (global unity in implementation) versus how they might play out in a more divided world with greater disparities between regions. The most positive assumptions of an ecologically friendly and globally unified approach yield a scenario ("B1" in SRES lingo) in which carbon dioxide emissions double between today and 2050 and fall to just below today's levels by 2100 (in gigatons of carbon per year). In the worst case, with business as usual, fossil-fuel-intensive assumptions yield a scenario ("A1FI" in SRES lingo) in which carbon dioxide emissions triple between today and 2050 and then slow down but continue to rise until quadrupling today's levels by 2100. While carbon dioxide emitted in gigatons of carbon per year is not exactly the same as its concentration in the atmosphere, under even the most optimistic scenarios the human-induced carbon output will double by the time the youngest of the post–World War II baby boomers reach the age of 90. [12]

watts. A gigawatt (GW) is 1 billion (10^9) watts, or enough electricity for 45,454 homes. With electrostatic precipitator ("smokestack scrubber") technology that removes particles from the effluent, the air pollutants can be reduced even further. But no smokestack scrubber yet exists to "unburn" the CO_2, H_2O, and nitrogen vapor. [29]

Together, electric utilities as energy suppliers and industrial facilities account for about half of all carbon dioxide emissions. Capture of carbon dioxide emissions is possible for electric power plants and many larger industrial facilities. Indeed, rules currently in place in the European Union target the lowering of emissions at thousands of these installations.

Even if every household plugged in a fully electric hybrid car each night as a replacement for a gasoline-powered car, we may see no net reduction in carbon dioxide emissions if the electricity the hybrid consumes is generated by

a power plant fueled by coal, oil, or natural gas. Therefore, examining the energy mix in electricity production is a major task that we will explore in later chapters.

While traditional particulate pollution has a local or regional impact (e.g., smog in Beijing, Los Angeles, or Mexico City), it rarely reaches high enough into the troposphere (lower atmosphere) to travel very far. One malicious impact of the carbon dioxide gas is that—once airborne at high altitude—it travels around the globe. Hence, carbon dioxide emitted anywhere becomes everyone's carbon dioxide.

Is a Doubling of Atmospheric CO_2 Concentrations Likely?

Unfortunately, yes. The US National Oceanic and Atmospheric Administration (NOAA) makes headlines every year when it announces the annual atmospheric carbon dioxide con-

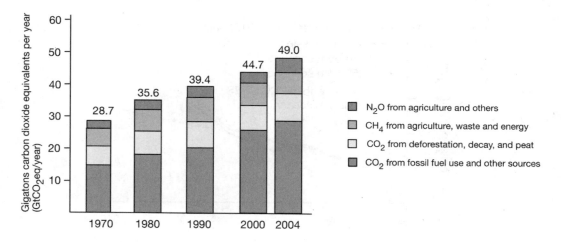

FIGURE 3.9 Global rise of greenhouse gas emissions: 1970–2004

In this bar graphic, we see the increase in human induced greenhouse gases from 1970 to 2004 expressed as gigatons of carbon dioxide equivalent per year. A gigaton is one billion (10^9) tons. Source: Adapted from [1: fig. SPM]

centrations at new record highs. Atmospheric carbon dioxide has been rising without a break since the industrial era required the burning of massive amounts of coal, initially. Within 100 years, another fossil fuel, petroleum oil, which was easier to extract and ideal for use in motor vehicles, was added as a fuel. Shortly thereafter, natural gas use for heat and electricity generation became widespread, as underground pipeline networks made its delivery to homes and industry even easier than transporting either coal or petroleum. [25]

Not only are we continuing to increase our emissions, we are doing it faster every year—exactly the opposite of what is needed for climate stabilization. Since 2000, the increase in atmospheric carbon dioxide concentrations has jumped between 1.5 and 3 parts per million each year (see Figure 3.9). The rate of the increase since 2000 is increasing as well. Human-induced carbon dioxide emissions have been growing about four times faster since 2000 than during the previous decade, despite efforts to curb emissions in a number of countries (Figure 3.10). Emissions from the combustion of fossil fuel and land use change reached 10 billion tons of carbon in 2007. Natural carbon diox-

ide sinks are growing (good news) but doing so slower than growth in atmospheric carbon dioxide, which has been increasing at 2 ppm since 2000, or 33% faster than the previous 20 years. [7]

If this pattern continues, we will add another 200 ppm of atmospheric carbon dioxide concentration above our current 385 ppm (see Figure 3.11). That means we would be reaching almost 600 ppm within the lifetime of today's younger Americans, which would set in motion catastrophic climate disruptions. Since the IPCC issued Climate Change 2007, ample actual observations show that matters are worsening faster than projected: sea-level rise is more rapid, global ice extent is shrinking more quickly, and biological systems are altering themselves more rapidly in direct response to upwardly creeping atmospheric temperatures from changes in its chemical composition. The Earth has experienced that level of greenhouse gas in our planet's atmosphere before—about 23 million years ago. But, of course, the world looked very different then: Humans were not present and did not emerge until the Earth began a long gradual cooling period. [3] Dr. Pep Canadell, executive director of the Global Carbon Project, puts it this way:

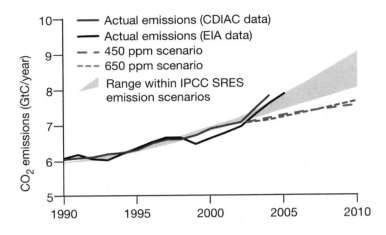

FIGURE 3.10 Global emissions growth accelerating since 2000

The growth trajectory in global fossil fuel emissions has accelerated since 2003 ahead of all projections. This is profoundly bad news for the prospect of stabilizing atmospheric conditions, and thereby climate. The solid lines here record the actual emission growth rates in percent per year for 1990 through 2007, as measured by the Carbon Dioxide Information Analysis Center and Energy Information Agency. Emissions had grown at about 2.2% between 2000 and 2003. But the growth trajectory has become even steeper from 2004 onward. In 2007, the growth rate was even higher than 2006, so the pattern has not abated. The straight lines on this graph represent a series of more and less optimistic emission scenarios that the IPCC projected in 2000. The most pessimistic of these (A1F1) was not pessimistic enough and is now outpaced by reality, as actual emission growth tracks higher than it. Source: Adapted from [7, 12, 22]

This new update of the carbon budget shows the acceleration of both CO_2 emissions and atmospheric accumulation are unprecedented and most astonishing during a decade of intense international developments to address climate change. [7]

Dr. Pep Canadell, 2008

How Much Warming Does Carbon Dioxide Doubling Cause?

By all accounts, we have a 60% likelihood that reaching a 580 to 600 ppm atmospheric carbon dioxide level will result in a rise of 2°C to 4°C (3.5°F to 7°F) in average surface temperature by the end of this century. Any rise in aver-age temperature leads to interlocking impacts. These impacts, already on the march, include ice and snow cover reduction, seasonal temperature changes, weather changes, species range changes, ocean circulation changes, and ocean acidification.

As ice and snow covers melt, not only do they release more freshwater into the world's oceans, but the smaller remaining white reflective surface area actually means that less heat is bounced up into space, and more heat is absorbed into the now newly exposed soil and open seawater, further heating up the surface and thus melting the edges of ice and snow fields even more rapidly. The rise of sea level associated with melting ice caps leads to the loss of coastal lands.

This alone is potentially very bad news for billions of people, as we will examine more deeply in Chapter 4. Eight of the world's 10 largest cities stand on ocean coasts. The rate of coastal population growth is increasing. Half of the world's people—more than were alive on the entire globe in 1950—live within 200 km (125 mi) of the ocean. Many of the world's largest cities and populations concentrate on coasts because ports are where goods and services are traded most heavily. [27]

Seasonal temperatures are likely to shift but are not distributed evenly in space or time. For example, the Arctic has warmed an average of 3°C to 4°C in the time that the globe as a whole has warmed 0.74°C. In addition, some widespread evidence shows that some regions now have fewer extremely cold days. That may not seem bad, but these cold snaps usually curb new, invasive insect and weed species from encroaching into such regions. Evidence also shows that the number of extremely hot days is on the rise. As sea level rises, more evaporation leads to more moisture in the atmosphere in the form of clouds and precipitation, and more heat in the form of energy for storms. Existing slow-moving weather patterns such as El Niño of the eastern Pacific are likely to happen more frequently and with more intensity. Warmer air holds more water. So it will rain less frequently, leading to more droughts. But when it does rain, it will rain harder, leading to more flooding.

The overall warming of the planet disrupts the distribution of plant and animal species. Most species—with the major exception of highly mobile *Homo sapiens*—have been relatively fixed in their locations for 10,000 years. Species occupy the ecological zones that maximize the opportunity for their nourishment and procreation.

FIGURE 3.11 online at ncse.org/climate solutions

Graphs depict the projected carbon dioxide emissions growth, and the consequent temperature increase, though the year 2100.

TABLE 3.1 What Does Climate Change Put at Risk?

Climate governs and therefore climate change affects:
- Availability of water
- Productivity of farms, forests, and fisheries
- Prevalence of oppressive heat and humidity
- Formation and dispersion of air pollutants
- Geography of disease
- Damages from storms, floods, droughts, and wildfires
- Property losses from sea level rise
- Expenditures on engineered environments
- Distribution and abundance of species

Ecological zones have many characteristics, but principal among them is the range between temperature highs and lows and the pattern of moisture present. If chilling frosts no longer occur, then plants that need cold snaps will diminish and be replaced by those that do not tolerate frosts. If rivers fluctuate to slightly warmer water temperature ranges, the cold-water fish species (such as trout) will die off or leave and the warm-water-tolerant species will move in.

Scientists estimate that the ranges of species are now moving toward the Earth's poles at a rate of about 6 km (3.7 miles) per year. Tropical plants and animals are marching steadily northward up the North American continent. Cold-loving plants and animals are retreating to ever-smaller northern territories or higher elevations, until they run out of places to go. [21] See also Insight 5, "Running Out of Mountain," in Chapter 5.

What Is "Dangerous" Here?

In ratifying the UN Framework Convention on Climate Change, all the signatory nations agreed to prevent "dangerous anthropogenic interference in the climate system." In January 2008, John Holdren, then chair of the board of the American Association for the Advancement of Science, now science advisor to President Obama, told his audience of 1,000 environmental scientists and policymakers in Washington, DC at the 2008 National Conference on Science,

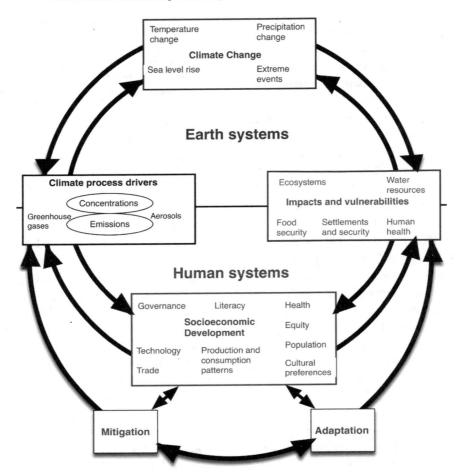

FIGURE 3.12 Human system and the Earth's climate process systems

Human systems interact with Earth systems in self-reinforcing feedback loops. Human drivers on climate change are numerous. These drivers cause physical changes in the Earth systems' climate process that have far-reaching effects on the life systems upon which human survival depends, from the natural systems that provide clean air and water to those that spark disease and famine. Source: Adapted from [2: fig. I.1]

Policy and the Environment, "The world is already experiencing 'dangerous anthropogenic interference in the climate system.' The question now is whether we can avoid catastrophic interference." [11]

National Academy of Sciences president Ralph Cicerone posed this question to the 1,000 attendees of the 2006 National Conference: What is "dangerous"? He offered the suggestion that *dangerous* may mean irreversible damages and a rate of disruption that is faster than the rate of adaptation. He also posed the question, Who should define *dangerous*? Cicerone pointed out that scientists and elected officials are both valid options; however, employing scientists in the task may not be appropriate, because of the value judgments that are implicit in the decision. Conversely, he stated, employing elected officials in the task may not be appropriate, because of the scientific knowledge necessary to make such a decision. [20]

National Aeronautics and Space Administration climate scientist James Hansen and his colleagues wrote the following, just 18 months earlier: "We conclude that global warming of more than about 1°C, relative to 2000, will constitute 'dangerous' climate change as judged from likely effects on sea level and extermination of species." The tropical Pacific is a primary driver of the global atmosphere and ocean. The tropical Pacific atmosphere-ocean system is the main source of heat transported by both the Pacific and Atlantic oceans. And the Pacific will be the primary recipient of warming Arctic water through the Bering Straits. Hansen and his team point out the warming detected in the vast and critical western Pacific appears to place the planet's thermometer within 1°C of as warm as it has been in about 1 million years. And, chillingly, Hansen points out that the warming trend we are on has already accelerated. [9]

A rise of more than 2 degrees above today's average temperature will put us in territory no human has seen: The last time the Earth was that warm was 3 million years ago, with sea levels 25 to 35 meters (80 to 130 feet) higher than today. If the polar ice caps reach a melting point of no return, then catastrophic physical changes will be set in motion, causing extinctions as plant and animal species hit their ecological temperature limits. [9] Even far short of that temperature, we can expect an increase in extinction rates well above the already unusually high rate of worldwide extinctions underway today. [26]

The interlocking effects of all the Earth system components—from cloud dynamics to jet stream alterations and feedback loops between melting ice, heat absorption, and ocean currents—make constructing a model challenging. But nonetheless, we can begin to see the fuller picture now of how the natural and human systems act upon each other in driving climate disruptions.

In sum, carbon in our atmosphere makes our planet hospitable to life. The Earth operates in an energy balance based on a very slow change—if any—in the natural sources of greenhouse gases. But human activities have put enormous quantities of additional greenhouse gases into the atmosphere, setting in motion an unprecedented disruption of the climate systems upon which human society depends for food, shelter, and general well-being.

CONNECT THE DOTS

- Preindustrial atmospheric CO_2 concentrations were 280 ppm versus current atmospheric CO_2 concentrations of 385 ppm.

- A strong positive feedback loop exists between atmospheric concentrations of greenhouse gases and temperature.

- Global average temperatures are rising now at more than 10 times the rate that they have risen since humans began farming and building cities after the last ice age ended.

- The Earth is warmer now than it has been in 650,000 years.

- The correlation between the rising CO_2 concentration in the atmosphere and the rising mean surface temperature of the planet is so strong, and the greenhouse warming potential of carbon in the atmosphere is so well established, that the globe's recent accelerated warming has no other explanation.

- We do not have a clear threshold for the precise tipping point at which CO_2 concentrations become dangerous. But, . . .

- We can only slow down the rate of greenhouse gas accumulations if we take unprecedented, permanent, and radical actions.

- Even under the most optimistic scenarios, the human-induced carbon output will double from preindustrial levels by the time the youngest of the post–World War II baby boomers reach the age of 90.

- Doubling atmospheric levels of carbon dioxide to 580 to 600 ppm would result in a rise of 2°C to 4°C in average surface temperature by AD 2100. The last time the Earth was that warm—3 million years ago—sea level was 25 to 35 meters (80 to 130 feet) higher than today.

Online Resources

www.eoearth.org/article/Carbon
www.eoearth.org/article/Carbon_dioxide
www.eoearth.org/article/Carbon_cycle
www.eoearth.org/article/Coal
www.eoearth.org/article/Atmosphere_layers
www.eoearth.org/article/Ocean_acidification
www.eoearth.org/article/Carbon_capture_and_storage

See also extra content for Chapter 3 online at http://ncseonline.org/climatesolutions

Climate Solution Action Item

Action Item: 15 Can Stabilizing Population Help Stabilize Climate?

Works Cited and Consulted

[1]   Bernstein L, Bosch P, Canziani O, Chen Z, Christ R, Davidson O, Hare W, Huq S, Karoly D, Kattsov V, et al. (2007) Summary for Policymakers of the Synthesis Report (in *Climate Change 2007: Fourth Assessment Report of the Intergovernmental Panel on Climate Change*, 22 pp, eds Allali A, Bojariu R, Diaz S, Elgizouli I, Griggs D, Hawkins D, Hohmeyer O, Pateh Jallow BP, Kajfež-Bogataj L, Leary N, Lee H, Wratt D) ar4_syr_spm.pdf: www.ipcc.ch

[2]   Bernstein L, Bosch P, Canziani O, Chen Z, Christ R, Davidson O, Hare W, Huq S, Karoly D, Kattsov V, et al. (2007) Synthesis Report (in *Climate Change 2007: Fourth Assessment Report of the Intergovernmental Panel on Climate Change*, 74 pp, eds Allali A, Bojariu R, Diaz S, Elgizouli I, Griggs D, Hawkins D, Hohmeyer O, Pateh Jallow BP, Kajfež-Bogataj L, Leary N, Lee H, Wratt D) ar4_syr.pdf: www.ipcc.ch

[3]   Canadell JG, Le Quéré C, Raupach MR, Field CB, Buitenhuis ET, Ciais P, Conway TJ, Gillett NP, Houghton RA, Marland G (2007) *Contributions to accelerating atmospheric CO_2 growth from economic activity, carbon intensity, and efficiency of natural sinks*. Proceedings of the National Academy of Sciences 104(47):18866. www.pnas.org/cgi/content/abstract/104/47/18866

[4]   Denman KL, Brasseur G, Chidthaisong A, Ciais P, Cox PM, Dickinson RE, Hauglustaine D, Heinze C, Holland E, Jacob D (2007) Couplings Between Changes in the Climate System and Biogeochemistry (in *Climate Change 2007: The Physical Science Basis. Contribution of Working Group I to the Fourth Assessment Report of the Intergovernmental Panel on Climate Change*, 541–584, eds Solomon S, Qin D, Manning M, Chen Z, Marquis M, Averyt KB, Tignor MH, Miller L) www.ipcc.ch

[5]   El T (2008) Population Curve. *English Wikipedia.* http://commons.wikimedia.org/wiki/File:Population_curve.svg

[6]   Forster P, Ramaswamy V, Artaxo P, Berntsen T, Betts R, Fahey DW, Haywood J, Lean J, Lowe DC, Myhre G, et al. (2007) Changes in Atmospheric Constituents and in Radiative Forcing (in *Climate Change 2007: The Physical Science Basis. Contribution of Working Group I to the Fourth Assessment Report of the Intergovernmental Panel on Climate Change*, eds Solomon S, Qin D, Manning M, Chen Z, Marquis M, Averyt KB, Tignor MH, Miller L) www.ipcc.ch

[7]   Global Carbon Project (2008) Carbon Budget and Trends 2007 (read September 26, 2008). www.globalcarbonproject.org

[8]   Hansen J, Sato M, Kharecha P, Beerling D, Berner R, Masson-Delmotte V, Pagani M, Raymo M, Royer DL, Zachos JC (2008) *Target atmospheric CO_2: where should humanity aim?* physics.ao-ph http://arxiv.org/abs/0804.1126

[9]   Hansen J, Sato M, Ruedy R, Lo K, Lea DW, Medina-Elizade M (2006) *Global temperature change*. Proceedings of the National Academy of Sciences 103(39):14288. www.pnas.org/cgi/content/abstract/103/39/14288

[10] Hegerl GC, Zwiers FW, Braconnot P, Gillett NP, Luo Y, Marengo Orsini JA, Nicholls N, Penner JE, Stott PA (2007) Understanding and Attributing Climate Change (in *Climate Change 2007: The Physical Science Basis. Contribution of Working Group I to the Fourth Assessment Report of the Intergovernmental Panel on Climate Change*, eds Solomon S, Qin D, Manning M, Chen Z, Marquis M, Averyt KB, Tignor M, and Miller HL) www.ipcc.ch

[11] Holdren J (2008) Meeting the Climate Change Challenge: Eighth Annual John H. Chafee Memorial Lecture on Science and the Environment. *Climate Change Science and Solutions: Eighth National Conference on Science, Policy, and the Environment.* http://ncseonline.org/climatesolutions

[12] IPCC (2000) Emissions Scenarios (SRES) (in *Special Report: A Special Report of Working Group III of the Intergovernmental Panel on Climate Change*, 1–22, eds Nakicenovic N, Swart R) sres-en.pdf: www.ipcc.ch

[13] IPCC (2007) FAQ (WG1) (in *Climate Change 2007: The Physical Science Basis. Contribution of Working Group I to the Fourth Assessment Report of the Intergovernmental Panel on Climate Change*, eds Solomon S, Qin D, Manning M, Chen Z,

Marquis M, Averyt KB, Tignor M, Miller HL) AR4-WG1_FAQ-Brochure_HiRes.pdf: www.ipcc .ch

[14] Jansen E, Overpeck J, Briffa KR, Duplessy J-C, Joos F, Masson-Delmotte V, Olago D, Otto-Bliesner B, Peltier WR, Rahmstorf S, et al. (2007) Palaeoclimate (in *Climate Change 2007: The Physical Science Basis. Contribution of Working Group I to the Fourth Assessment Report of the Intergovernmental Panel on Climate Change*, eds Solomon S, Qin D, Manning M, Chen Z, Marquis M, Averyt KB, Tignor MH, Miller L) ar4-wg1-chapter6.pdf: www.ipcc.ch

[15] Loulergue L, Schilt A, Spahni R, Masson-Delmotte V, Blunier T, Lemieux B, Barnola JM, Raynaud D, Stocker TF, Chappellaz J (2008) *Orbital and millennial-scale features of atmospheric CH_4 over the past 800,000 years*. Nature 453(7193):383–386. www.nature.com/nature/journal/v453/n7193/abs/ nature06950.html

[16] Lüthi D, Le Floch M, Bereiter B, Blunier T, Barnola JM, Siegenthaler U, Raynaud D, Jouzel J, Fischer H, Kawamura K (2008) *High-resolution carbon dioxide concentration record 650,000–800,000 years before present*. Nature 453(7193):379–382. www .nature.com/nature/journal/v453/n7193/abs/ nature06949.html

[17] Mann ME, Zhang Z, Hughes MK, Bradley RS, Miller SK, Rutherford S, Ni F (2008) *Proxy-based reconstructions of hemispheric and global surface temperature variations over the past two millennia*. Proceedings of the National Academy of Sciences 105(36):13252–13257. www.pnas.org/content/105/ 36/13252.full

[18] McKibben B (2008) 350.org (read August 15, 2008). www.350.org

[19] Meehl GA, Stocker TF, Collins WD, Friedlingstein P, Gaye AT, Gregory JM, Kitoh A, Knutti R, Murphy JM, Noda A, et al. (2007) Global Climate Projections (in *Climate Change 2007: The Physical Science Basis. Contribution of Working Group I to the Fourth Assessment Report of the Intergovernmental Panel on Climate Change*, eds Solomon S, Qin D, Manning M, Chen Z, Marquis M, Averyt KB, Tignor MH, Miller L) ar4-wg1-chapter10.pdf: www .ipcc.ch

[20] NCSE (2008) Climate Change Science and Solutions. *National Conference on Science, Policy,*

and the Environment. http://ncseonline.org/ climatesolutions/

[21] Parmesan C, Yohe G (2003) *A globally coherent fingerprint of climate change impacts across natural systems*. Nature 421:doi:10.1038/nature01286. www .nature.com/nature/journal/v421/n6918/abs/ nature01286.html

[22] Raupach MR, Marland G, Ciais P, Le Quéré C, Canadell JG, Klepper G, Field CB (2007) *Global and regional drivers of accelerating CO_2 emissions*. Proceedings of the National Academy of Sciences 104(24):10288. www.pnas.org/cgi/content/ long/104/24/10288

[23] Segar D (2007) *Introduction to Ocean Sciences*. (W. W. Norton & Company, New York). www .wwnorton.com/college

[24] Stern N (2006) The Stern Review Report on the Economics of Climate Change. HM Treasury, London. www.hm-treasury.gov.uk/independent_ reviews/stern_review_economics_climate_ change/stern_review_report.cfm

[25] Tans P (2008) Trends in Atmospheric Carbon Dioxide NOAA/ESRL. *Global Monitoring Division*. US Department of Commerce, National Oceanic and Atmospheric Administration Earth System Research Laboratory (ESRL), Boulder, CO. www .esrl.noaa.gov/gmd/ccgg/trends/

[26] Thuiller W, Lavorel S, Araujo MB, Sykes MT, Prentice IC (2005) *Climate change threats to plant diversity in Europe*. Proceedings of the National Academy of Sciences 102(23):8245–8250. www .pnas.org/cgi/content/abstract/102/23/8245

[27] UN Oceans (2008) United Nations Atlas of the Oceans (read August 24, 2008). www.oceansatlas .org/cds_static/en/human_settlements_coast__ en_1877_all_1.html

[28] USGS (2005) Caribbean Tsunami and Earthquake Hazards Studies. *The Caribbean Program*. Woods Hole Science Center (read October 24, 2008). http://woodshole.er.usgs.gov/project-pages/ caribbean/atlantic+trench_Large.html

[29] Wolfson R (2008) *Energy, Environment, and Climate*, chap 5 (W. W. Norton & Company, New York). www.wwnorton.com/college/physics

CHAPTER 4

Rising Carbon, Rising Oceans

Sea level will rise if the ocean warms. . . . Analysis of the last half
century of temperature observations indicates that the ocean has
warmed in all basins. [2]

NATHANIEL BINDOFF and his IPCC colleagues, 2007

The eight inhabited islands of Tuvalu lie midway between Hawaii and Australia. Like thousands of other islands in Polynesia and Micronesia, Tuvalu's islands are low lying, formed by sediment trapped in coral atolls. It is one of the smallest and most remote countries on Earth. Although people have been living on the islands for thousand of years, Tuvalu became a fully independent nation within the British Commonwealth just 30 years ago and has a population today just under 12,000. In addition to income from the sale of beautiful stamps and coins, Tuvalu derives royalties from the lease of its ".tv" Internet domain name, with revenue of more than $2 million in 2006. [6]

Tuvaluans depend on coral lagoons for fish. But the coral reefs are under attack from the spread of the crown-of-thorns starfish. Tuvalu already relies on rain cachements for drinking water, as rising saltwater renders most of its groundwater undrinkable. Heavy storms already erode its beaches and flood its homes and roads. In 2000, the Tuvalu government

appealed to Australia and New Zealand to take in Tuvaluans if rising sea levels should make evacuation necessary. As young as Tuvalu is, it may be among the first nations whose very soils disappear under a rising sea. [6]

A Warmer Sea Is a Rising Sea

In the prior chapter, we examined the changes in the physical world that climate change will set in motion, such as more intense storms, and more frequent droughts. Many of us may not be alarmed yet about a warming atmosphere. But a rising sea is different. [35]

A sea rising for any reason—thermal expansion, storm surge, ice melt—is bad news for everyone everywhere. Places like Bangladesh's delta or New Orleans that are already at or below sea level will be inundated first, but the impacts are far reaching, as we will see in this chapter.

Most of us may not realize that sea levels will rise much more along some coasts than others, owing to the complex interactions of tides, water volume, and currents. [28] The Permanent Service for Mean Sea Level at the Proudman Oceanographic Laboratory in Liverpool, UK, has been collecting sea level change information from tide gauges since 1933 and has the world's most extensive records on sea level history. This research center reports, "Global-average sea level is believed to have risen by between 10-20 cm during the past century and best estimates are that it will rise by approximately 50 cm in the next 100 years." In short, sea level between now and 2100 will rise three times faster than in the prior century. [32] Other scientists, such as James Hansen and his colleagues, project a steeper rise in sea level of 1 meter (m) or more. [17] The precise projections will continue to be debated by specialists who are continually refining the very complex models used in such work. Figure 4.1 takes the sea level data in Figure 1.3 and adds the continued rise projected to 2100, showing the range of uncertainty. But, again, the research community unanimously agrees that the ocean level is rising and will continue to rise for the foreseeable future.

So where is this water coming from? Believe it or not, most of the actual rise in sea level is thought to be due to the rise in temperature of the water in the top surface layer of the ocean. When fluid is heated, it expands. Scientists call this effect thermal expansion.When the volume of water in the oceans increases, sea level rises. When the ocean warms up, the water level rises because the atmosphere above it offers less resistance than the seafloor. [2] Water is a very unusual liquid in that it expands both when it is heated and when it reaches a freezing point as ice. So as atmospheric warming is causing saltwater levels around the world to rise due to thermal expansion, the same

---

FIGURE 4.1 online at ncse.org/climate solutions

Satellite altimetry data show that global average sea levels have been rising at an increased rate in recent decades.

---

warming is also speeding up the melting of polar and glacial ice.

The Polar Express Is Slowing Down

We think of the Arctic polar cap as a frozen landscape. But it is not land at all. It is an ocean covered with a surface of ice that is constantly reshaping itself. The polar ice cap is the permanent mass of sea ice that forms a jagged circle around the North Pole, which used to cover about 70% of the Arctic Ocean. Driven by easterly currents and winds, the floating polar ice cap rotates in a clockwise motion with the North Pole at its center. It takes about 4 years to complete a single rotation. The ocean at the North Pole in the Arctic basin is deep, ranging from 2,500 to 4,400 m, or 8,200 to 14,400 feet (ft). [18] About 12 to 13 million square kilometers (km^2), or 4.6 to 5 million square miles (mi^2), of the Arctic basin are covered by ice in winter. Nearly all of the Arctic and sub-Arctic shores are icebound throughout the winter. By late summer, the amount of Arctic sea ice usually shrinks to about 9 million km^2 (3.5 million mi^2). Lately, the ice-free zone has been expanding. In the summer of 2007, the minimum ice extent dropped by half to about 4.2 million km^2 (1.6 million mi^2). [14] (See online Figure 4.2.) As the National Snow and Ice Data Center reported in 2008, "The extent of arctic ice in September, when extent is at its annual minimum, is decreasing at a rate of 7.7 percent per decade, which corresponds to approximately 1.4 million square kilometers (540,543 square miles). The Septembers of 2002 to 2004 showed dramatically lower arctic ice extent. This trend is a major sign of climate change in the polar regions and may be an indicator of the effects of global warming." The Arctic ice cover is now down to far less than 70% and falling fast, and this permanent ice is also becoming thinner very fast. [30] And that change has huge consequences for life on Earth.

In areas of the Arctic basin where surface currents are strong, open ice-free expanses called polynyas appear. The largest polynya,

North Water, is located at the head of Baffin Bay and usually has a surface area as large as Lake Superior. Compared with the surrounding air, the waters of a polynya are so warm that steam billows up. The average temperature of the waters within a polynya is about 0 degrees Celsius (°C), and it never dips below −2°C, the freezing point of seawater. Polynyas allow heat from warm ocean currents to escape; this exchange is important in the regulation of the Earth's temperature. The release of heat from the polynya water mirrors the energy release that happens when the warmer Atlantic waters arrive off Greenland. Paul Hebert, director of the Biodiversity Institute of Ontario, writes, "Polynyas teem with animal and plant life. It is only here, where the sea ice is absent, that the sun's energy directly reaches the waters. . . . Phytoplankton in the polynya use this energy to produce a nutrient rich grazing area for zooplankton. Feeding on these small animals are whales and fishes that in turn feed seals, walruses and polar bears." [18]

The North Atlantic Current provides about 60% of the inflow to the Arctic Ocean bringing warmer water from the Atlantic Ocean. This warmer water, part of what we call the Gulf Stream, moderates the temperature of eastern North America, northern Europe, and Greenland. Once the heat escapes this water, the steady conveyor of colder water sinks to the seafloor in the North Atlantic and returns to the south. [18]

When cold Arctic air freezes seawater, the ice on the surface contains only freshwater, and there is a layer of fresher, less salty water

---

FIGURE 4.2 online at ncse.org/climate solutions

(A) Arctic sea ice extent reached record lows in 2007. The extent of summer melting is accelerating faster than at any period since tracking this phenomenon began. For the latest data see http://nsidc.org/arcticseaicenews/
(B) Seawater exposed in a polyna heats up and creates steam in the cold Arctic air above.

about 50 m deep just below the ice. Below this fresh surface water layer lies a denser, saltier water layer that may be over 200 m thick. In the critical North Atlantic area, this colder, saltier middle layer acts as insulation for the surface ice, keeping the warmer, saltier layer arriving from the Atlantic from melting the pack ice at the surface. [27] So the pack ice, which moderates our temperatures, depends on both stable air temperatures and stable deep ocean temperatures. While we know that the average global air temperatures have been rising steadily (and rising more at the poles than the equator), we have also recently discovered that the North Atlantic currents may be becoming less salty.

This continuous loop of circulating ocean water was last disrupted about 8,500 years ago, possibly by the sudden influx of freshwater behind ice dams that reached a breaking point. At that point, some parts of the world become much drier, others much colder, and all in all, climate becomes more variable.

By examining deep seafloor sediment, researchers are able to establish the likely patterns of past temperatures and currents in the oceans. And that research points to some alarming data. The seas bordering the North Atlantic have become noticeably less salty in the past 40 years, and especially since 1990. "This is the largest and most dramatic oceanic change ever measured in the era of modern instruments," declares Bob Dickson of the Centre for Environment, Fisheries, and Aquaculture Science, "This has resulted in a freshening of the deep ocean in the North Atlantic, which in the past disrupted the Ocean Conveyor and caused abrupt climate changes." [8] Woods Hole Oceanographic Institution sci-

---

FIGURE 4.3 online at ncse.org/climate solutions

A global system of currents called the ocean conveyor carries warm surface waters from the tropics to the high latitudes, where the water cools. There cold water sinks and flows back toward the equator in the deep ocean.

entists have launched a variety of missions to explore how global climate change is affecting the Arctic and how changes in the Arctic, in turn, could spill out and cause further climate change well beyond the polar region through the ocean conveyor, shown in Figure 4.3.

> If too much fresh water enters the North Atlantic, its waters could stop sinking. The Conveyor would cease. Heat-bearing Gulf Stream waters would no longer flow into the North Atlantic, and European and North American winters would become more severe. [12]
>
> *Robert Gagosian, President,*
> *Woods Hole Oceanographic Institution*

The summer of 2007 set a record for the minimum sea ice extent in the Arctic. Michael C. MacCracken, chief scientist for climate change programs at the Climate Institute, reports that reductions in Arctic Sea ice are already having and will have significant effects inside the region. Access to the region will increase, especially for shipping and mineral rights, leading to sovereignty claims* and challenges for ensuring safety and environmental quality. Adverse impacts will threaten Arctic ecosystems and species (e.g., polar bears). Sea ice loss allows increased coastal erosion, which will force relocation of about 150 indigenous communities. The melting of permafrost is weakening soils and foundations for buildings and pipelines. [1]

It Is Not Just the Polar Bears

While polar bears have become the iconic symbol of the threat posed by climatic disruption, the Arctic is also home to approximately 4 million people. Sarah James, a member of the

---

*All parties to the United Nations Convention on the Law of the Sea (LOS) must file claims by 2009. The United States is the only major nation not to have ratified the LOS, as of December 2008. The United States cannot make legitimate claims under the treaty unless it ratifies the treaty. (http://untreaty.un.org)

Gwich'in Steering Committee, described her people's dependence upon nature:

> The Gwich'in are the northernmost Indian Nation living in fifteen small villages scattered across a vast area extending from northeast Alaska in the U.S. to the northern Yukon and Northwest Territories in Canada. There are about nine thousand Gwich'in people who currently make their home on or near the migratory route of the Porcupine River Caribou Herd in communities in Alaska, Yukon, and the Northwest Territories. The word "Gwich'in" means "people of the land", and it refers to a people who have lived in the Arctic since before the political boundaries that now transect the Gwich'in homelands were drawn on maps dividing Alaska and Canada. Oral tradition indicates that the Gwich'in have occupied this area since time immemorial, or, according to conventional belief, for as long as 20,000 years. [19]

The Gwich'in are "caribou people." About 75% of their food derives from caribou; their parkas and huts are made of caribou; their songs and dances are about caribou. They use the term "caribou skin hut" to refer to their home, their village, the Earth, the universe. According to Ms. James, "We believe that the Creator put us where we are today to take care of this part of the world. We did well by keeping it the way it is." [20]

Yet, people in other parts of the world have not taken care of their parts of the world, and climate change means that the people of the North are unable to keep the world the way it is. Ms. James told the Eighth National Conference on Science, Policy, and the Environment:

> Global warming and climate change are real new to the Arctic. Last year it didn't get cold like it used to (only 40 degrees F below zero!). There were fast changes in the weather. Last year there was no snow until almost Christmas. It displaced the animals. We had problems with wolves coming into

Global water cycle

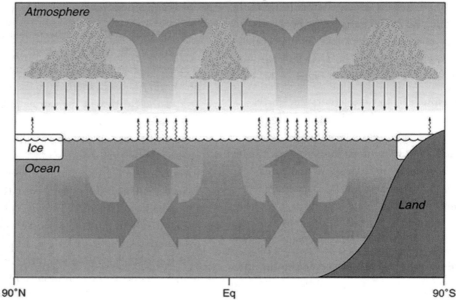

FIGURE 4.4 New view of the global water cycle

This diagram is an oceanographer's view of the global water cycle. It is drawn as a north-south section and shows the atmosphere carrying water vapor from evaporation to precipitation regions. In contact with both the oceans and the continents, the thin, transparent atmosphere acts as a conduit connecting the terrestrial and oceanic components of the water cycle. This movement of water and the energy it carries is called flux. The oceans function as a reservoir and buffer in the planetary circulation of water. Storing 23 times more water than is stored on land and a million times the water in the atmosphere, the ocean has air-sea fluxes that are many times larger than the terrestrial equivalents. The temperature of the ocean's surface is a very important driver of global climate. This diagram accurately depicts the pattern of evaporation at mid latitudes and precipitation at high and low latitudes accurately depicts the dominance of the ocean-atmosphere processes, and it delegates land processes to a more suitable minor role. It also shows the complementary return flows induced in the ocean. These flows redistribute water in the ocean, moderating the rise and fall of sea levels on our coasts. Source: [33], illustration by Jack Cook

Arctic Village. When the snow is late, the ice and ground freeze deeper, because snow provides insulation. Without snow, it freezes out the animals and disturbs the peace. The wolves can't keep up with their prey. They hurt their paws on the hard ground. They are attracted to the village and kill sled dogs. We have to be careful when we have to get meat. [20]

Ms. James notes that the problem is "addiction to greed, waste, and all that. We need to get back to a simple way. Clean air, water, land, and life is the only way to have peace. Maybe we'll do a better job of housekeeping the whole universe." [20]

The world system is interconnected. A warmer Arctic will also have significant impacts on mid-latitude climate. Up to now, the Arctic

I: What We Know About Climate

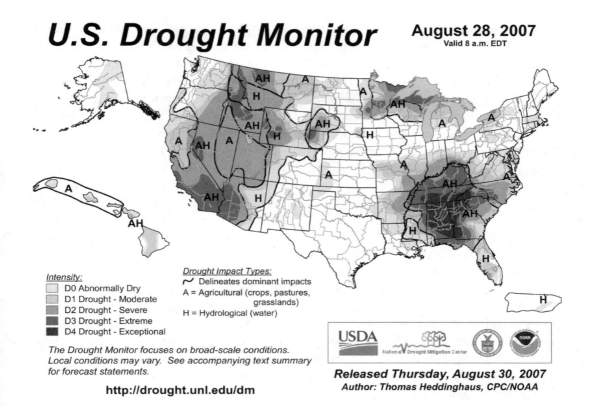

U.S. Drought Monitor

August 28, 2007
Valid 8 a.m. EDT

Intensity:
- D0 Abnormally Dry
- D1 Drought - Moderate
- D2 Drought - Severe
- D3 Drought - Extreme
- D4 Drought - Exceptional

Drought Impact Types:
~ Delineates dominant impacts
A = Agricultural (crops, pastures, grasslands)
H = Hydrological (water)

The Drought Monitor focuses on broad-scale conditions. Local conditions may vary. See accompanying text summary for forecast statements.

http://drought.unl.edu/dm

USDA National Drought Mitigation Center

Released Thursday, August 30, 2007
Author: Thomas Heddinghaus, CPC/NOAA

FIGURE 4.5 Persistent drought in the southeastern United States

As the cold air masses from Canada diminish, the cold fronts are no longer strong enough to push over the Appalachian Mountains in the south where they used to meet the moist warm air from the Gulf of Mexico. Hence, fewer rain events occur and drought ensues. Source: [9]

generated enough cold air that the intersection of cold air meeting warmer air from the tropics hovered near the coast of the Gulf of Mexico. "That no longer seems to be the case," explains MacCracken, "through the fall and early winter. With less cold air coming out of the Arctic and northern Canada, tropical air pushes north." Until Arctic sea ice 1–2 m thick insulates the air from the ocean, really cold winter air masses cannot form, and warm, moist air pushes north into the United States; the resulting clash can yield violent weather. [26] For decades, American school children were shown the water cycle in which the picture was largely land based. That is where the people are. But that is not where most of the water is. We can see how these water cycle fluxes work in Figure 4.4 by using Woods Hole

oceanographer Ray Schmitt's reconceptualization of the traditional land-based water cycle to an ocean-based water cycle.

But what does the water cycle mean for us on land? Let's look at what has been happening along the eastern seaboard of the United States in the past few years. Warm-season thunderstorms require the presence of warm, moist air arriving from the Gulf of Mexico and Caribbean, plus a trigger such as a cold front arriving from northern Canada. As the Arctic warms, fewer cold fronts build up to move south. Weaker cold fronts get blocked by the Appalachians. This prevents the two air streams from converging, so fewer thunderstorms develop over the land just to the south and east of the Appalachians, leaving their southeastern side drier and hoping for rain. This disruption triggered by a warming Arctic air mass may explain the prolonged drought that Georgia and much of the southeastern United States experienced in 2007 and 2008. The University of Nebraska hosts the US Drought Monitor website at which browsers may visit online archives to pull up maps of past conditions, as Figure 4.5 shows for drought faced in 2007 by residents, plants, and animals of the Southeast.

Greenland's Uncertain Future Prospects

In addition to the sea ice that is diminishing, the land-bound ice in the Arctic is also in trouble. The area of Greenland experiencing summer melting has been increasing significantly. The melt area on Greenland in 2007 was about 10% larger than in 2005, as part of the long-term trend of expanding melt areas. The 2007 report of the Intergovernmental Panel on Climate Change (IPCC) projected that sea level would rise by about 0.2 to 0.5 m, or 8 to 20 inches, by 2100, excluding rapid dynamical changes in ice flow and not including meltwater from mountain glaciers that are not in the polar zones. Tide gauge and satellite data presented by IPCC showed in 2007 that the rate of sea level rise is increasing. Newer satellite imagery since 2007 shows a further acceleration, that the ice is disappearing even faster than the upper range of the 2007 IPCC projections (see Figure 4.1). [29]

Sea levels are rising now at about twice the average rate of rise for the 20th century. "If this rate of rise continues, with no greater additions from Greenland and Antarctic Ice Sheets," warns Michael MacCracken, "We are at midpoint of IPCC projection." [26] That begs a question: How well do the IPCC projections of sea level rise compare to observations? Sea level is now rising at the very upper limits of the IPCC estimates made just a few years ago in 2001, and these estimates are now under a state of revision.

The Greenland ice sheet is massive and extremely vulnerable. At its thickest point, Greenland's ice is about 3,000 m (almost 2

miles) above the land underneath. If it melted in its entirety, sea level would rise about 6 to 7 m (20 to 23 ft). Most maps give the impression that Greenland is very big. But Greenland is only about one-fourth the size of Brazil or Australia and close in size to Libya or Mexico. Greenland's annual average temperature is projected to rise by more than the global average. By the late 21st century, Greenland may be 3°C to 6°C warmer than in the late 20th century. Greenland's underlying topography is quite vulnerable. Contrary to the prevailing view, much of the Greenland ice sheet in interior areas is below sea level as a vast shallow bathtub. The land has been depressed by the weight of the ice. So ocean waters can carry heat underneath the ice and help lift the ice sheet. In addition, fjords connect the interior ice sheet to the surrounding seas along the west and northern coasts. These fjords offer seawater a path toward the "bathtub," where the water enables more rapid movement of the ice from the interior to the ocean.

In other words, vast swaths of Greenland are currently below sea level but for the ice on top of them. National Aeronautics and Space Administration satellite data show that the most rapidly melting areas are located in the extreme northern reaches of Greenland. Why? No one knows for sure, but a simple reason may be that those northern reaches include areas where the underlying land is at sea level. So the days during which ice melts increase in number due to both a warmer atmosphere above and intruding sea water below. [36] Bindoff and his IPCC coauthors say:

All of these observations taken together give high confidence that the ocean state has changed, that the spatial distribution of the changes is consistent with the large-scale ocean circulation and that these changes are in response to changed ocean surface conditions. [2]

Nathaniel Bindoff and IPCC, 2007

The IPCC 2007 report contains the helpful schematic representation shown in Figure 4.6, of the ocean undergoing temperature rise, as well as a decrease in pH (a rise in acidity). In the North Atlantic, the warming is penetrating deeper than in the Pacific, Indian, and Southern oceans. The main reason is that the Arctic Ocean has deeper seafloor channels connecting to the North Atlantic on either side of Greenland than it does connecting to the Pacific through the relatively shallow Bering Straits.

Historically, about 20,000 years ago with the last glacial maximum extent, the Earth was about 6°C (11°F) cooler and sea level was about 120 m lower than today. About 125,000 years ago (in the Eemian period), the global mean temperature was about 1°C (1.8°F) warmer than today and Greenland was about 3°C to 4°C (5°F to 6°F) warmer than at present. At that time, reductions in polar ice volume led to a sea level rise of about 3 to 5 m (13 to 20 ft). In plain English, for every degree that average global temperature rises, sea level will rise by 10 to 20 m (33 to 66 ft).

Climatic evidence suggests that the equilibrium sensitivity of sea level to global average temperature is between 10 and 20 meters per degree Celsius. [26]
Michael MacCracken of The Climate Institute

How Does the Ocean Store So Much Carbon?

The ocean covers over 70% of planet Earth's surface. While we continue to make new discoveries about our ocean each year, we already know a great deal about the critical role the ocean plays in creating the planet's climate. The ocean affects climate by transporting vast volumes of warmer surface water and colder deep water around the globe. The vast surface of the ocean is

---

FIGURE 4.6 online at ncse.org/climate solutions

Observed changes in the state of the Atlantic Ocean include changes in temperature, salinity, sea level, sea ice, and biogeochemical cycles.

TABLE 4.1 Relative Abundances of Organic Carbon in the Oceans in Living and Nonliving Forms

Organic matter form	Percent of total organic carbon in oceans
Dissolved organic matter	94.9%
Nonliving particulate organic matter	5.0%
Phytoplankton	0.1%
Zooplankton	0.01%
Fishes	0.001%

Source: [34: table 14.2]

continually interacting with the air above it. The ocean's surface water absorbs heat in the tropical zones and transports this heat with its currents toward the poles, where it releases the heat to the atmosphere in creating water vapor, which falls elsewhere as rain or snow. This transport of heat energy—in conjunction with the prevailing wind and current patterns—shapes the planet's climate zones. Wind and water determine, for example, where it will be colder or warmer or drier or wetter, in short, where monsoons will be common and where deserts will form. As the atmosphere's composition changes, with rising carbon dioxide as a principal culprit, so too the ocean below it is changing.

The ocean does far more than move heat around. Nature stores the building block of life, carbon, everywhere. But it is the ocean's intermediate and deep water that stores the vast majority of the planet's carbon (84%).[†] In fact, it is the tiniest creatures of the sea, the mighty phytoplankton, that do the most carbon storage work. As we see in Table 4.1, almost all of the carbon found in the ocean is dissolved as

---

[†]But it may surprise you to learn that underground oil and gas deposits contain only a tiny fraction of the planet's carbon (1%), and coal contains not that much more (6%). Living vegetation also represents a tiny fraction (1%). Slightly more is dissolved in the ocean's surface water (2%), is airborne as gas in the atmosphere (2%), and is found in solid compounds in soil and organic matter (3%). There most carbon is locked away as dissolved organic matter. Scientists call such storage a carbon sink.

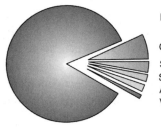

Deep ocean 85%

Coal, oil, and gas 7%
Soil 3%
Surface ocean 2%
Atmosphere 2%
Vegetation 1%

FIGURE 4.7 Earth carbon reservoirs

A comparison of the Earth carbon reservoirs shows the deep ocean reservoir dwarfs all other carbon reservoirs. The biological pump is the primary mechanism for sequestering carbon in the deep ocean. (A petagram is 10 to the 15th grams, or 10,000 billion kilograms—a lot! 1 PgC = 1 GtC = 3.66 GtCO$_2$e.) Source: [38]

organic matter, that is, formerly living matter dissolved into the seawater.

The Mighty Phyto

As a large carbon sink, or repository for excess carbon, ocean water dissolves carbon dioxide and houses the living creatures that convert carbon dioxide to other useful compounds. All the marine organisms that do not swim or do not live on the ocean floor are called plankton. The biggest plankton are animals called zooplankton, which include herbivores, carnivores, and omnivores. The smallest and most abundant plankton are plants (algae) called phytoplankton, which photosynthesize. Photosynthesis is the chemical process that plants use to convert light energy, water, and carbon dioxide into chemical energy for growing and living. Phytoplankton are too small to be seen by the unassisted eye—much smaller than 1 millimeter in diameter. In the ocean's surface water where the sunlight for photosynthesis is strongest, there may be as many as a billion individual phytoplankton per liter. Scientists have identified tens of thousands of species of phytoplankton, but more are discovered each year. The ultraphytoplankton species are so tiny that they escaped discovery until the 1990s. [34]

Plankton have a larger effect on climate than any single other process or group of organisms. Plankton are critical foodstock for larger creatures such as the sea butterfly, which in turn are food for larger species such as fish. Plankton grow, molt, excrete, and die—taking carbon as

dioxide (two oxygen atoms to one of carbon) and as carbonate (three oxygen atoms to one of carbon) with them to the deep ocean. In deep water, marine bacteria and fungi continue the chemical conversion process until producing water and carbon dioxide. As dead plankton fall toward the ocean floor, they eventually form sedimentary layers of shale, storing this carbon for millions of years, until the movement of the Earth's crustal plates might push the ocean floor skyward. For example, the striking White Cliffs of Dover are composed of chalk (pure white calcium carbonate) from the skeletons of coral, sponges, and other small marine species that accumulated on the ocean floor over 130 million years ago. We can see in Figure 4.7, that 85% of all the carbon on Earth is sequestered safely in the deep ocean waters and floor.

The biological pump, represented by the mighty phytoplankton, efficiently transfers carbon to the deep ocean. Within the global carbon cycle, 45% of the annual carbon turnover of 750 billion tons is driven by the primary productivity of ocean phytoplankton. As these organisms bloom and mature, they can be eaten or die. A significant fraction of the dead organisms or fecal pellets aggregate into falling particles and sink into the deep ocean. Plankton also form the base of the marine food chain. At least, 90% of all marine life relies on plankton, including all the creatures that absorb carbon as carbonate to build their shells. But the plankton's biological life cycle is highly dependent on the physics of water chemistry and temperature. [38] Figure 4.8 offers a simple schematic of the multi-

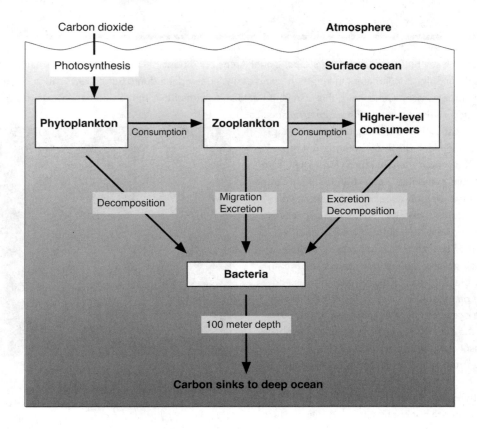

Ocean's biological pump

FIGURE 4.8 Ocean's biological pump

The biological pump schematic shows pathways for carbon into the deep ocean. This is a natural process by which plankton grow at the surface and then lose buoyancy after they die. Carbon is exported to the deep ocean in the "marine snow" composed of sinking plankton bodies and the fecal pellets from higher-level consumers that eat plankton. Source: [38]

faceted, biologically driven action of the ocean's natural carbon sink.

Margaret Leinen, chief scientific officer of Climos, describes the importance of the biological pump, as follows: "It has been operating very efficiently since the beginning of photosynthesis." But Leinen notes, "Even if we eliminate all of the carbon dioxide emissions, in other words if we become carbon neutral, we will continue to observe impacts upon calcification." [25] More carbon emission means more acid in the ocean, and that means less calcium for the ocean's inhabitants.

How is the ocean reacting to an ever-increasing load of carbon? Alarming data from the Global Carbon Project show that carbon emissions appear to be increasing on a tripling track, as opposed to the doubling track that the IPCC had used for many of their predictions. Data on the uptake of carbon by land indicate that land is a poorer carbon sink than once believed (as deforestation spreads) and that the ocean has been declining in its ability as a carbon sink as a result of human causes.

The Global Carbon Projects puts it plainly: "The efficiency of natural sinks has decreased by

INSIGHT 4: THE SEA BUTTERFLY

Meet the sea butterfly, *Limacina helicina*. Under Arctic conditions, this creature is quite small, reaching only 15 millimeters in length. It could easily hide under a dime. But it grows much larger in warmer waters. Genetically, it is a snail. It floats and swims freely in the water and is carried along with the currents. Some *Limacina* species have lost their shells and gills through adaptation to life in the water column. But most rely on seawater chemistry to form shell from calcium carbonate. Active hunters, they can deploy a web of mucus that gathers mostly plankton but also bacteria, small crustaceans, gastropod larvae, dinoflagellates, and diatoms. The web, which appears as wings in photographs, can be many times larger than the creature itself. These small animals and the plankton they feed on form the basis for the ocean's capacity to absorb carbon. But as the ocean becomes more acidic, the shells dissolve or fail to form. Researcher Gretchen Hoffman of the University of California, Santa Barbara, calls the sea butterfly the potato chip of the ocean for its role in the food web. "These animals are not charismatic, but they are talking to us just as much as penguins or polar bears," explains Hoffman. "They are harbingers of change. It's possible that by 2050 they may not be able to make a shell anymore. If we lose these organisms, the impact on the food chain will be catastrophic." [13]

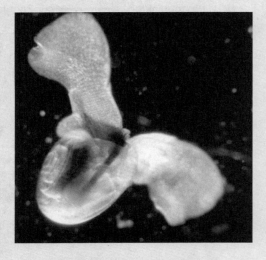

FIGURE 4.9 **Arctic sea butterfly**

The Arctic waters are surprisingly rich in plant and animal life, including this sea butterfly. Mats of plants form on the underside of ice floes. Planktonic species, while partly endemic to the Arctic, are believed to be mostly derived from Atlantic origins rather than Pacific, despite considerable inflow of Pacific species through the Bering Strait. The shallow Bering Strait bars any deep-water Pacific species from entering the Arctic Ocean. On the Atlantic front, both deep-water and shallow-water species are able to migrate to the Arctic. Source: [3]

From sea butterflies to sea urchins, as marine invertebrates adapt to increasing acidity in the water, their larvae recalibrate their metabolism in order to still be able to make shells from the remaining calcium. But the price they pay is significant. This physiological adjustment means the adult animals are both smaller and less able to withstand the warmer waters. "These observations suggest that the 'double jeopardy' situation—warming and acidifying seas—will be a complex environment for future marine organisms," concludes Hoffman. [13]

10% over the last 50 years (and will continue to do so in the future), implying that the longer we wait to reduce emissions, the larger the cuts needed to stabilize atmospheric carbon dioxide. Fifty years ago, for every ton of carbon dioxide emitted to the atmosphere, natural sinks removed 600 kilograms. Currently, the sinks are removing only 550 kilograms and this amount is falling. All of these changes characterize a carbon cycle that is generating stronger climate forcing and sooner than expected." The longer we wait to reduce emissions, the larger the cuts needed to stabilize atmospheric carbon dioxide. [16]

Why Is the Ocean Losing Its Capacity as a Carbon Sink?

Although biology plays a leading role in retaining carbon in the oceans, the reasons for the declining efficiency of the ocean sink lie primarily in the realm of physics. The Southern Ocean is where large quantities of carbon enter the ocean. In Leinen's words, "The various climate changes that we've seen, driven by carbon dioxide emissions and temperature change have strengthened winds around Antarctica, deepening the amount of winter circulation and winter mixing." These changes have permitted the ocean to release carbon, rather than hold it, by bringing up some of this carbon that the biological pump has pushed into deep water. The strengthening of the winds is attributed to both global warming at the poles and the ozone hole hovering over Antarctica. [24]

The result of the decline in the ocean's efficiency as a sink is significant. Global Carbon Project research shows that although most (65%) of the increase in carbon in the atmosphere can be attributed to the growing global economy, about one-fifth (18%) is due to reductions in the ability of natural sinks, including the ocean, and slightly less (17%) is due to a recent disturbing rise in the carbon intensity of the world's economies—a topic covered in Chapter 9. [5]

Increased carbon emissions affect how acidic the ocean becomes. As levels of dissolved carbon increase in the ocean water, the ratio of bicarbonate to carbonate changes, making the ocean water more acidic. When carbon dioxide (CO_2) dissolves in seawater, an unstable compound, carbonic acid (H_2CO_3), is formed.‡ Ocean organisms that build calcium-based shells convert the resulting bicarbonate into calcium carbonate. More hydrogen ions mean less salty and more acidic ocean water. As ocean salinity levels fall, so does the concentration of the carbonate ion. The carbonate concentrations drop, and the carbonate molecules then become vulnerable to dissolving and not being available for shell building. The ocean has not been able to equilibrate to a normal pH.§

These changes have happened in the past. But the current atmospheric concentrations of carbon dioxide are rising much more rapidly than historically was the case. An acidifying ocean reduces the calcium available for the most critical creatures that rely on precipitated calcium carbonate and other materials to form their skeletons. Corals are included in this group, and observations of "bleached" and dying coral have already become widespread in the world's oceans. Calcification of coral is projected to decrease by 10% to 30% under doubled carbon dioxide concentrations. [15, 22] Laboratory studies of corals in artificially doubled atmospheric carbon dioxide show this same trend. [23]

Leinen emphasizes that these changes are already underway and will inevitably continue.

---

‡Chemically, adding carbon dioxide (CO_2) to water (H_2O) yields carbonic acid: $CO_2 + H_2O \rightarrow \rightarrow H_2CO_3$.

§Science uses "pH" as the term for the acid-base level of a solution. An acid is a compound that donates a hydrogen ion (H+) to another compound, known as a base. You may recall "acid plus base equals salt plus water" from high school. At normal ocean pH, 90% of the carbon is in bicarbonate, 9% is in carbonate, 1% is in carbon dioxide. Specifically, CO_2 plus H_2O yields H_2CO_3 and dissociates into bicarbonate, HCO_3, and carbonate, CO_3^{-2}. These join with calcium to form shells of calcium carbonate, $CaCO_3$, or further dissociate back into water, H_2O, and carbon dioxide, CO_2, to start the cycle again.

One possible solution to stem this increase in ocean acidification is the direct removal of carbon dioxide from the atmosphere. The most widely discussed technique is called ocean fertilization, which would permanently sequester carbon dioxide in the deep ocean by stimulating phytoplankton growth. If we sprinkle iron dust onto the ocean, it stimulates plankton growth that in turn take up more carbon from the atmosphere. The plankton bloom ends within approximately 60 days of the first application of iron. But the technique is controversial because of possible ecological consequences and requires a great deal more study (see Part IV). [38]

Though sequestering carbon is straightforward from a technical perspective, Leinen cautioned attendees at the Eighth National Conference on Science, Policy, and the Environment that it will not be easy: "This will be one of the most difficult tasks that we have to address in the future." [38]

What Is Causing the Changes in the Ocean?

A major driver accelerating the amount of carbon in the atmosphere, and hence the acidification of our oceans, is the carbon intensity of our global economy.** Since 2000, the rate of growth of annual carbon emissions from human fossil fuel combustion has tripled compared with the prior decade: Annual emission growth is observed at 3.3% for 2000–2006 and was at about 1.1% for the 1990s. The current emission growth exceeds the predictions of the highest IPCC emission scenarios. Atmospheric CO_2 has grown at 1.9 parts per million (ppm) per year since 2000, compared with an average growth of 1.5 ppm growth from 1970 to 1999. This acceleration can be seen in the steeper climb in emissions after 2000 in Figure 3.10.

Of all the carbon emitted by human activity, 45% remains airborne in the atmosphere. Airborne as carbon dioxide and other carbon compounds, the human contribution acts as a powerful greenhouse gas that we discussed in Chapter 2. Natural systems on land and sea remove 55% of the human-caused carbon. Of this amount, the ocean takes up 24% of the carbon, and forests 30%. [5]

As carbon dioxide emissions rise, they have to go somewhere. As we can see in Figure 4.10, the trend line for carbon in the atmosphere has been rising steadily. While the amount of carbon absorbed by land has remained nearly constant since 1960, carbon absorbed by the ocean chemically or through plankton photosynthesis has been in steady decline.

Seeing the Forest for the Carbon

Many of the same factors affecting the ocean also affect land. As Inez Fung and her collaborators note, "A series of experiments . . . [shows] that

---

**We will examine carbon emission more closely in later chapters. Since 2000, the carbon intensity of the world's economy has stopped decreasing. This is bad news. For the prior 100 years, economies became ever more efficient at producing more wealth per unit of fossil fuel emission. The carbon intensity of the economy is the amount of carbon emissions required to produce $1 of wealth—defined as gross domestic product (GDP) at the country level or as gross world product at the global level. The lack of improvement has been maintained from 2000 to 2006. More global wealth is now being produced by using more carbon-intensive energy systems than ever before. For example, for the same unit of energy produced, burning coal emits twice as much carbon dioxide as burning a biofuel derived from oilseed rape plants and 45% more than burning natural gas.

FIGURE 4.10 online at ncse.org/climate solutions

Carbon flux and carbon dioxide flux in the carbon budget for 1956–2000.

FIGURE 4.11 online at ncse.org/climate solutions

Carbon dioxide emissions remain in the atmosphere, on land, and, in declining amounts, in the ocean.

carbon sink strengths vary with the rate of fossil fuel emissions, so that carbon storage capacities of the land and oceans decrease and climate warming accelerates with faster CO_2 emissions." [11] So, let's examine forests as a critical asset in the fight against climate change. American forests sequester major quantities of carbon, offsetting approximately 10% of US emissions. US Forest Service Chief Abigail Kimbell says, "With the fourth largest forest estate of any country in the world, the U.S. has the potential to do more to wield this resource as a weapon against warming. The health of America's forests is key to the ability of forests to take up carbon." [21]

Around the world, when forests are destroyed or degraded, they are a source of carbon dioxide in the atmosphere. When conserved, managed, or planted sustainably, forests are a carbon sink. Almost 40% of all carbon stored in terrestrial ecosystems is held within forest vegetation and soils, much of this in the boreal forests of the northern hemisphere and in the tropical forests of South America and Asia. The IUCN points out a major dichotomy between the hemispheres: "Forest re-growth in the northern hemisphere currently absorbs carbon dioxide from the atmosphere, creating a 'net sink'. However, in the tropics, forest clearance and degradation are together acting as a 'net source' of carbon emissions." [39]

But the reforestation trend north of the equator is hardly sufficient and requires active management with sustainable practices. Thirteen US Forest Service scientists joined with others on the IPCC in accepting the Nobel Peace Prize in 2008. The Forest Service has taken an active role in studying the effects of global change on the nation's forests. It has 81 experimental forest areas where research is conducted. A key part of the Forest Service's strategy is developing partnerships with stakeholders. Because over half of US forest land is privately owned, it is critical to give landowners an incentive to properly manage and hold onto their forests. Research conducted in the Mendocino National Forest shows that removal of excess woody material

could help reduce carbon emissions in three ways: (1) avoiding wildfire emissions, (2) gains in carbon sequestration, and (3) avoiding the burning of fossil fuels. However, businesses need a steady supply of small-diameter woody materials to justify investments in biomass utilization. In central Oregon, the Forest Service worked with partners to develop a "coordinating resource protocol," or CROP—a way of coordinating delivery to furnish a steady stream of small-diameter materials to businesses. [37]

Yet several other issues would have to be emphasized within the Forest Service and elsewhere to deal effectively with and communicate about climate change. These include handling uncertainty in forecasting the effects of climate change on ecosystems and species, understanding how to increase public acceptance of active management, helping rural communities adapt, taking into account demographic changes and globalization in climate change mitigation and adaptation strategies, boosting environmental education, and sorting out the technical aspects of biofuels and other alternative energies. Suburban sprawl is an additional significant factor that obliterates thousands of acres of forested land each year. The Forest Service projections for loss of forest habitat on private land in critical watersheds is sobering. (See the watershed map in Figure 4.12.)

While ocean waters are shared by all nations, the trees in a forest are the sovereign property of the nation within which they grow. Hence, forest management policies at the national level are critical and can have a very strong effect. Philip Fearnside of the National Institute for Research in the Amazon in Brazil tells us, "Deforestation sacrifices environmental services such as maintenance of biodiversity, water cycling and carbon stocks. The substantial impact of this

---

FIGURE 4.12 online at ncse.org/climate solutions

The US Forest Service projects where sprawl will replace forested areas by 2030.

---

deforestation on loss of environmental services has so far not entered into decision-making on infrastructure projects, making strengthening of the environmental assessment and licensing system a high priority for containing future loss of forest." [10]

> There is a positive feedback between the carbon and climate systems, so that climate warming acts to increase the airborne fraction of anthropogenic CO_2 and amplify the climate change itself. [11]
>
> *Inez Fung and colleagues, 2005*

Put differently, the industrial age since 1850 has greatly expanded the carbon in the atmosphere, but neither Earth's land nor the oceans have expanded. The system is now overloaded, and recovery of ocean chemistry will take tens of thousands of years. [4] The only way to prevent ocean acidification and sea level rise is to greatly reduce carbon dioxide emissions starting today.

CONNECT THE DOTS

- The Arctic Ocean is undergoing rapid changes as sea ice diminishes.
- These changes in the North bring added drought and climate disruption to the United States by altering the location of weather fronts.
- The 4 million people of the Arctic are already feeling the impact of climate change.
- Land-based changes in the polar regions, such as Greenland, may unleash additional freshwater that will disrupt ocean circulation patterns and accelerate sea level rise.
- In addition to the climate effects of warming at the poles, the ocean is failing to keep up with humanity's carbon output.
- Global change impacts physical processes that can lead to feedback that will reduce the effectiveness of the oceanic carbon sink.
- Rising atmospheric concentrations of carbon dioxide affect the physical solubility of CO_2 but not its uptake by biological processes that transport CO_2 to deep water.
- The high rate of CO_2 increase has led to an out-of-equilibrium condition that is causing the increasing acidity (decreasing pH) of the ocean.
- The increasing CO_2 greatly weakens the capacity of calcifying organisms at the base of the food web to survive.
- These destructive impacts cascade along the food web, ultimately destroying the food sources in the ocean upon which humans rely.
- Even if we eliminate CO_2 emissions now, we will observe further ocean acidification from the current atmospheric CO_2 concentrations.
- The only way to avoid ocean acidification is to directly remove CO_2 from the atmosphere by ocean fertilization—a very controversial process.
- Land use is reducing the capacity of forests to sink carbon as well, especially in South America.

Online Resources

www.eoearth.org/article/Deforestation_in_Amazonia
www.eoearth.org/article/Forest_environmental_services
www.eoearth.org/article/Land-use_and_land-cover_change
www.eoearth.org/article/Marine_carbonate_chemistry
www.eoearth.org/article/Ocean
www.eoearth.org/article/Ocean_acidification
www.eoearth.org/article/Plankton
www.eoearth.org/article/Sea_ice_in_the_Arctic
www.eoearth.org/article/Sea-level_rise_and_coastal_stability_in_the_Arctic
Arctic Climate Impact Assessment, www.acia.uaf.edu/
Communication Partnership for Science and the Sea, www.compassonline.org
Drought Monitor, http://drought.unl.edu/DM/monitor.html
Gwich'in Steering Committee, www.gwichinsteering committee.org/gwichinnation.html
International Union for Conservation of Nature, www.iucn.org
NASA Earth Observatory, http://earthobservatory.nasa.gov/Features/CarbonCycle/
National Snow and Ice Data Center, http://nsidc.org

Ocean Thermal Energy Conversion, www.otecnews
.org
Sea Level CCAR, http://sealevel.colorado.edu/
United Nations Atlas of the Oceans, www.oceansatlas
.org
Woods Hole Oceanographic Institution, www.whoi.edu

Climate Solution Actions

Action 18: Coastal Management and Climate Change
Action 19: Forest Management and Climate Change
Action 25: Ocean Fertilization for Carbon
Sequestration

Works Cited and Consulted

[1] Arctic Climate Impact Assessment, Hassol SJ
(2004) in Impacts of Arctic Warming: Arctic
Climate Assessment Report. Arctic Council and
the International Arctic Science Committee. www
.acia.uaf.edu

[2] Bindoff NL, Willebrand J, Artale V, Cazenave
A, Gregory J, Gulev S, Hanawa K, Le Quéré C,
Levitus S, Nojiri Y, et al. (2007) Observations:
Oceanic Climate Change and Sea Level (in *Climate
Change 2007: The Physical Science Basis. Contribu-
tion of Working Group I to the Fourth Assessment
Report of the Intergovernmental Panel on Climate
Change*, eds Solomon S, Qin D, Manning M, Chen
Z, Marquis M, Averyt KB, Tignor MH, Miller L)
ar4-wg1-chapter5.pdf: www.ipcc.ch

[3] Bluhm B, Hopcroft R (2006) Arctic Biodiversity.
Ocean Explorer. NOAA (read November 13, 2008).
http://oceanexplorer.noaa.gov/explorations/05
arctic/background/biodiversity/biodiversity.html

[4] Caldiera K, Kleypas J (2006) COMPASS Briefings
"It's Not Just the Heat, It's the Acidity (Caldeira)."
COMPASS 2006 (May 23)(read September 18,
2008). www.compassonline.org/meetings/
briefings_hoc.asp

[5] Canadell JG, Le Quéré C, Raupach MR, Field CB,
Buitenhuis ET, Ciais P, Conway TJ, Gillett NP,
Houghton RA, Marland G (2007) *Contributions to
accelerating atmospheric CO$_2$ growth from economic
activity, carbon intensity, and efficiency of natural
sinks*. Proceedings of the National Academy of Sciences 104(47):18866. www.pnas.org/cgi/content/
abstract/104/47/18866

[6] CIA (2007) The World Factbook. (Central Intelligence Agency, Washington, DC). www.cia.gov/
library/publications/download/index.html

[7] Cook J (2005) "The Ocean Conveyor" in Illustration Gallery in Images & Multimedia. *Oceanus
Magazine*. (Woods Hole Oceanographic Institution, Woods Hole, MA) (read November 17, 2008).
www.whoi.edu/oceanus

[8] Dickson B, Yashayaev I, Meincke J, Turrell B, Dye
S, Holfort J (2002) *Rapid freshening of the deep
North Atlantic Ocean over the past four decades.*
Nature 416 (April 25, 2002). www.nature.com/
nature/journal/v416/n6883/abs/416832a.html

[9] Drought Monitor (2008) *National Drought Mitigation Center*. Lincoln, NE (read November 16, 2008).
http://drought.unl.edu/dm/

[10] Fearnside P (2007) "Deforestation in Amazonia,"
(topic ed) Hall-Beyer M, in *Encyclopedia of Earth*,
(ed) Cleveland CJ. (Environmental Information
Coalition, National Council for Science and the
Environment, Washington, DC). www.eoearth.
org/article/Deforestation_in_Amazonia

[11] Fung IY, Doney SC, Lindsay K, John J (2005)
Evolution of carbon sinks in a changing climate.
Proceedings of the National Academy of Sciences
102(32):11201–11206. www.pnas.org/cgi/content/
abstract/102/32/11201

[12] Gagosian RB (2003) Abrupt Climate Change:
Should We Be Worried? *World Economic Forum*.
Davos, Switzerland. www.whoi.edu/page.do?pid
=12455&tid=282&cid=9986

[13] Gallessich G (2008) Climate Change Seen Turning
Seas Acidic. *931060 UCSB* March 3, 2008. www.ia
.ucsb.edu/93106/2008/March3/Climate.html

[14] Gallessich G (2008) "Sea Ice," (topic ed) Duffy
JE, in *Encyclopedia of Earth*, (ed) Cleveland, CJ.
(Environmental Information Coalition, National
Council for Science and the Environment,
Washington, DC). www.eoearth.org/article/
Ilulissat_Icefjord%2C_Denmark-Greenland

[15] Gattuso JP, et al. (2007) "Ocean acidification," in
Encyclopedia of Earth, (ed) Cleveland, CJ. (Environmental Information Coalition, National Council
for Science and the Environment, Washington,
DC). www.eoearth.org/article/Ocean_acidification

[16] Global Carbon Project (2008) *Carbon Reductions
and Offsets*, in *Earth System Science Partnership
Report No. 5*, eds Coulter L, Canadell JG, Dhakal S.
www.globalcarbonproject.org

[17] Hansen J, Sato M, Ruedy R, Lo K, Lea DW,
Medina-Elizade M (2006) *Global temperature
change*. Proceedings of the National Academy
of Sciences 103(39):14288. www.pnas.org/cgi/
content/abstract/103/39/14288

[18] Hebert P, et al. (2008) "Arctic Ocean," (topic
ed) Duffy JETE, in *Encyclopedia of Earth*, (ed)
Cleveland CJ. Biodiversity Institute of Ontario.
(Environmental Information Coalition, National
Council for Science and the Environment,
Washington DC). www.eoearth.org/article/
Arctic_Ocean

[19] James S (2008) Gwich'in Culture. *Gwich'in Steering Committee* (read November 15, 2008). www
.gwichinsteeringcommittee.org/gwichinnation
.html

[20] James S (2008) Plenary Presentation: Summarizing Global Change Science and the Likely Implications of Global Climate Change. *National Conference on Science, Policy, and the Environment: Climate Science and Solutions.* http://ncseonline.org/2008conference

[21] Kimbell G (2008) Forest Management and Climate Change Response. *National Conference on Science, Policy, and the Environment.* http://ncseonline.org/2008conference

[22] Kleypas JA, Feely RA, Fabry VJ, Langdon C, Sabine CL, Robbins LL (2006) Impacts of Ocean Acidification on Coral Reefs and Other Marine Calcifiers: A Guide for Future Research. Report of a workshop held April 18–20, 2005, St. Petersburg, FL, sponsored by NSF, NOAA, and the US Geological Survey. 88 pp (read November 1, 2008). www.isse.ucar.edu/florida

[23] Langdon C, Takahashi T, Sweeney C, Chipman D, Goddard J, Marubini F, Aceves H, Barnett H, Atkinson MJ (2000) *Effect of calcium carbonate saturation state on the calcification rate of an experimental coral reef.* Global Biogeochemical Cycles 14(2):639–654. www.ncoremiami.org/members/personnel/GBC2000.pdf&oi=ggp

[24] Le Quéré C, Rödenbeck C, Buitenhuis ET, Conway TJ, Langenfelds R, Gomez A, Labuschagne C, Ramonet M, Nakazawa T, Metzl N (2007) *Saturation of the Southern Ocean CO_2 sink due to recent climate change.* Science 316(5832):1735. www.sciencemag.org/cgi/content/abstract/sci;316/5832/1735

[25] Leinen M (2008) Oceans: A Carbon Sink or Sinking Ecosystems? *National Conference on Science, Policy, and the Environment: Climate Science and Solutions.* http://ncseonline.org/2008conference/

[26] MacCracken MC, Moore F, Topping JC (2007) *Sudden and Disruptive Climate Change: Exploring the Real Risks and How We Can Avoid Them.* (Earthscan Publications). www.climate.org/publications/sudden-disruptive-climate-change.html

[27] McManus J, Oppo D (2006) "The Once and Future Circulation of the Ocean: Clues in Seafloor Sediments Link Ocean Shifts and Climate Changes." *Oceanus Magazine,* November 16, 2006. www.whoi.edu/page.do?pid=12455&tid=282&cid=17906

[28] Meehl GA, Stocker TF, Collins WD, Friedlingstein P, Gaye AT, Gregory JM, Kitoh A, Knutti R, Murphy JM, Noda A, et al. (2007) Global Climate Projections (in *Climate Change 2007: The Physical Science Basis. Contribution of Working Group I to the Fourth Assessment Report of the Intergovernmental Panel on Climate Change,* eds Solomon S, Qin D, Manning M, Chen Z, Marquis M, Averyt KB, Tignor MH, Miller L) ar4-wg1-chapter10.pdf: www.ipcc.ch

[29] Nerem RS, Dorsi S, Willis JK, Chambers DP, Mitchum GT (2007) Satellite and In Situ Observations of Regional Sea Level Change: What Can They Tell Us about Future Changes? *American Geophysical Union.* www.agu.org/cgi-bin/wais?jj=G44A-01

[30] NSIDC (2008) Arctic Sea Ice Down to Second-Lowest Extent; Likely Record-Low Volume Arctic Sea Ice Extent during the 2008 Melt. *National Snow and Ice Data Center.* Images & Multimedia, Boulder, CO (read November 1, 2008). www.nsidc.org/pubs/notes/

[31] OTEC (2008) Ocean Thermal Energy Conversion (read November 31, 2008). www.otec.org

[32] Permanent Service for Mean Sea Level (PSMSL) (2008) Home Page. Proudman Oceanographic Laboratory, Liverpool, UK (read October 1, 2008). www.pol.ac.uk/psmsl

[33] Schmitt RW (1992) "Mysteries of Planetary Plumbing." *Oceanus Magazine* 35(2):38–45. (Woods Hole Oceanographic Institution, Woods Hole, MA). www.whoi.edu/oceanus/

[34] Segar D (2007) *Introduction to Ocean Sciences.* (W. W. Norton & Company, New York). www.wwnorton.com/college

[35] UN Oceans (2008) United Nations Atlas of the Oceans (read August 24, 2008). www.oceansatlas.org/cds_static/en/human_settlements_coast__en_1877_all_1.html

[36] UNEP (2008) "Ilulissat Icefjord, Denmark-Greenland," (topic ed) McGinley M, in *Encyclopedia of Earth,* (ed) Cleveland CJ. (Environmental Information Coalition, National Council for Science and the Environment, Washington, DC). www.eoearth.org/article/Ilulissat_Icefjord%2C_Denmark-Greenland

[37] USFS and BLM (2008) Coordinated Resource Offering Protocol (CROP). US Forest Service and Bureau of Land Management (read December 17, 2008). www.forestsandrangelands.gov/Woody_Biomass/supply/CROP/index.shtml

[38] Whilden K, Margaret Leinen M, Whaley D, Grant B (2007) *Ocean Fertilization as an Effective Tool for Climate Change Mitigation.* Greenhouse Gas Market Report 2007. www.climos.com/publication.php

[39] World Conservation Union (2006) "Forest Environmental Services," in *Encyclopedia of Earth,* (ed) Cleveland CJ. (Environmental Information Coalition, National Council for Science and the Environment, Washington, DC) (read August 29, 2008). www.eoearth.org/article/Forest_environmental_services

PART II

How to Think About Climate Solutions

CHAPTER 5

The Five Horsemen of Extinction

Extinction by habitat destruction is like death in an automobile accident: easy to see and assess. Extinction by the invasion of exotic species is like death by disease: gradual, insidious, requiring scientific methods to diagnose. [65]

E. O. WILSON, 1997

The nation's Kennedy Space Center occupies most of Merritt Island, on Florida's Atlantic coast. Thousands of visitors know this beautiful site from watching rocket launches along the Banana River. The land there that is not devoted to aeronautics is for the birds, quite literally. The island and its surrounding estuaries form Merritt Island National Wildlife Refuge, encompassing more than 140,000 acres (57,000 hectares). The entire site is 30 miles, or 48 kilometers (km), from north to south and 10 miles (16 km) from east to west at its widest. National Aeronautics and Space Administration (NASA) science and technology coexist on the refuge with more than 1,000 plant and 500 wildlife species, including 300 species of birds. These waterfowl, shorebirds, and songbirds, including 93 federally or state-listed endangered or threatened species, include important populations of southern bald eagles, brown pelicans, wood storks, and mottled ducks. The largest mammal on Merritt is the manatee, which grazes the island's watery channels. The plants and animals inhabit a surprisingly diverse mix of coastal habitats, including the beach and dune system, estuarine waters, forested and nonforested wetlands, impounded wetlands, and upland shrublands and forests. [58]

Even on a federally controlled island where visitors need passes and every shipment in and out is examined carefully, the invasion of exotic species is rampant. Florida, with its subtropical environment, is being overrun at an alarming rate. Invasives on Merritt today include Australian pine, paperbark melaleuca, Brazilian pepper, water hyacinth, and the southern pine bark beetle, just to name a few unwelcome newcomer species.

But invasives are not the only threat to native species on the island. To quote the US Department of the Interior, others come from "feral animals, free roaming pets, recreational boating, elevated nutrient loading, and pollution, as well as from the increased demand for public use activities that are not directly linked to fish and wildlife goals." [58] Climate change

FIGURE 5.1 **Kennedy Space Center: science amidst nature**

(A) *(right)* The Kennedy Space Center sits on a coastal island bursting with ecologically rich habitat for hundreds of plant and animal species. But rockets are not the only exotics to invade the island. Thousands of humans visit each month, as Adrian, Lisa, and Sarah did recently.

(B) *(above)* The waterfowl are mere specks hugging the shore when compared with the rocket towers. The New Horizons rocket seen on the left in the picture blasted off in January 2006 and will not reach Pluto until 2015. Will these birds still be able to dabble for dinner at the Merritt Island refuge even under modest sea-level rise projections? Source: [64]

adds a whole new dimension to these already serious threats. As rising sea turns wetlands into open water, and freshwater marshes into saltier water, pressure increases on plants and animals that have evolved to conditions now fast eroding.

When disruptions caused by climate change—including changes in temperatures, seasonality, rainfall, and storm frequency and intensity—occur in the context of already fragmented landscapes and stressed seascapes, the net effect is a biologically impoverished planet—a world of weeds. Combined with warming temperatures and rising seas, the introduction of invasive species, unfettered catch of fish and other species, pollution, and habitat destruction all diminish the diversity of life on Earth—and the health of existing ecosystems.

This chapter will examine some of the connections between life for plants and animals and life for the species reading this book, *Homo sapiens.*

All the problems we just mentioned are the direct result of human activity, pushed relentlessly forward especially in the industrialized world and by rising human populations. The consequence of providing more than 6.5 billion people with food, shelter, and consumer goods is that we consume more than the entire Earth's annual output of renewable resources within the first 10 months of each year. This ecological overshoot is clearly unsustainable. [61]

TABLE 5.1 The Extinction History of Life on Earth

First major extinction, about 440 million years ago (mya): 19% of families* were lost (there was little or no life on land at this time) because of severe and sudden cooling.

Second major extinction, about 370 mya: 19% of families were lost because of global climate change.

Third major extinction, about. 245 mya: 54% of families were lost because of plate tectonic movements and bolide (meteor) impact.

Fourth major extinction, about 210 mya: 23% of families were lost, with causes unknown.

Fifth major extinction, about 65 mya: 17% of families were lost because of meteor impact.

Sixth major extinction, today: Potentially, nearly 50 % of all species across the entire globe will disappear by 2100.

*A family may consist of a few to thousands of species.
Source: [16]

TABLE 5.2 Benefits of Biodiversity

Biological resources that provide goods for human use include:

- Food: species that are hunted, fished, and gathered, as well as those cultivated for agriculture, forestry, and aquaculture
- Shelter and warmth: timber and other forest products and fibers such as wool and cotton
- Medicine: both traditional medicines and those synthesized from biological resources and processes

Source: ESA [18]. The Ecological Society of America has excellent short fact sheets on key topics such as acid deposition, acid rain, biodiversity, biocomplexity, carbon sequestration in soils, coral reefs, ecosystem services, environmental justice, fire ecology, floods, hypoxia, and invasion at www.esa.org/education_diversity/factsheets .php.

Even before humans were an ecological factor, the natural evolution of land and life led to both extinctions and evolution of new species. Each generally occurred on a slow, geologic timescale (with species lifetimes of millions of years). During the more than 3.5 billion–year history of life on the planet, Earth has experienced at least five periods of mass extinction. We are currently creating the sixth, with rates of extinction many hundreds of times above historical levels. Potential losses include nearly half of the world's more than 10 million species over the next century. [2]

The current mass extinction differs from the five prior episodes by its cause (human), speed (within decades), and extent (global). All prior extinction episodes were set in motion by physical changes on Earth. Human activities have destroyed forests, wetlands, grasslands, rivers, lakes, coral reefs, and other ecosystems at alarming levels and have increased extinctions perhaps more than 1,000 times the natural background rate of roughly 10 extinctions per year. [48]

Cultivated systems now cover one-quarter of Earth's terrestrial surface. Areas that were formerly forests, grasslands, wetlands, and other habitats are now plowed over, paved over, drained, desertified, or otherwise degraded. [37] The species that once occupied these habitats often have no other place to go. [34]

Benefits of Biodiversity

The word *biodiversity* is a contracted version of *biological diversity*. In a prologue to Eric Chivian's book *Sustaining Life*, Former UN Secretary General Kofi Annan eloquently writes the folowing:

Biological diversity—the variety of life on Earth—is at the heart of our efforts to relieve suffering, raise standards of living, and achieve the UN Millennium Development Goals. We cannot do without the countless services provided by biodiversity: pollinating our crops; fertilizing our soils with nitrogen, phosphorus, and other nutrients; providing millions of people with livelihoods, medicine, and much else. Advances in medicine, including treatments for currently untreatable diseases, would not be possible without the powerful pharmaceuticals derived from plants, animals and microbes or without the knowledge gained from other species in biomedical research. We must conserve and sustainably use this pillar of human life. Yet biodiversity is declining at an unprecedented rate and is woefully underappreciated as a resource and as an issue meriting high-level attention. [6]

Humans have always depended on the Earth's biodiversity for food, shelter, health, and wealth. (See Table 5.2 for a short summary of those

TABLE 5.3 Four Services That Insects Provide the Nation

Ecological service		Cost of replacing this service ($ billions/year)
Dung burial	Insects remove feces from livestock ranges.	0.380
Pollination	Insects, such as native bees, pollinate agricultural products.	3.07
Pest control	Insects are predators on the 6,000 species that eat agricultural products.	4.49
Recreation	Insects provide nutrition for animals valuable for hunting, fishing, and observing.	49.96
Total cost		$57 billion/year

Source: [31]

benefits.) The services delivered to us by flora and fauna and microbes are critical to our well-being. Let's look at a few examples by starting in our own backyards. Take the dung beetle, for example. Among 90,000 described insect species in the United States, the unglamorous dung beetle family is easy to overlook. Fifty carrion or dung beetle species are found in the New York metropolitan area. In New York City, however, only about a dozen species are usually found in traps placed in city parks, and only four species are found in downtown Manhattan, most of which are nonnative beetles. [8] Carrion beetles feed on remains of dead animals. Dung beetles feed on feces that would otherwise contaminate water supplies or render forage grasses unpalatable to livestock. Dung beetles are rollers, rolling dung into small balls, or tunnelers, burying dung underground, or dwellers, living within dung on the ground. Dung beetles are common everywhere except Antarctica. Dung provides all the necessary nutrients a dung beetle needs.

By burying feces, dung beetles save the US cattle industry alone an estimated $380 million annually, as burial reduces habitat for pest flies and recycles the nitrogen and carbon from the dung into the soil, instead of releasing carbon into the air as methane. [31] In more-urban areas, beetles feed their larvae and help protect human health by burying bacteria-laden wastes that otherwise might be spread to people by houseflies or might contaminate the runoff that fills our water reservoirs. When the humble beetles are not around to remove carrion and dung, flies take over. But to do what they do, the beetles need soft soil in which they can dig. The hard, compacted turf of many eastern woodlands, especially in urban and suburban areas, appears to hinder the beetles. Native dung beetles reportedly are declining and being replaced by nonnative, exotic, and invasive dung beetles.

Four services provided by wild insects—dung burial, pest control, pollination, and wildlife nutrition—are sufficiently quantifiable for researchers to estimate the financial value (see Table 5.3). Other services are less easily quantified. As John Losey and Mace Vaughn write, "We base our estimations of the value of each service on projections of losses that would accrue if insects were not functioning at their current level.... We estimate the value of those insect services we address to be almost $60 billion a year in the United States, which is only a fraction of the value for all the services insects provide." [31]

The bad news is we have already observed a steady decline in these beneficial insects, such as native bees, as part of an overall decline in biodiversity, especially in localized environments heavily degraded by human impacts. [26]

The initial four horsemen of the extinction apocalypse are habitat loss and degradation, invasive exotic species, direct take of wild plants and animals, and pollution. [1] Now a fifth horseman—global climate disruption—magnifies the already dangerous effects of the first four (see Table 5.4 for a summary). The effect of more than one horseman is a case of multiplication, not addition. These five horsemen of extinction are not a case of pouring different tinted dyes into a cup to mix a paint color. This is more like adding multiple explosives to a barrel already on fire.

TABLE 5.4 The Five Horsemen of Extinction: How Nature Copes

C	Competition arrives to claim your day job.	Invasive species are making huge inroads and displacing key species with unknown long-term consequences.
O	Overharvesting reduces the number of your family left to support each other.	The clearest examples of overharvested species are some types of fish.
P	Pollution is poisoning your groundwater.	Water, soil, and air pollution are no more healthy for plants and animals than they are for humans.
E	Eviction notices appear on your home.	Humans are building homes and roads in ways that fragment what little plant and animal habitat is left in many regions and remove habitat entirely in others.
S	System-wide disruption from global climate change magnifies all of the above threats to life on Earth.	Human-induced climate disruption acidifies oceans, changes rainfall patterns, brings spring sooner, and makes storms more severe.

Innumerable species are subject to the loss of their homes. Many of these same species are also subjected to competition from invasive and exotic species that never before co-occured with them. Many species are undergoing a reduction in population numbers due to human exploitation (overharvesting) and deleterious effects of pollution. Species under such stress are much more vulnerable to the impacts of climate disruption. These extinction forces act in synergistic ways—for example, as the present locations of already stressed species become unsuitable for their future existence because of changed climate, those plants or animals will have to move to any remaining locations or else go extinct. [37]

Horseman #1:
Habitat Loss and Fragmentation, Leaving No Place Left to Hide

Thomas Lovejoy, president of the H. John Heinz III Center for Science, Economics and the Environment, explains that we face biological impoverishment due to a few key facts. Very few places on the planet escape the work of human hands. Human intervention has destroyed and fragmented the habitat upon which flora and fauna rely—through clearing forests, building road networks, tilling agricultural fields, channeling rivers, removing mountaintops, dredging harbors, trawling seafloors, and overfishing oceans, all to serve expanding human popula-

tions. Roadways also carve up the habitat available to native species, effectively fencing them into ever-smaller boxes, and speed up the invasion pathways for the arrival of exotics. Species do not exist in isolation. There are historical co-evolved relationships, such as between flowers and their pollinators, between predators and their prey, and even between trees and specific fungi in the soil. Therefore, when one species disappears—for example, the polar bear or sugar maple—its absence has deep repercussions on the surrounding ecosystem of complex interactions among plant and animal species. [33] The new ecosystem is biologically impoverished and less stable.

Also due to climate disruption, areas that we have established as protected areas may not continue to provide the proper conditions for the species they were set up to protect. As Camille Parmesan and Hector Galbraith observe, "In particular, such shifts in [species'] composition are likely to alter important competitive and predatory/prey relationships, which can reduce local or regional biodiversity. A particularly compelling example of this is the change observed over more than 60 years in the intertidal communities of Monterey, California, where [an ecological] community previously dominated by northern colder-water species has been 'infiltrated' by southern warmer-water species in response to oceanic warming. . . . Thus, many protected areas such as the [legally established] marine reserve in Monterey Bay, are experiencing a shift in the communities that they protect." [42]

In short, changes in temperature, salinity, acidity, and a myriad of other parameters are occurring in damaged environments in which species are already threatened by encroachment, dwindling gene pools, or a host of other forces. The net result is that there are literally few places to hide. This is particularly tragic for endemic species—those that live only in one place and who have no other place to go.

For example, the ancient scrub and beach dunes of Lake Wales Ridge formed 1 to 3 million years ago and managed now as a satellite of the larger Merritt Island refuge are home to some 40 plant species found nowhere else in the world, including the Florida ziziphus, one of the rarest and most endangered plants in the state. A flowering shrub that can grow 2 meters tall, the ziziphus has yellow-orange fruit that develop in spring. It was thought to be extinct in the wild until rediscovered in 2007 in its natural sandhill habitat on Lake Wales Ridge. Florida's agribusiness, commercial, and residential development—combined with inadequate conservation measures—have pushed hundreds of plant and animal species to the brink of extinction. [59] Unchecked, climate change will push many of them over the brink.

Horseman #2: Invasive Species from the Last Port of Call

Life is more diverse nearer the Earth's equator than farther away from it. For over two centuries, biologists have been trying to understand why this is so. Plant and animal species rely on a host of conditions to which they have adapted genetically and behaviorally. Temperature, water, elevation, sunlight, soil composition, and the presence of other species each play roles in determining which species live in a particular location. Now there is a new more powerful force. Regardless of where we live, humans have been moving species around at a pace that far outstrips any natural process.

Economic trade introduces many species accidentally. Invasive species may cost the US economy as much as $137 billion per year. About 90% of the transport of goods globally occurs by oceangoing freight ship, often taking organisms from one ecosystem into another far away. [46] Before humans, there was no way for those ecosystems to be connected. As cargo ships unload, it is common practice to take aboard a vast quantity of water to balance the ship. Ships keep a great volume of this ballast water on board on their way to the next port and pump it out when new cargo comes on board. Harbor water is full of life. On an average day, cargo ships around the world may be transporting more than 7,000 species to new homes. [7] The vast majority perish, but a few thrive in their new environment. For example, first seen in the United States in 1988, tiny zebra mussels arrived in ballast water from freight ships visiting the Great Lakes from Asia. These mussels proliferated, crowded out native shellfish, and clogged underwater drains for municipal water systems and even blocked the intake and outflow valves of power plants that used lake water for cooling their machinery. [30] The incredibly rapid spread and economic impacts of the invasive zebra mussels led in 1990 to the first US legislation to regulate ballast water.

FIGURE 5.2 online at ncse.org/climate solutions

Data from Florida and Ohio show the increasing numbers of exotic species invading US waters in recent years.

FIGURE 5.3 online at ncse.org/climate solutions

Pie charts show a stunning variety of ways in which invasive species are introduced to US waters.

FIGURE 5.4 online at ncse.org/climate solutions

Pie charts show the areas of origin of invasive species transplanted into Florida and Ohio.

Of all the continental states, Florida is the most impacted by invasives, as we learned in the Merritt Island case. But all states are affected. Even northern states like Ohio have been hard hit by invasive plants and animals. Invasives common in Ohio—and many northern states—include aquatic or wetland plants like Eurasian watermilfoil and purple loosestrife. [66] Of the 3,000 plant species known to occur in Ohio, approximately 25% are not native to the state. These invaders often outcompete the indigenous species and provide inferior homes and poorer food to native animals. Ohio's soybean crop is worth $1 billion to the state. But invasive plants that infest soy fields, such as johnsongrass, threaten the soy crop. [40] Of course, the soybean is not from Ohio but instead is a species of legume native to East Asia, where it has been cultivated for 5,000 years. But soybeans got to Ohio at least a century ago, so they can claim squatter's rights to complain about all the newer invasives. [40]

Climate change is already exacerbating the pressure, either directly or indirectly, favoring highly opportunistic species that thrive in disturbed environments. [39, 56]

Horseman #3: Overtaking: What the Grim Reaper Soweth

Over 8,000 plant species worldwide are threatened with extinction, according to the IUCN. The number grows daily. Researchers recently estimated that between 22% and 47% of the world's flora is in serious decline. In the United States alone, 744 plant species are federally listed as threatened or endangered by the US Fish and Wildlife Service, comprising over half of all listed species. * Overconsumption of wild plants for medicine and food endangers some species. [20] The taking of natural resources for private gain or public good has been a subject of heated and long debate in Washington, DC, and in state capitols. The Endangered Species Act

has long been a flash point in the property rights debate. [9] But for the most egregious example of overtaking of natural resources, we need to look to the sea.

The story of declining marine life becomes crystal clear if we look at the long-term effect of the killing of the ocean's largest creatures, its whales. Removing the largest animals from an ecosystem can cause strong enduring changes throughout the entire ecosystem, as predator-prey balances are upset and begin to shift to affect other species. Alan Springer of the Institute for Marine Science in Fairbanks, Alaska, and colleagues studied 50 years of data on mammal take from the Pacific Ocean. Commercial whaling appears to have had a debilitating impact, greatly reducing the number of whales by the mid-1970s. In succession, harbor seal populations plummeted a decade later, possibly because killer whales (also known as orcas) were being forced to vary their diet in the absence of whales—their former prey. [55]

After harbor seals became too few, orcas turned to even smaller pinnipeds, such as fur seals and sea otters, as prey. In the 1990s, fur seal, sea lion, and sea otter populations plummeted.

Marine mammals, like the otter, play a critical role in the overall ecosystems they inhabit. Kelp are large macroalgae that form underwater forests along the Pacific coast to create much of the offshore habitat of many other aquatic species. Sea urchins eat kelp. Sea otters eat urchins, keeping the sea urchin population in check, when sufficient otters are present. But, as the National Marine Sanctuaries warns, "when sea otters decline, urchin numbers explode and grab onto kelp like flies on honey. The urchins chew off the anchors that keep the kelp in place, causing them to die and float away, setting off a chain reaction that depletes the food supply for other marine animals causing their numbers to decline."†

---

*Many species that are imperiled have not yet been listed as threatened or endangered.

†For more on the sea otter as a keystone species, visit http://sanctuaries.noaa.gov/about/ecosystems/kelpdesc.html.

Industrial whaling set in motion a chain reaction in which orcas had no choice but to "fish down" to smaller animals of the marine mammal food web. [44] "We propose that decimation of the great whales during the modern era of industrial whaling ultimately caused the declines by forcing the great whales' foremost natural predators, killer whales, to turn elsewhere for food." Springer continues, "If our hypothesis is correct, either wholly or in significant part, commercial whaling in the North Pacific Ocean set off one of the longest and most complex ecological chain reactions ever described, beginning in the open ocean more than 50 years ago and leading to altered interactions between sea urchins and kelp on shallow coastal reefs." [55]

Managing fish resources on the high seas, that is, beyond the 200-mile (300 km) limits of individual nations' exclusionary economic zones, is proving vexingly difficult. Even as the mammals of the sea suffer population collapses, fishing boats continue to catch the "chicken of the sea" in escalating volume (see Figure 5.6). While there has been some success in establishing marine reserves for protecting species and habitat, Mark Spalding, president of the Ocean Foundation, points out, "In the high seas, beyond the jurisdiction of individual nations, marine protection remains virtually non-existent." [54] Marine food chains are being shortened from both ends. The loss of the multitudinous microscopic plankton base erodes the

capacity for smaller fish. And the loss of larger, less numerous apex species at the top, as we have just seen, places marine life in "hot water double jeopardy."

More than a billion people worldwide rely on fish as their main source of protein. [11] Bluefin tuna have been severely overfished, and some scientists believe they are in danger of extinction. Researchers have shown that marine ecosystems with naturally low diversity had lower fishery productivity, more frequent "collapses" (defined as strong reductions in fishery yield), and lower tendency to recover after overfishing than naturally species-rich systems. [12]

Since 1900, 123 freshwater animal species have been recorded as extinct in North America. Hundreds of additional species of fishes, mollusks, crayfishes, and amphibians are considered imperiled. As Anthony Ricciardi and Joseph Rasmussen wrote almost a decade ago, "Assuming that imperiled freshwater species will not survive throughout the next century, our model projects a future extinction rate of 4% per decade, which suggests that North America's temperate freshwater ecosystems are being depleted of species as rapidly as tropical forests." [50] Changing hydrological and temperature patterns will rapidly further impoverish biotic diversity of freshwater communities.

Horseman #4: Pollution

The fourth horseman of extinction is pollution. Human activity alters the chemical composition of air, soil, and water. Pollution of each of these natural components hinders ecosystems' ability to maintain the diversity of species they may have previously enjoyed. Pollution can take the form of adding toxins to the environment, such as toxic mercury, a contaminant of coal that becomes airborne from smokestack emissions. Pollution can take the more insidious form of simply overloading natural systems with ingredients that appear harmless but become killers as they overwhelm a system. For example, runoff from rainstorms may contain perfectly

---

FIGURE 5.5 online at ncse.org/climate solutions

Graph shows the decline of smaller prey populations when predators are forced to eat lower on the food chain after killer whales' (orcas') usual prey were taken in industrial whale harvests.

FIGURE 5.6 online at ncse.org/climate solutions

Fish catches on the high seas: 1950–2004

TABLE 5.5 Some Ecological Consequences of
Human Activity on Ecosystem Processes

Ecosystem structure
- Loss of biodiversity
- Structural asymmetry and downsizing of communities
- Loss of keystone species and functional groups

Ecosystem processes
- Low internal regulation
- High nutrient turnover
- High resilience
- Low resistance
- Low variability
- Low adaptability

Ecosystem functions
- High porosity of nutrients and sediments
- Loss of productivity
- Loss of reflectance (reflection of sunlight upward)

Global processes
- Modified biogeochemical cycles
- Atmospheric change
- Accelerated climatic change

Source: [63]

natural items, from leaf detritus to suspended soil particles. This runoff collects in rivers that flow to the ocean, where it reduces water clarity, robs water of needed oxygen as it decomposes, reduces light that reaches the sea bottom, and often has catastrophic impacts on life in the ocean. Land-based pollution has dramatic effects on coral species diversity, live coral cover, composition of the coral fauna, coral growth rates, erosion intensity, and reef carbonate budgets. [14] Stressed species and ecosystems have lowered resilience and are less able to withstand the added impacts of climate disruption (see Table 5.5).

A simple example of the impact of airborne and waterborne pollution on biodiversity is the effect of nitrogen. As excess nitrogen from fertilizer use, animal waste releases, and many other human sources settles in water far removed from its origin, it causes harmful algal blooms that kill other plants and animals. This in turn reduces the diversity within each group, a problem also caused by excess nitrogen on land. Half of the world's diversity of flowering

plants is restricted to 34 biodiversity hot spots. The National Geographic Society has prepared an excellent world map of biodiversity hot spots, including details of the individual endangered fauna in each hot spot, which is available from Conservation International. [10] Researchers have found increased nitrogen entering biologically diverse hot spots. The average deposition rate across these areas was 50% greater than the global terrestrial average in the mid-1990s and could more than double by 2050. And that means a high likelihood of plant extinctions within these "protected" zones within our lifetimes. [45]

Horseman #5:
Climate Change, Disrupting the
Mother Nature We Knew

Everywhere around us, nature is changing her behavior. Yellow-bellied marmots are ending hibernation earlier. Many flowers are blooming earlier. Frogs are spawning earlier. Compared with 50 years ago in Mediterranean orchards, leaf unfolding is 16 days earlier and fruit flower opening is 6 days earlier. Many bird species, from flycatchers to jays, are nesting earlier. Tree swallows are nesting 9 days earlier than 40 years ago. Why? While we do not know with certainty, the earlier egg-laying date is highly correlated with a rise in temperatures in the month of May since 1960. Biologists Camille Parmesan and Gary Yohe examined the change in behavior of more than 1,700 species. They found that such changes could be predicted by simultaneous shifts in climate. [43]

We who welcome an early spring may like this, but as with all of these rapid changes in the natural order of life, there are negative consequences. Not every species is shifting at the same rate. Orchard keepers rely on the free pollination services of butterflies and other insects. But since the mid-1970s, the butterflies that pollinate the fruit have been arriving 11 days earlier. Around the world, plants and animals that coevolved based on their mutual dependence are falling out of synch. The consequences of this uncoupling could be quite damaging and far-reaching. [41]

If the gap between these natural patterns continues to widen, the fruit trees may bloom out before pollinating butterflies, bees, and other insects emerge. Because they're missing the flower nectar, the butterflies' populations may collapse before the birds that feed on them arrive. Laboratory experiments with species endemic to North America's tallgrass prairies show that when plant species that usually flower before the peak of summer are grown in warmer temperatures, their flowering and fruit production is advanced. But early warming delays reproduction in other, "late" species that flower after the peak of summer heat. [52] Climate disruption is uncoupling the timing of natural events. So how do the insects bridge the gap after the early plants bloom out and before the late plants flower? And what does that mean for the red-wing blackbird and other birds that rely on the insects?

In addition to the timing of natural events, the geographic range for species is changing. Scientists from the National Audubon Society analyzed 40 years of Christmas Bird Count data—and found that nearly 60% of the 305 species found in North America in winter are on the move, shifting their ranges northward by an average of 35 miles. Their findings provide new and powerful evidence that global warming is already having a serious impact on natural systems. Northward movement was detected among species of every type, including more than 70% of highly adaptable forest and feeder birds.

Similar impacts are being found in the polar regions. The breeding range for the Adélie penguins on the Antarctic coast has shrunk by 3 km in just 10 years. That may not sound like much, but it increases the distance adults have to travel for food during the critical egg-hatching period. In addition, the krill and small fish that the penguins rely on are disappearing from coastal waters. Why?

Sea butterflies, which we met in Chapter 4, are about the size of a pea. They are popular dining fare for many marine species, including a wide variety of fishes that are, in turn, consumed by penguins in the Antarctic and polar bears in the Arctic and by almost every fish species that humans harvest for dinner, as well. These small creatures come in many varieties, many of which are still being documented. All over the world's ocean, sea butterflies feed on the smallest plankton and build their shells by converting carbon dioxide in the water into calcium carbonate. We often use *ocean* as a singular noun, as you may have noticed in Chapter 4. Ocean scientists regard the collective ocean as one biophysical unit, albeit complex, in which the components are all connected. This is analogous to using the phrase *the land*.

A higher level of carbon dioxide in the atmosphere above the ocean means more carbon dioxide enters the ocean. The ocean currently absorbs on average of about 1 metric ton of carbon dioxide produced by each person every year. As a point of comparison, in the United States the average per capita production of carbon dioxide is 20 metric tons per year. As Jean-Pierre Gattuso reports, "It is estimated that the surface waters of the oceans have taken up over 500 thousand million metric ton of carbon dioxide (500 Gt CO_2), about half of all that generated by human activities since 1800." [22]

In addition to temperature changes, climate disruption is already starting to alter precipitation and wind patterns, all of which are driven by the vast oceans as heat sinks. The world's oceans, which will warm less quickly than land, are undergoing both warming and acidification. These impacts are already being seen on coral reefs, which are the ocean's most biodiverse type of ecosystem. At least three different but compounding mechanisms brought on by climate change kill living corals and threaten coral populations: (1) Temperature rise forces coral bleaching, which is the expulsion of tiny plant-like organisms (zooxanthellae) that live within the coral tissue and provide the host with food and oxygen; (2) warming temperatures magnify the effect of infectious diseases on coral, leading to more coral loss; (3) acidification of ocean water makes it more difficult and more costly in terms of energy for corals to secrete their calcium carbonate skeleton. We have already lowered the pH level of the ocean by about 0.1 unit. Lower pH slows coral growth, which compounds the problems brought on by bleaching and disease. Slow growth for coral means that coral loses its ability to compete with other species such as sponges and seaweeds and to keep up with sea level rise. As the concentration of atmospheric carbon dioxide increases, ocean warming and acidification will accelerate. Even conservative forecasts suggest the planet could lose coral reef systems on a large scale by 2100.

What kind of marine ecosystem would be left if coral died out in a warmer, more acid ocean that also threatens the sea butterfly? The fewer little creatures, like sea butterflies, there are in the sea, the fewer big creatures there will be. We do not really know how species that have coevolved over millions of years will co-disappear, but we are certain that the consequences will be severe (see Figure 5.7 for the future of coral as carbonate concentrations and water temperatures rise).

Returning to the land, climate shifts that warm our winters may seem pleasant in the short term, but they are also pleasant for mosquitoes and other insects that normally would be killed off in the winter. These include insects that can damage trees and other plants. Insect populations increase and outbreaks follow. Also, the physical changes in timing and quantity of temperature and moisture put enormous pressure on native plants. For example, trees like the sugar maple that turns such vibrant colors in the fall foliage season need snow-covered ground in the winters to protect their extensive root systems. A mid-winter thaw in Quebec in 1981 melted surface snow. A later freeze cycle left the shallower roots damaged, especially those of the maples. For a decade afterward, maples grew in stunted fashion. Trees in northern forests are more likely to show stunted growth and dieback where snow cover is prevented from developing over a winter. Also, insect-induced defoliation during the growing season weakens the trees and makes them more susceptible to secondary pathogens. In short, as winter temperatures rise, the trees fare less well, weaken, and die. They are replaced by more heat-tolerant species, such as oak and hickory, arriving from the south or lower elevations.

Replacing one species with another may seem inconsequential. But deep-rooted trees contribute much more to the moderation of our climate than scientists previously thought. During a drought in 1990, biologist Todd Dawson and his colleagues noticed that smaller plants closest to the trunks of large maples wilted far less than plants a short distance farther from the trees. Not every tree species establishes deep roots. Sugar maples have both deep taproots and extensive shallow roots. In times of drought, sugar maples draw water from deep lower soil layers through plant root systems into the drier upper soil layers in a process called hydraulic lift. This hydraulic lift benefits both the tree and many other plants growing around it. Neighboring plants take advantage of the "free" water supply during drought periods. Dawson and colleagues found that a large sugar maple can lift as much as 100 liters of deep-stored water overnight, about a quarter of the total amount of water it needs for daily photosynthesis.

But how did the deep water get to the taproot in the first place? More recently, Dawson and his colleagues have uncovered direct links between plant root functioning and climate. In a wet season, maples actually transport excess water to deep reservoirs underground, to be drawn up when needed.

The energy that drives this hydraulic redistribution comes from the leaves' transpiring water to the atmosphere in the photosynthetic process. The evaporation of water in dry times helps lower the surrounding air temperature. Scientists had wondered why, during a drought in the Amazon, the air temperature remained

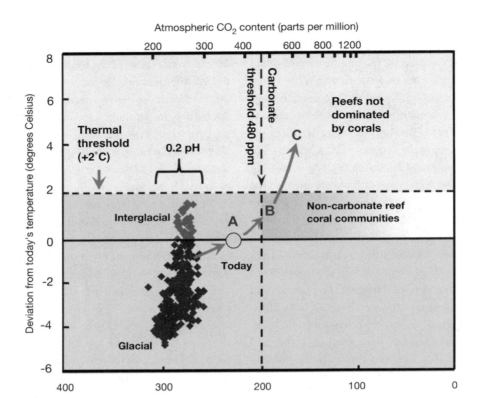

Atmospheric CO$_2$ content (parts per million)

FIGURE 5.7 Outlook for coral reefs, hosts of ocean biodiversity

Within 20 years, from 1980 to 2000, the sea surface warming accelerated to the current upward trajectory. Warm water peaks cause the animals that form coral to suffer bleaching as their shells disintegrate. As water approaches 28 degrees Celsius in summers, coral no longer has enough time to recover, and bleaching accelerates. The unit of measure in this figure is micromoles per kilogram. Source: [32] using [24]

cooler than would be expected. The release by the trees of deep water stored during the wet season explains this cooler air. This water redistribution helps the trees photosynthesize and therefore store more carbon during drought conditions than otherwise possible. [27]

There's this skin on the Earth—plants— that has an effect on a global scale, pulling carbon dioxide out of the atmosphere and letting water go, in a dynamic way that has climatic implications. [51]

Todd Dawson, UC Berkeley

Forests That Bring the Rain: A Case Example of All of the Above

The most significant forests on the planet are those in the Amazon. The vast Amazonian rainforests contain more tons of plant matter per square kilometer than almost any other area. This biomass stores enormous amounts of carbon and water. Deforestation in Brazil in particular has been driven by government policies that favor large agricultural firms. Philip Fearnside, an ecologist at Brazil's National Institute for Research in the Amazon, identi-

II: How to Think About Climate Solutions

fies the impact of forest loss on climate change: "Over three-fourths of Brazil's contribution to this global problem is the result of Amazonian deforestation. Half of the dry weight of the trees is carbon, and when forest is cut this carbon is released to the atmosphere either as carbon dioxide or as methane, both from burning and from decomposition of wood that fails to burn." [21] The tropical forests of the Amazon have suffered from uncontrolled logging that has reduced the area of forest and left fragmented remnants, from pollution from runoff and fertilizer use, and now from climate change. But here is where this story of the Amazon gets more interesting.

Recent analysis of satellite images of the Amazon shows that it may be the rain forest that makes the rain, rather than the other way around.* Contrary to conventional wisdom, the Amazon has both wet and dry seasons. The trees stay green year-round as tree roots redistribute water from deeper underground up to drier soils, as we examined earlier in the chapter. Most of the Amazon has greater leaf area in the dry season than the wet season. In other words, adequate sunlight appears more important to spurring leaf growth than is rainfall, as had been previously thought. In turn, leaf surface area plays a critical role in carbon, water, and climate cycles on local and global scales. The trees use the sunlight of the drier season to grow new leaves, using water stored during the wetter season. In fact, new leaf growth peaks just before the monsoon season arrives. The growth increases the release by the foliage of water vapor (evapotranspiration) into the air above the forest. The water vapor makes the air above the forest more buoyant, allowing it to rise and initiate the season's first thunderstorms, which in turn help change the wind patterns and bring more moist ocean air ashore. Hence, the leaf

---

*Visit these NASA Earth Observatory feature articles for great visuals on the Amazon forest: http://earth observatory.nasa.gov/study/AmazoneEVI and http://earthobservatory.nasa.gov/Study/AmazonLAI.

growth in the Amazon appears to help bring on the rainy season. So deforestation could alter the climate system enough to remove the triggers for the tropical monsoon season. [36]

The big picture here is simple: Reducing deforestation (and encouraging reforestation) will moderate climate change and thereby help protect biodiversity. A positive feedback loop exists between protecting rainforests worldwide and the impact they have on moderating climate extremes and reducing change, through their impact on the water cycle and their removal of greenhouse gases such as carbon dioxide from the atmosphere.

Health Consequences for Humans: Disruption Triggers Disease

The most vulnerable human and ecological systems are not difficult to find. One third to one half of the world's population already lacks adequate clean water, and climate change—involving increased temperature and droughts in many areas—is already adding to the severity of these issues. [25]

Roger E. Kasperson and Kirstin Dow, 2007

Another sobering aspect of the impact of climate disruption is the introduction of invasive disease carriers to new populations of humans and wildlife. Warmer, wetter climate increases the population of disease carriers. We are witnessing the globalization of what health professionals call vector-borne diseases, that is, contagious illnesses spread by certain carriers (vectors) such as insects and ticks, which are themselves not the direct causes of disease.

Global warming and extreme weather affect the breeding and survival of disease vectors such as mosquitoes responsible for malaria and West Nile virus. Malaria currently kills 3,000 African children a day. West Nile virus expands in regions suffering drought conditions that concentrate the mosquitoes.

Lyme disease, the most widespread vector-borne disease in the United States, is currently

TABLE 5.6 Ten Health Impacts of Climate Disruption

Infectious and respiratory diseases

1. *Malaria* is the deadliest, most disabling, and most economically damaging mosquito-borne disease worldwide. Warming affects its range, and extreme weather events can precipitate large outbreaks. Malaria underwent a fivefold increase in illness following a 6-week flood in Mozambique and is expanding in the highlands of Zimbabwe as mosquitoes find warming higher elevations more tolerable.

2. *West Nile virus* is an urban-based, mosquito-borne infection afflicting humans, horses, and more than 138 species of birds. It is already present in the United States, Europe, the Middle East, and Africa, and warm winters and spring droughts play roles in amplifying this disease. To date, there have been over 17,000 human cases and over 650 deaths from West Nile virus in North America since it arrived in 1999, probably with an infected traveler.

3. *Lyme disease* is the most widespread vector-borne disease in the United States and can cause long-term disability. Lyme disease is spreading in North America and Europe as winters warm. Models project that warming will continue to shift the suitable range for the deer ticks that carry this infection.

4. *Asthma* prevalence has quadrupled in the United States since 1980, and this condition is increasing in developed and underdeveloped nations. New drivers include rising CO_2, which increases the allergenic plant pollens and some soil fungi, and dust clouds containing particles and microbes coming from expanding deserts, compounding the effects of air pollutants and smog from the burning of fossil fuels.

Extreme weather events

5. *Heat waves* are becoming more common and more intense throughout the world. The outcome of the deadly 2003 summer heat wave in Europe and the potential impact of such "outlier" events elsewhere are stark for human health, forests, agricultural yields, mountain glaciers, and utility grids.

6. *Floods* inundated large parts of central Europe in 2002 and had consequences for human health and infrastructure. Serious floods occurred again in central Europe in 2005 and in the midwestern United States in 2008. The return times for such inundations are projected to decrease in developed and developing nations. Climate change is expected to result in more heavy rainfall events in such areas.

Natural and managed systems

7. *Water*, life's essential ingredient, faces enormous threats. Underground stores are being over-drawn and underfed. As weather patterns shift and mountain ice fields disappear, changes in water quality and availability will pose growth limitations on human settlements, agriculture, and hydropower. Flooding can lead to water contamination with toxic chemicals and microbes, and natural disasters routinely damage water-delivery infrastructure.

8. *Forests* are experiencing numerous pest infestations. Warming increases the range, reproductive rates, and activity of pests, such as spruce bark beetles, while drought makes trees more susceptible to the pests and wildfire. Large-scale forest diebacks are happening (e.g., the devastating bark beetle infestations in Colorado underway now) and have severe consequences for human health, property, wildlife, timber, and the carbon cycle.

9. *Agriculture* faces warming, more extremes, and more diseases. More drought and flooding under the new climate disruption, and accompanying outbreaks of crop pests and diseases, can affect yields, nutrition, food prices, and political stability. Chemical measures to limit infestations are costly and unhealthy.

10. *Marine ecosystems* are under increasing pressure from overfishing, excess wastes, loss of wetlands, and diseases of bivalves that normally filter and clean bays and estuaries. Even slightly elevated ocean temperatures can destroy the symbiotic relationship between algae and animal polyps that make up coral reefs, which buffer shores, harbor fish, and act as nurseries to juvenile fish populations and contain organisms with powerful chemicals useful to medicine. Warming seas and diseases may cause coral reefs to collapse.

Source: [17]

II: How to Think About Climate Solutions

increasing in North America as winters warm and the vectors—deer ticks—proliferate without harsh winters to cause diebacks. Under moderate climate change, the area suitable for deer ticks may increase by 213% by the 2080s.

Disease carriers are called vectors for two reasons. They have both a direction-of-spread and a speed-of-spread component. In the 1700s, outbreaks of dengue fever were few and far between. In the preindustrial days of sailing vessels, infrequent outbreaks occurred and were local. Only when viruses and their mosquito vectors could survive the slow transport on a sailing ship between population centers did a new strain wreak havoc. As cargo transport has sped up with motorized craft, dengue fever has now spread steadily northward from its origins in tropical Africa and South America, because winter conditions that previously killed its mosquito vectors are less common. By 1975 dengue fever had become a frequent cause of hospitalization and death among children in many countries in Southeast Asia. It is estimated that there are over 100 million cases of dengue worldwide each year. Dengue virus was accidentally reintroduced into Central America in 1994 and is now found in several countries in that region. Because this type of dengue has been absent from the Americas for almost 20 years, the human population has a low level of immunity. The virus is expected to spread rapidly in the coming decade throughout both North and South America, aided by the warming climate that allows their insect vectors to spread.

Increases in ragweed pollen, stimulated by increasing levels of carbon dioxide, may be contributing to the rising incidence of asthma. Microbes and other living organisms tend to increase their numbers exponentially, as their population levels reflect environmental conditions and resource availability. Mold, pollen, mushroom spores, and airborne particulates will all rise in incidence and thereby increase asthma rates. Indeed, they already are doing so. Asthma rates have climbed fourfold in the United States since 1980. [17]

Physician Paul Epstein and his colleagues at the Harvard Medical School compiled 10 case studies, summarized in Table 5.6, of actual events to explore the role climate disruption may have played in each. The results are sobering. As we begin to place monetary values on the costs of treating malaria or asthma and on social costs of diminished health, the case for taking immediate action to minimize climatic disruption becomes clearer.

Another aspect of the impact of climate disruption deserves sober reflection. Climate protection is a social justice issue at its most fundamental core. Let's call this aspect of climate disruption "the disproportionality factor." A disproportionately small number of wealthier, industrialized nations contribute the largest volumes of atmospheric greenhouse gases. The corollary to the disproportion of cause is the disproportion of impact. A disproportionately large number of less-affluent nations will suffer the heaviest and most immediate health burdens of climate disruption. Disproportionality also applies to people within a country. Those who contribute the fewest per capita emissions of carbon are ironically most vulnerable to the impacts of climate change. In essence, the poor, the young, the elderly, and the already ill (and future generations) will bear the heaviest burden as diseases spread, as water shortages occur, as food supply shocks set in motion supply-and-demand chains beyond their reach. Where one is born or whether one is wealthy or not should not be the determining factor of how quickly one might succumb to disease or famine or drought. Thus climate change is ultimately a matter of justice.

Whither Nature?
Tipping Elements We Should Fear!

Thomas Lovejoy points out that we are seeing the first signs of distressing positive feedback in the climate-ecosystem connection. Neither the climate nor the ecological changes we face may be gradual or linear. The disruptions may

Going up a mountain is like moving towards the poles in terms of climate and habitat. Each thousand feet of elevation is roughly equivalent to moving 300 miles northward. The higher you go, the colder the winters and the shorter the growing season. Ecologists identify life zones or ecological communities along an altitudinal gradient, largely as a function of colder temperatures at higher elevation. Mountaintops that reach 8,000–11,000 feet in Arizona often are covered with coniferous forests. These alpine forests are remnants from a cooler climate of the past and are now geographically isolated from each other, so they are known as "sky islands." Alpine plants and animals, such as the endangered Mount Graham red squirrel (*Tamiasciurus hudsonicus grahamensis*) have no way to connect with others of their species on other mountaintops.

Atop the highest mountains (above 13,000 feet), such as in the Rockies and Sierras, rock and ice dominate in conditions so severe that no trees can grow. Yet these habitats are far from barren. Among the species that live above tree line is a small mammal known as the American pika (*Ochotona princes*). Hikers may be familiar with the cute animal whose high-pitched whistle alerts their family members of danger. Pikas have a short active season during which they gather grass and raise their families. They literally make hay while the sun shines. In the winter, they feed on the hay they gathered and on small plants under the snow (http://www.eol.org/pages/133021).

On first glance, one would think that global warming would be good for pikas. A longer growing season would allow more time to feed and gather hay. However with rapid warming, the ideal conditions for pikas are fast disappearing. In the mountains of the Great Basin in Nevada, seven of 25 populations of pikas that were present in the 1930s had become extinct by the 1990s (Beever et al, 2003 cited by Parmesan and Galbraith 2004) [42]. The seven former populations were at significantly lower elevations than the survivors. Apparently warming conditions made it impossible for pikas to survive in these lower locations.

As the present locations of species become unsuitable for their future existence due to changed climate, the plants or animals will have to move to a more suitable location or else go extinct. Evidence of the loss of low elevation populations has been found in butterflies and plants. The Edith's checkerspot butterfly (Euphydrays editha) on the west coast studied by population biologist Paul Ehrlich and others has experienced nearly 80% extinction among its southernmost and lowest elevation populations. They have been able to shift to newly favorable conditions in northern and higher elevations. Alpine species are not so fortunate. There is great concern that as the climate warms, species that require cooler conditions will literally run out of mountaintop habitat in which to live.

be quite swift and exponential. In early 2002, the massive Larsen B ice shelf on the Antarctic Peninsula collapsed with a swiftness that stunned polar observers. This ice shelf had been stable for 12,000 years as a 220-meter (720-foot) layer of ice on top of the seawater underneath. The Larsen shelf chunk, comparable in size to the state of Rhode Island (3,250 square km), fractured and fell into the sea in a matter of weeks. Just as a vast ice sheet may suddenly break up when the stresses on it reach a tipping point, ecosystems may go haywire, to use a vernacular term for nonlinear dynamics.

For example, in the Rocky Mountains, pine

bark beetles may be spreading rapidly because milder winters are not killing them off. Native to the Rocky Mountains, the beetles are spreading quickly while killing their most common host, the lodgepole pine. Just west of Denver in Grand County, a surging human population and a warming climate are all connected. As large stands of lodgepole have been weakened or killed by the beetles, these dried stands have become much more susceptible to fire. Fire frequency is increasing in the West because of a combination of drier, hotter weather and land mismanagement. The loss of vast swaths of mature forest in just a decade changes the water retention capacity of the land and alters the habitat of species that rely on the thermal shelter and forage opportunities such forests offer. [53]

Up through the 1980s, Alaska was a net sink for carbon. Now, as the Arctic warms, its melting permafrost releases methane and carbon. This release induced by warming increases greenhouse gases that in turn amplify warming. Since the early 1990s, Alaska's tundra has become a carbon source of sizable and increasing dimension. Siberia is also a net source of methane and carbon, so the problem is not confined to North America. Arctic tundra may reach a point of warming where a sudden massive outgassing of methane will be unleashed, previously locked up in permafrost as frozen decomposing plant matter. [38] This will have a stunning impact on the climate.

Throughout the globe we find specific locations where a small change in a system may have strong effects on interactions, feedbacks, or connections of nearby systems. Such bottlenecks or switch elements will either cause a return to past climate or trigger a shift to a new mode of operation for the climate system.

For example, the only connection between the vast Pacific Ocean and the deep cold Arctic Ocean is the narrow and shallow Bering Strait (about 85 km wide and 50 meters deep) at the far northern end of the Pacific. As ice floes that bottleneck in the Bering Sea break up and as open water above the Arctic Circle expands in the summer, a huge volume of saltier water may be released into the northern Pacific. This release would have unknown but likely negative impacts on Pacific fisheries. Other important salinity valves exist in the Straits of Gibraltar, holding back the Mediterranean's saltier water from the eastern Atlantic, and in the Baltic, holding its brackish water back from the much saltier North Sea. The El Niño–Southern Oscillation (ENSO) cycles of the central Pacific, which bring, among other weather, the annual dry Santa Ana winds to fire-prone California, are also vulnerable to shifts in moisture content of air that climate disruption may tip one way or another. We mentioned the deforestation impacts in the Amazon on climate timing already.

Scientists are most concerned about the tipping of elements in the polar regions. University of East Anglia scientist Timothy Lenton and his collaborators introduced the term *tipping element* to "describe subsystems of the Earth system that are at least subcontinental in scale and can be switched—under certain circumstances—into a qualitatively different state by small perturbations. The tipping point is the corresponding critical point—in forcing and a feature of the system—at which the future state of the system is qualitatively altered." [28, 29] Once these tipping points are reached, there is no turning back.

What Can Be Done? Coevolution and Co-Disappearance

Thomas Lovejoy expresses our options succinctly: We need to revise our conservation strategies. For example, we need to (1) increase the natural connectivity between pieces of habitat with natural borders and corridors to facilitate the movement of plants and animals; (2)

FIGURE 5.8 online at ncse.org/climate solutions

Global map shows potential tipping elements in the climate system.

TABLE 5.7 Ecological Principles for Conserving Ecosystem Processes

Maintain or replicate

- Species richness, structural symmetry, and keystone processes
- Internal regulatory processes (e.g., predator-prey interactions)
- External diversifying forces
- Large habitat areas and spatial linkages between ecosystems
- Ecological gradients and ecotones

Minimize

- Erosion, nutrient leaching, and pollution emissions
- Landscape simplification
- Landscape homogenization

Mimic

- Natural processes in production cycles of materials

Source: [63]

minimize climate change impacts by reducing other stresses, such as siltation on coral reefs and deforestation; and (3) downscale climate projections from global to local in order to better understand the problem. Translation: Based on the existing global models, we should create localized projections more useful to regional policymakers and planners. In short, either we save what we can of the habitats in which all living things have coevolved or we all co-disappear.

Human-dominated marine ecosystems are experiencing accelerating loss of populations and species, with largely unknown, but certainly negative, consequences. On the other hand, according to Boris Worm and colleagues, "Restoration of biodiversity, in contrast, increased productivity fourfold and decreased variability by 21%, on average. We conclude that marine biodiversity loss is increasingly impairing the ocean's capacity to provide food, maintain water quality, and recover from perturbations. Yet available data suggest that at this point, these trends are still reversible." [67] Ecologists often describe strategies for conserving ecosystem functions as a choice to maintain extant systems, minimize further degradation, or mimic natural processes (see Table 5.7).

To protect currently threatened areas, we

need to engage in forward-looking biogeography. Climate change is shifting the location of suitable habitat, and these movements need to be considered in order to protect the parcels critical to future generations of affected species, including humans. In acknowledging that we will not be able to protect every parcel, we should recognize that some key parcels can foster larger safe havens. For example, marine reserves support population recovery and growth in neighboring areas. Fish species have increased in biomass inside the closed areas and are spilling over into surrounding waters, as has been documented in the Gulf of Maine, where cod and haddock stocks nose-dived in the 1990s. [23, 49] Additionally, due to local geography and topography, some locations may be less impacted by climate change. These areas should be protected as buffer zones against climatic disruption and as future reserves (see online Figure 5.9).

These conservation measures may help some species to get through the transition to a new climate change regime, but at the 2 degrees Celsius warming that is unavoidable, we still face catastrophic effects on biodiversity. Even below that temperature, some species, such as high-elevation and high-latitude species, may be irreversibly diminished and lost. Above that temperature, very little adaptation will be possible for species. Therefore, mitigating the forces of extinction is essential. The impacts of urbanization and the resulting air and water pollution and habitat destruction and fragmen-

FIGURE 5.9 online at ncse.org/climate solutions

Graph shows the maximum distances that fish from marine reserves traveled into surrounding unprotected waters.

FIGURE 5.10 online at ncse.org/climate solutions

Map of the Chesapeake Bay shows areas of loss and gain of its crucial aquatic grasses.

tation on native species have been poorly studied. But many researchers conclude that educating a highly urbanized human population about these impacts can greatly improve species conservation in all ecosystems. [35] Ultimately, we are facing an unprecedented situation with global climatic disruption on top of the forces already diminishing life on Earth. Therefore there is a need for new thinking, new science, and new approaches.

We cannot understand, let alone seek to improve, what we have not measured. Our scientific knowledge of life on Earth has only just scratched the surface of how diverse life really is. The reality of human-induced global climate disruption is that unless we radically mitigate—starting yesterday—we are pretty deep into irreversible and substantial biological losses already. Given the breadth and depth of the sixth mass extinction underway today, we have to act now based on the best available data to conserve what we can.

CONNECT THE DOTS

- We are losing species and other elements of biodiversity at an unprecedented rate.

- Very small changes can create thresholds beyond which conditions are irreversibly altered.

- The timing of natural events is affected differently for different species.

- Climate disruption is decoupling basic cycles in nature that we rely on for food, health, and wealth.

- Physical tipping elements around the globe exist and may be sensitive to small local changes.

- The extinction crisis we face is exacerbated by the four prior horsemen of the apocalypse: habitat loss and degradation, invasive exotic species, direct take of wild plants and animals, and pollution.

- Climate disruption is the fifth horseman that

makes the extinction crisis arrive faster and harder.

- Human-dominated marine ecosystems are experiencing accelerating loss of populations and species, with largely unknown consequences.

- The loss of biodiversity reduces the productivity of ecosystems and reduces their ability to recover from environmental shocks.

- The rise of ongoing extinctions may be more severe for marine and freshwater species than terrestrial species.

- In either case, the extinction rise and loss of biodiversity is unprecedented, worldwide, and potentially disastrous from the tropics to the temperate and polar zones, on land, in freshwater, and in the ocean.

- The consequences for human health include the spread of infectious and respiratory diseases, extreme weather events of life-threatening drought and flood, and erosion of life-supporting natural systems that supply water, natural products, and food from agriculture and marine ecosystems.

- The burdens of these consequences fall hardest on the weak, poor, elderly, young, and sick and future generations—in humans and in other living species.

- It is critical that we undertake conservation actions to blunt the worst effects of the forces of extinction, before the tipping elements are breached.

- We have a number of actions we can take, including reducing non-climate stresses on our ecosystems, increasing natural connections between existing natural areas to allow species to move, and downscaling global climate projections to local levels for local policymakers. Reducing deforestation will moderate climate change and thereby help protect biodiversity.

- But we also need new thinking, new science, and new approaches to help nature respond to a new situation.

Online Resources

Conservation International, www.conservation.org

Convention on Biological Diversity, www.cbd.int

Earth Observatory, http://earthobservatory.nasa.gov

Ecological Society of America, www.esa.org

International Union for Conservation of Nature, www.iucn.org

Millennium Ecosystem Assessment, www.millenniumassessment.org

National Academy of Sciences, http://nationalacademies.org/evolution

National Audubon Society, www.audubon.org/globalWarming/

National Audubon Society: http://birdsandclimate.audubon.org/

National Ecological Observatory Network, www.neoninc.org

National Marine Sanctuaries, http://sanctuaries.noaa.gov/

UN FAO Fishery Data, www.fao.org/fishery

Wildlife Habitat Policy Research Program, http://ncseonline.org/WHPRP/

World Database on Protected Areas, www.wdpa.org

www.eoearth.org/article/Biodiversity

www.eoearth.org/article/Biodiversity_fact_sheet

www.eoearth.org/article/Causes_of_forest_land_use_change

www.eoearth.org/article/Conservation_and_management_of_rare_plant_species

www.eoearth.org/article/Coral_reefs_and_climate_change

www.eoearth.org/article/Deforestation_in_Amazonia

www.eoearth.org/article/Dengue_and_Dengue_Hemorrhagic_Fever

www.eoearth.org/article/Evapotranspiration

www.eoearth.org/article/Human_development_and_climate_change

www.eoearth.org/article/Human_vulnerability_to_global_environmental_change

www.eoearth.org/article/Invasive_species

www.eoearth.org/article/Marine_biodiversity_and_food_security

www.eoearth.org/article/Marine_ecosystem_services

www.eoearth.org/article/Nonpoint_source_pollution

www.eoearth.org/article/Ocean_acidification

www.eoearth.org/article/Species_richness

www.eoearth.org/article/Terrestrial_biome

Climate Solution Actions

Action 18: Coastal Management and Climate Change

Action 19: Forest Management and Climate Change

Action 20: Climate Change, Wildlife Populations, and Disease Dynamics

Action 35: Climate Change and Human Health—Engaging the Public Health Community

Works Cited and Consulted

[1] Blockstein DE (1989) *A federal policy is needed to conserve biological diversity.* Issues in Science and Technology 5(4):63–67. www.issues.org/

[2] Cadotte MW, Cardinale BJ, Oakley TH (2008) *Evolutionary history and the effect of biodiversity on plant productivity.* Proceedings of the National Academy of Sciences. www.pnas.org/content/105/44/17012.abstract

[3] Cardinale BJ, Wright JP, Cadotte MW, Carroll IT, Hector A, Srivastava DS, Loreau M, Weis JJ (2007) *Impacts of plant diversity on biomass production increase through time because of species complementarity.* Proceedings of the National Academy of Sciences 104(46):18123. www.pnas.org/cgi/content/abstract/104/46/18123

[4] CBD (2008) Convention on Biological Diversity. Montreal (read December 18, 2008). www.cbd.int

[5] Chesapeake EcoCheck (2007) Chesapeake Bay Health Report Card. Partnership between NOAA, IAN, and UMCES (read October 4, 2008). www.eco-check.org/reportcard/chesapeake/2007/indicators/aquatic_grasses/

[6] Chivian E, Bernstein A (2008) *Sustaining Life: How Human Health Depends on Biodiversity.* (Oxford University Press, Oxford). www.oup.com/us/catalog/general/subject/LifeSciences/Ecology/?view=usa&ci=9780195175097

[7] Cohen AN, Carlton JT (1998) *Accelerating invasion rate in a highly invaded estuary.* Science 279:555–558. www.sciencemag.org/cgi/content/abstract/279/

[8] Cohn JP (2005) *Urban wildlife.* BioScience 55(3):201–205. http://www.tufts.edu/vet/seanet/pdf/bioscience_cohn_2005.pdf

[9] Congressional Research Service (CRS) (2008) "The Endangered Species Act and Claims of Property Rights 'Takings,'" (topic ed) McGinley M, in *Encyclopedia of Earth,* (ed) Cleveland CJ. (Environmental Information Coalition, National Council for Science and the Environment, Washington, DC). www.eoearth.org/article/The_Endangered_Species_Act_and_Claims_of_Property_Rights_%E2%80%9CTakings%E2%80%9D

[10] Conservation International (2007) Biodiversity Hotspots. Glossary (read November 23, 2008). www.biodiversityhotspots.org/xp/Hotspots/resources/pages/maps.aspx

[11] Duffy E (2006) "Marine Ecosystem Services,"(topic ed) Smith W, in *Encyclopedia of Earth,* (ed) Cleveland CJ. (Environmental Information Coalition, National Council for Science and the Environment, Washington, DC). www.eoearth.org/article/Marine_ecosystem_services

[12] Duffy E (2007) "Marine Biodiversity and Food Security," (topic ed) Cleveland CJ, in

Encyclopedia of Earth, (ed) Cleveland CJ. (Environmental Information Coalition, National Council for Science and the Environment, Washington, DC). www.eoearth.org/article/Marine_biodiversity_and_food_security

[13] Dunn RR, Danoff-Burg JA (2007) *Road size and carrion beetle assemblages in a New York forest.* Journal of Insect Conservation 11(4):325–332. www.springerlink.com/index/64645T21H777138L.pdf

[14] Edlinger E (1998) Effects of land-based pollution on Indonesian coral reefs: biodiversity, growth rates, bioerosion, and applications to the fossil record. Paper AAINQ42843. http://digitalcommons.mcmaster.ca/dissertations/AAINQ42843

[15] Ehrlich PR, Pringle RM (2008) *Where does biodiversity go from here? A grim business-as-usual forecast and a hopeful portfolio of partial solutions.* Proceedings of the National Academy of Sciences. www.pnas.org/cgi/content/abstract/105/Supplement_1/11579

[16] Eldredge N (2005) The Sixth Extinction. *American Institute of Biological Sciences.* www.actionbioscience.org/newfrontiers/eldredge2.html

[17] Epstein P (2005) Climate Change Futures: Health, Ecological and Economic Dimensions. *Harvard Medical School.* http://chge.med.harvard.edu/programs/ccf/

[18] ESA (2008) Education & Diversity Student Resources Fact Sheets. *Ecological Society of America.* Washington, DC (read September 15, 2008). www.esa.org/education_diversity/factsheets.php

[19] FAO (2007) The State of World Fisheries and Aquaculture 2006. Figure A. *Fisheries and Aquaculture Department.* www.fao.org/fishery

[20] Farnsworth E (2007) "Conservation and Management of Rare Plant Species," (topic ed) Sarkar S, in *Encyclopedia of Earth*, (ed) Cleveland CJ. (Environmental Information Coalition, National Council for Science and the Environment, Washington, DC). www.eoearth.org/article/Conservation_and_management_of_rare_plant_species

[21] Fearnside P (2007) "Deforestation in Amazonia," (topic ed) Hall-Beyer M, in *Encyclopedia of Earth*, (ed) Cleveland CJ. (Environmental Information Coalition, National Council for Science and the Environment, Washington, DC). www.eoearth.org/article/Deforestation_in_Amazonia

[22] Gattuso JP, et al. (2007) "Ocean Acidification," (topic ed) Duffy JE, in *Encyclopedia of Earth*, (ed) Cleveland CJ. (Environmental Information Coalition, National Council for Science and the Environment, Washington, DC). www.eoearth.org/article/Ocean_acidification

[23] Hannah L (2008) *Protected areas and climate change.* Annals of the New York Academy of Sciences 1134(1 The Year in Ecology and Conservation Biology 2008):201–212. www.blackwell-synergy.com/doi/abs/10.1196/annals.1439.009

[24] Hoegh-Guldberg O, Fine M, Skirving W, Johnstone R, Dove S, Strong A (2005) *Coral bleaching following wintry weather.* Limnology and Oceanography 50(1):265–271. http://aslo.org/lo/toc/vol_50/issue_1/index.html

[25] Kasperson RE, Dow K (2007) *Vulnerable Peoples and Places.* (Island Press, Washington, DC). www.millenniumassessment.org/en/Condition.aspx#download

[26] Kremen C, Williams NW, Thorp RW (2002) *Crop pollination from native bees at risk from agricultural intensification.* Proceedings of the National Academy of Sciences 99(26):16812–16816. www.pnas.org/cgi/doi/10.1073/pnas.262413599

[27] Lee JE, Oliveira RS, Dawson TE, Fung I (2005) *Root functioning modifies seasonal climate.* Proceedings of the National Academy of Sciences 102(49):17576–17581. www.pnas.org/cgi/content/abstract/102/49/17576

[28] Lenton TM, Held H, Kriegler E, Hall JW, Lucht W, Rahmstorf S, Schellnhuber HJ (2008) *Inaugural article: tipping elements in the Earth's climate system.* Proceedings of the National Academy of Sciences 105(6):1786. www.pnas.org/cgi/content/abstract/105/6/1786

[29] Lenton TM, Schellnhuber HJ (2007) Tipping the Scales. *Nature Reports: Climate Change.* www.nature.com/climate/2007/0712/full/climate.2007.65.html

[30] Leung B (2002) *An ounce of prevention or a pound of cure: bioeconomic risk analysis of invasive species.* Proceedings of the Royal Society B: Biological Sciences 269(1508):2407–2413. http://journals.royalsociety.org/index/A02NHXJ6QJ4FU45Q.pdf

[31] Losey JE, Mace V (2006) *The economic value of ecological services provided by insects.* BioScience 56(4):311–323. www.bioone.org/

[32] Lovejoy TE, Flannery T, Steiner A (2008) "Climate Change: We Did It, We Can Undo It." *International Herald Tribune*, October 27, 2008. www.iht.com/articles/2008/10/27/opinion/edlovejoy.php

[33] Lovejoy TE, Hannah L (2005) *Climate Change and Biodiversity.* http://yalepress.yale.edu/yupbooks/book.asp?isbn=0300104251

[34] Mace G, Masundire H, Baillie J (2007) Biodiversity. Ecosystems and Human Well-Being: Current State and Trends: Findings of the Condition and Trends Working Group. www.millenniumassessment.org/en/Condition.aspx#download

[35] McKinney ML (2002) *Urbanization, biodiversity, and conservation.* BioScience 52(10):883–890. www.bioone.org/

[36] Myneni RB, Yang W, Nemani RR, Huete AR, Dickinson RE, Knyazikhin Y, Didan K, Fu R, Negron Juarez RI, Saatchi SS (2007) *Large seasonal swings in leaf area of Amazon rainforests.* Proceedings of the National Academy of Sciences 104(12):4820. www.pnas.org/cgi/content/abstract/104/12/4820

[37] Nelson GC (2007) Drivers of Ecosystem Change: Summary Chapter. Ecosystems and Human Well-Being: Current State and Trends: Findings of the Condition and Trends Working Group. www.millenniumassessment.org/en/Condition.aspx#download

[38] Oechel WC, Vourlitis GL, Verfaillie J, Crawford T, Brooks S, Dumas E, Hope A, Stow D, Boynton B, Nosov V (2000) *A scaling approach for quantifying the net CO_2 flux of the Kuparuk River Basin, Alaska.* Global Change Biology 6(S1):160–173. www.blackwell-synergy.com/doi/abs/10.1046/j.1365-2486.2000.06018.x

[39] Ogden J, Davis S, Jacobs K, Barnes T, Fling H (2005) *The use of conceptual ecological models to guide ecosystem restoration in South Florida.* Wetlands 25(4):doi:10.1672:795–809. www.bioone.org/

[40] Ohio Department of Natural Resources (ODNR) (2006) Invasive Species: Aliens Among Us. *Division of Wildlife* (read November 23, 2008). www.dnr.state.oh.us/tabid/5825/default.aspx

[41] Parmesan C (2007) *Influences of species, latitudes and methodologies on estimates of phenological response to global warming.* Global Change Biology 13(9):1860–1872. www.blackwell-synergy.com/doi/abs/10.1111/j.1365-2486.2007.01404.x?ai=tx&mi=2rb8z&af=R

[42] Parmesan C, Galbraith H (2004) *Observed Impacts of Global Climate Change in the US.* (Pew Center on Global Climate Change). www.worldcat.org/wcpa/oclc/56941852&oi=institution

[43] Parmesan C, Yohe G (2003) *A globally coherent fingerprint of climate change impacts across natural systems.* Nature 421:doi:10.1038/nature01286. www.nature.com/nature/journal/v421/n6918/abs/nature01286.html

[44] Pauly D, Christensen V, Dalsgaard J, Froese R, Torres Jr F (1998) *Fishing down marine food webs.* Science 279(5352):860. www.sciencemag.org/cgi/content/abstract/sci;279/5352/860

[45] Phoenix GK, Hicks WK, Cinderby S, Kuylenstierna JCI, Stock WD, Dentener FJ, Giller KE, Austin AT, Lefroy RB, Gimeno BS (2006) *Atmospheric nitrogen deposition in world biodiversity hotspots: the need for a greater global perspective in assessing N deposition impacts.* Global Change Biology 12(3):470–476. www.blackwell-synergy.com/doi/abs/10.1111/j.1365-2486.2006.01104.x

[46] Pimentel D, Lach L, Zuniga R, Morrison D (2000) *Environmental and economic costs of nonindigenous species in the United States.* Bioscience 50:53–65. www.bioone.org/

[47] Pimm SL, Raven P (2000) *Biodiversity: extinction by numbers.* Nature 403(6772):843–844. www.nature.com/nature/journal/v403/n6772/full/403843a0.html

[48] Pimm SL, Russell GJ, Gittleman JL, Brooks TM (1995) *The future of biodiversity.* Science 269(5222):347–350. www.sciencemag.org/cgi/content/abstract/269/5222/347

[49] PISCO (2007) The Science of Marine Reserves. (eds) Lubchenco J, Gaines S, Grorud-Colvert K, Airamé S, Palumbi S, Warner R, Smith BS. www.piscoweb.org

[50] Ricciardi A, Rasmussen JB (1999) *Extinction rates of North American freshwater fauna.* Conservation Biology 13(5):1220–1222. www.blackwell-synergy.com/doi/abs/10.1046/j.1523-1739.1999.98380.x

[51] Sanders R (2006) *Deep-rooted trees contribute to the Earth's climate much more than scientists thought.* Medical News Today 2006 (13 January). www.medicalnewstoday.com/articles/36105.php

[52] Sherry RA, Zhou X, Gu S, Arnone III JA, Schimel DS, Verburg PS, Wallace LL, Luo Y (2007) *Divergence of reproductive phenology under climate warming.* Proceedings of the National Academy of Sciences 104(1):198. www.pnas.org/cgi/content/abstract/104/1/198

[53] Smith LC, MacDonald GM, Velichko AA, Beilman DW, Borisova OK, Frey KE, Kremenetski KV, Sheng Y (2004) *Siberian peatlands a net carbon sink and global methane source since the early Holocene.* Science 303(5656):353–356. www.sciencemag.org/cgi/content/abstract/303/5656/353

[54] Spalding M, Fish L, Wood LJ (2008) *Toward representative protection of the world's coasts and oceans—progress, gaps, and opportunities.* Conservation Letters 1:1–10. www3.interscience.wiley.com/cgi-bin/fulltext/121427657/

[55] Springer AM, Estes JA, van Vliet GB, Williams TM, Doak DF, Danner EM, Forney KA, Pfister B (2003) *Sequential megafaunal collapse in the North Pacific Ocean: an ongoing legacy of industrial whaling?* Proceedings of the National Academy of Sciences 100(21):12223–12228. www.pnas.org/cgi/content/abstract/100/21/12223

[56] Twilley RR (2001) *Confronting Climate Change in the Gulf Coast Region: Prospects for Sustaining Our Ecological Heritage.* (Union of Concerned Scientists; Ecological Society of America, Cambridge, MA). www.ucsusa.org/gulf/gcchallengereport.html

[57] US Geological Survey (USGS) (2008) Nonindigenous Aquatic Species. http://nas.er.usgs.gov/

[58] US Fish and Wildlife Service (USFWS) (2006)

Draft Comprehensive Conservation Plan and Environmental Assessment: Merritt Island National Wildlife Refuge. www.fws.gov/southeast/planning/

[59] US Fish and Wildlife Service (USFWS) (2008) Lake Wales Ridge NWR (read November 26, 2008). www.fws.gov/merrittisland/subrefuges/LWR.html

[60] USDA ARS (2008) "Improving Honey Bee Health: Coordinated Areawide Program Is Under Way." *Agricultural Research*, February:7. http://www.ars.usda.gov/is/AR/archive/key.htm

[61] Wackernagel M (2008) Ecological Footprint. *Global Footprint Network* (read August 20, 2008). www.footprintstandards.org

[62] Walther GR, Post E, Convey P, Menzel A, Parmesan C, Beebee TJC, Fromentin JM, Hoegh-Guldberg O, Bairlein F (2002) *Ecological responses to recent climate change*. Nature 416:389–395. http://www.nature.com/nature/journal/v416/n6879/abs/416389a.html

[63] Western D (2001) *Human-modified ecosystems and future evolution*. Proceedings of the National Academy of Sciences 98(10):5458–5465. http://www.pnas.org/cgi/content/abstract/98/10/5458

[64] Wiegman L (2006) Images from Kennedy Space Center. January. http://etothefourth.com

[65] Wilson EO (1997) *Strangers in Paradise: Impact and Management of Nonindigenous Species in Florida*, ix–x, (eds) Simberloff D, Schmitz DC, Brown TC. (Island Press, Washington, DC). http://islandpress.org

[66] Windle PN, Kranz RH, La M (2008) *Invasive Species in Ohio: Pathways, Policies, and Costs.* (Union of Concerned Scientists, Cambridge, MA). www.ucsusa.org/publicationswww.ucsusa.org/publications

[67] Worm B, Barbier EB, Beaumont N, Duffy JE, Folke C, Halpern BS, Jackson JBC, Lotze HK, Micheli F, Palumbi SR (2006) *Impacts of biodiversity loss on ocean ecosystem services*. Science 314(5800):787–790. http://www.sciencemag.org/cgi/content/abstract/314/5800/787

CHAPTER 6

The Cheapest Carbon

Doing nothing about climate change is far more expensive and
risk than taking strong pro-active and immediate measures. [10]
SIR NICHOLAS STERN, 2006

This chapter is short, but important. If the rise in atmospheric greenhouse gases is the problem, how do we fix it? Anyone can sequester carbon by planting a tree. But we cannot "plant our way" out of the current steep rise in carbon dioxide emissions. Trees need time to grow and we would need to find, buy, or convert space for thousands of square kilometers of new forests. Once carbon becomes airborne as an emission, removing carbon from the atmosphere is very expensive. As Thomas Dietz of Michigan State University says, "The cheapest carbon to eliminate from the atmosphere is carbon that we don't put there in the first place." [2]

Amory Lovins of the Rocky Mountain Institute talks about "negawatts." By that, Lovins means investing to reduce electricity demand instead of investing to increase electricity generation capacity. For example, enhancing the energy efficiency of our buildings—such as by painting our roofs white and adding insulation—and installing local renewable energy sources—such as solar panels but also passive design features—so buildings actually produce more power than they use, can be much less expensive than building more power plants. [10]

To stabilize emissions in 50 years (by 2054), Princeton University's Stephen Pacala and Robert Socolow identify 15 different strategies and technologies that could each reduce greenhouse gas emissions by 1 billion tons annually. These greenhouse gas reduction methods include increasing biofuel production 50-fold over the current levels in Brazil and the United States, capturing carbon at the source and sequestering it with new and developing technologies, and doubling nuclear power capacity to replace fossil fuel–powered plants. [13] Building safe new nuclear power plants and subterranean carbon injection systems are big, very expensive projects, even for the most ambitious and well-financed do-it-yourselfers. So what meaningful emission reduction actions are available for individuals, families, and mom-and-pop businesses?

What Can a Family Do?

A lot! Each gallon (3.7 liters) of gasoline burned produces 8.8 kilograms (19.4 pounds) of carbon dioxide emissions. The average US automobile emits 5,200 kilograms (5.735 tons) of carbon dioxide per year. American households and transportation they rely on produce somewhere between 32% and 41% of all US greenhouse gas emissions each year. That is 2.1 billion tons, or 8% of the world's total emissions. [16]

The individual and household sector is a potential source of prompt and large emissions reductions (as we can see in Figure 6.1). As individuals, we have at least four ways in which our behavior can make a difference: environmental activism (we can be part of the movement and individually commit to activism); community activism (we can join groups that are trying to effect change, petition civic leaders, and vote); organizational behavior (we can seek to change behaviors of the organizations in our lives); and finally, private-sphere behaviors (we can do things differently at home, as consumers, and as maintainers and operators of our homes and vehicles). [14] Specifically, many private behaviors carry large potential public benefits. These include consumer purchasing behavior, maintenance of homes, household equipment and vehicles, changes in equipment use (lifestyle, curtailment), waste disposal behavior, and "green consumerism." Our diet plays a huge role here too. Animal agriculture is responsible for 18% of global greenhouse gas emissions, especially in the form of methane gas. Total greenhouse gas emissions worldwide from animal agriculture exceed that of transportation. For the entire agriculture sector, livestock constitute nearly 80% of all emissions. [6] (See also Figure 3.9, Concentrations of Greenhouse Gases from 0 to 2000, in Chapter 3.) Thus reducing the amount of red meat in our diet will improve our health and that of the planet.

American families have big feet. The ecological footprint of American households and their transportation is greater than the total

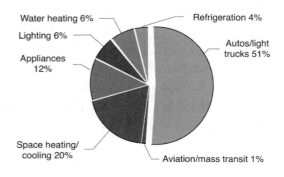

FIGURE 6.1 **Greenhouse gas emissions from US households and transportation**

Household motor vehicles are a very substantial source of greenhouse gases, accounting for over half of all annual household-related emissions. The heating and cooling of space in our homes is another big slice of the emission pie, at 20%. The 12% attributable to appliances is roughly equivalent to the emissions of the entire US chemical industry. The 6% of household emissions coming from heating water and the 6% from lighting are each roughly equivalent to the combined emissions of all iron and steel producers and paper mills in the United States. Source: Adapted from [17]

emissions footprint of any country in the world. Mathis Wackernagel, executive director of the Global Footprint Network who originally conceived of ecological footprints, with William Rees of the University of British Columbia, defines the footprint concept as follows: "a measure of how much biologically productive land and water an individual, population or activity requires to produce all the resources it consumes and to absorb the waste it generates using prevailing technology and resource management practices." [18] Carbon footprints are an important subset of our overall ecological footprint, accounting for between half and three-quarters of humanity's total ecological impact. (See Figure 6.2.) A carbon footprint is the amount of greenhouse gases, measured as units of carbon dioxide, produced by a human activity. [17]

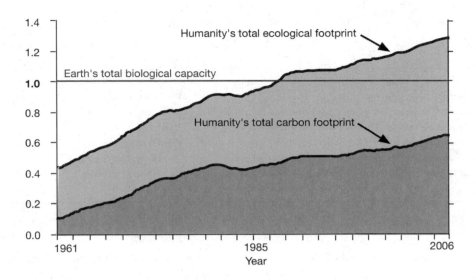

FIGURE 6.2 **Humanity's growing ecological and carbon footprints**

Carbon emissions are a significant contributor to the total human ecological footprint, that is, the impact human activities have on the planet's capacity to renew the resources consumed each year. In this figure, the entire capacity of our one Earth is 1.0. Source: [18]

Seven Low-Hanging Fruits for the Family Carbon Diet

In an important analysis subtitled "Low-Hanging Fruit," Michael Vandenbergh of Vanderbilt University's Law School and his colleagues have pointed to seven changes individual consumers can make that add up to very significant emissions reductions, even if just 10% to 33% of Americans undertake them. These "quick and easy" behavior changes are listed in Table 6.1.

Note that none of these changes, other than buying compact fluorescent lightbulbs (CFLs) to replace incandescent bulbs, affect consumption. Additional behavior changes such as buying less and driving less would save even more money and carbon dioxide emissions. Vandenbergh describes the seven "low-hanging fruit" as "actions that have the potential to achieve large reductions at less than half the cost of the leading current federal legislation, require limited up-front government expenditures, generate net savings for the individual, and do not confront other barriers." [16, p. i]

Taken together, these behavior changes alone would generate roughly 150 million tons in emissions reductions and several billion dollars in net social savings by 2014. That is the equivalent of removing 26 million automobiles from the road! These actions can be put into action immediately with some public education. They would generate a net social savings from lowered utility bills for households at the cost of energy at current prices.

Some of the Cheapest Carbon You're Going to Find

Now here is the good news: These changes will cost us very little—if anything. Efficiency actions usually have a positive financial impact, actually saving money. Vandenbergh estimates these seven emissions reductions can be achieved for a cost of $10 per ton of carbon avoided, or less. The government's out-of-pocket cost in this campaign would be $2 billion for public information campaigns, subsidies, and other activities. That sounds like a lot. But $2 billion is just 2% of what

TABLE 6.1 Seven Low-Hanging Family Fruits

Action	Plasticity (with this level of action)	Emissions saved (million tons of carbon per year)
Reduce motor vehicle idling.	10% of drivers complying	6–10
Reduce "standby power" electricity use.	33% reduction	16–22
Accelerate use of compact fluorescent lightbulbs (CFLs).	300 million bulbs	12–37
Adjust temperature settings 2°F in both summer and winter.	33% of households	18–36
Lower temperature settings on water heaters.	50% of households	28–38
Maintain recommended tire pressure in personal motor vehicles.	33% of drivers complying	14
Change air filters in personal motor vehicles at recommended intervals.	25% of drivers complying	24
Total		**118–181**

Source: [16]

the US government spent on the 2008 economic stimulus checks mailed to US taxpayers. Put differently, these changes can be assisted by the government's investing about $19 for each of America's roughly 105 million households.

The barriers to personal behavior change that reduce energy use are high, but not insurmountable. Americans are accustomed to a high-consumption lifestyle. We have not been very attentive to energy practices. We have limited trust in both government and our energy providers. We have high levels of built-in costs that discourage incurring new up-front costs for higher-efficiency technologies. We are often captives of energy decisions made by others on behalf of consumers. And we are used to relatively cheap energy. But as Vandenbergh points out, "Although each of these barriers can be overcome, to constitute a low-hanging fruit action these types of personal barriers must be minimal." So the seven actions that avoid these barriers are a very good place to start. All of these low-hanging fruit use existing off-the-shelf technologies. For example, to replace 300 million incandescent lightbulbs with CFLs—that use 75% less electricity for the same amount of light—each American household would install three such CFL bulbs. Or a third of American households would install nine CFL bulbs. While high-quality CFL bulbs used to be much more expensive up front than incandescent bulbs, the price has dropped dramatically to a few dol-

lars in the past 2 years as the CFL market has expanded. And now, we are developing less-expensive lighting technologies every year. We are literally building a better lightbulb.

> There can be behavioral ways that can be deployed quickly with low, zero, or in many cases, negative cost. This isn't always giving up your car to use public transportation or carpool—even though avoiding unnecessary driving does save money and reduce pollution. These can be very small things but cumulatively they can have a greater effect. [2]
>
> *Thomas Dietz*

To get to what Dietz called "some of the cheapest carbon you're going to find," we need a true commitment to capitalize on science to promote the need for these behavioral changes. Investment in research on the human dimensions aspects of climate change is woefully low and possibly declining; less than a page of the Intergovernmental Panel on Climate Change Working Group III's chapter on building energy was devoted to behavioral change. The percentage of the federal global change research budget devoted to human dimensions declined from 3% in 1991 to 2% in 2006. Despite that, we have very substantial social science research on which we can base campaigns to drive behavioral change. Hence we need to increase investment in social science research and increase use of social science findings in developing and implementing carbon reduction strategies. [3]

If a sufficient number of Americans take these actions, we can reduce our emissions by roughly 150 million tons of carbon dioxide per year—while saving money.

Cost-to-Benefit Details of the Seven Low-Hanging Fruit

Here are some details for each of the seven consumer actions. Since Michael Vandenbergh writes so clearly, his owns words below (the extracts) will do most of the talking. (Specific pages from his article are noted in the references.)

#1. Reduce Engine Idling
(Cost: $0, Benefit: immediate)

Modern car engines need almost no warm-up. Few of us realize that restarting a warm engine consumes less fuel and emits less pollution than idling for 5 to 10 seconds.

> If a vehicle will idle for more than 5 to 10 seconds, shutting the engine off and restarting it when the driver is ready to resume driving typically will not only reduce fuel consumption, but also will reduce wear and tear on the engine, improve fuel economy, and improve the performance of catalytic converters. For idle times of 45 seconds or more, the savings in fuel consumption and engine maintenance from shutting off the engine vastly exceed the minor wear-and-tear associated with restarting the engine. [16, p. 24]

#2. Reduce Standby Power
(Cost: $0, Benefit: immediate)

Many electronic devices use power all day and night, from cell phone chargers to televisions, whether anyone is home or in the room. Home computer and wireless networks and home entertainment centers with flat-screen televisions are big standby power hogs. So are most devices with remote controls. A quick solution

is to cut off power to any such systems at night by switching off the power strip into which they are plugged. Eventually, the government Energy Star program (www.energystar.gov) must mandate that manufacturers use lower-wattage settings for standby draw. Using energy-monitoring smart power strips ($30–$60) in a home or office could pay for itself within a year.

> Some large-screen televisions can use as much power in standby mode as a refrigerator. According to the U.S. Department of Energy, 40 percent of electricity consumption by home electronics occurs in standby mode. Certain appliances, such as microwave ovens and video recorders (VCRs), actually consume more electricity over the course of a year running their clock displays in standby mode than they do while in use. [16, p. 32]

#3. Install CFL Bulbs
(Cost: $5–$10, Benefit: immediate)

We can replace 10% of the nation's 3.1 billion incandescent bulbs with compact fluorescents (CFLs) if every American household swaps 3 CLF bulbs for the old ones. CFLs use 75% less electricity for the same amount of light and last two to four times longer.

> Unlike many other emissions-generating technologies, light bulb turnover is quite rapid. The common [incandescent bulb] has a life of only 1,000 hours, so CFLs can be substituted quickly, and they produce significant short-term emissions reductions. Further, CFL prices have dropped dramatically in the past few years, and consumers are now able to purchase these bulbs for less than $3 per bulb. This means the CFL payback to the consumer will occur within months after purchase. [16, p. 40]

#4. Lower Thermostat Settings
(Cost: $0, Benefit: immediate)

Lowering the thermostat slightly in winter and raising it slightly in summer could save at least $125 a year in costs per household and add up to big emission reductions. With proper insulation in our attics and sill plates, we may not notice any difference in comfort.

We believe a modest two degree Fahrenheit (F) change in ambient indoor temperatures, combined with a more significant reduction in overnight winter temperatures, does not constitute a significant lifestyle adjustment. The range of annual savings derived from a two-degree F change in summer and winter temperatures runs from 1,000 to 2,000 pounds of CO_2 per household, depending on the source of the energy used for home heating and cooling, the efficiency of existing equipment, current temperature settings, and other factors. [16, p. 45]

#5. Lower Water Temperatures
(Cost: $0, Benefit: immediate)

We may rarely think of our water heaters as overworking. Lowering the water heater setting by 20°F would make very little difference in our comfort level and yield an everyday savings in energy and emissions. With proper insulation around our hot-water delivery pipes, we may not notice any difference in comfort. Many basements may have 30 to 40 feet of new uninsulated hot-water pipe!

Many hot water heaters are installed with a default temperature setting of 140–150 degrees F, when in most cases temperatures of 115 or 120 degrees F will be perfectly adequate to meet household needs. Individuals can adjust the temperature settings by themselves with only a small time cost and without any financial cost. The financial savings from reducing temperatures by 20 degrees F would be about $24 to $40 per year per household. [16, p. 47]

#6. Maintain Tire Pressure
(Cost: $10, Benefit: immediate)

Our cars produce roughly half the total greenhouse gas emissions over which consumers have direct control. Given the potential for high gasoline prices, as we experienced in the summer of 2008, proper air pressure in tires makes good financial sense. A good tire gauge is less than $10.

The U.S. Department of Energy estimates that vehicle gas mileage improves an average of 3.3 percent by inflating tires regularly to proper pressures. Tire gauges are inexpensive, and routine oil changes often include tire inflation as a matter of course. The low-hanging fruit action is simply to get the U.S. public to check and maintain tire pressure on a consistent basis. A two-car family could save about $120 per year by taking this action. [16, p. 49]

#7. Replace Air Filters
(Cost: about $30, Benefit: immediate)

Most of us assume our car engines' air filters will be replaced at the regular tune-ups every 15,000 miles or so. But it pays to be sure.

Gasoline savings alone from changing an air filter at the recommended interval total about $240 per year. As a result, it is cost effective for the individual to maintain a regular schedule for changing filters. Periodic air filter changes can save the vehicle owner anywhere from 7 to 10 percent in fuel mileage. [16, p. 50–51]

The bottom line is, by adopting these low-hanging fruit, we can save money and reduce our emissions with very little up-front expense. We don't have to wait for massive government programs. No doubt, each of you can think of many other energy and material conservation steps that you can harvest for meaningful social good—as well as your own good—from avoiding plastic shopping bags to carpooling and bicycling more. We can take ownership over making a difference—one household at a time.

Efficiency-improving actions generally save more energy than curtailing use of intrinsically inefficient equipment. [7]

Gerald Gardner and Paul Stern

INSIGHT 6: CARBON AND ITS EQUIVALENTS

Each greenhouse gas—such as carbon dioxide, methane, or any synthetic fluorinated gas—has a different atmospheric concentration and a different strength as a greenhouse gas. The strength of the compound to force heat retention is called a "climate forcing." A potent greenhouse gas with a very small atmospheric concentration can contribute to the overall greenhouse effect just as much as a weaker greenhouse gas with a much larger atmospheric concentration. Methane is about 20 times more powerful in its global-warming potential than carbon dioxide in the short term; nitrous oxide is about 70 times more powerful than carbon dioxide in the short term; and many of the synthetic fluorinated gases are much more powerful than that and much longer lasting. Some compounds persist longer in the atmosphere than others. [11]

A carbon footprint is the measure of the amount of greenhouse gases, measured in units of carbon dioxide, produced by human activities. A carbon footprint can be measured for an individual or an organization, and it is typically given in tons of "carbon dioxide equivalent" (often abbreviated as CO_2e or CO_2eq) per year. For example, the average North American generates about 20 tons of CO_2e each year. The global average carbon footprint is about 4 tons of CO_2e per year (see online Figure 6.3). [18]

Stern and most other researchers use carbon dioxide equivalents in discussing greenhouse gas emissions, particularly when comparing emission impacts across many individual gases. The truth is that most human activities emit more than one greenhouse gas. Burning coal for electricity produces carbon dioxide but also nitrous oxide and sulfur oxide gases. The "carbon" footprint we have here bandied about often actually includes the impact of the several greenhouse gases combined that are produced by a given activity. If so, their impact should be measured in tons of CO_2e, or the tons of CO_2 that would cause the same level of radiative forcing. CO_2e is expressed in parts per million by volume (ppmv). For brevity in this book, we will shorten this unit to ppm, as in CO_2e ppm.

A related but distinct concept is "equivalent carbon dioxide." This is a quantity that describes for a given mixture and amount of greenhouse gas the amount of carbon dioxide that would have the same global-warming potential when measured over a long timescale, typically 100 years. Equivalent carbon dioxide is a time-factored measure obtained by multiplying the mass emitted by the global-warming potential of the gas. Methane is powerful in the short term, but its lifetime in the atmosphere is relatively brief (10–12 years) compared with some other greenhouse gases (such as carbon dioxide, nitrous oxide, or the synthetic fluorocarbons). So the equivalent carbon dioxide of a given mass of methane declines over the long term (see Table 6.2).

(continued)

FIGURE 6.3 online at ncse.org/climate solutions

Carbon dioxide emissions per capita for selected nations

TABLE 6.2 Relative Global Warming Potentials of Four Greenhouse Gases

| Gas | Atmospheric lifetime (years) | Global warming potential (relative to CO_2) time frames | | | Concentration levels (parts per billion) | | Main human activity source |
		20 years	100 years	500 years	Preindustrial (ppb)	2007 levels (ppb)	
Water (H_2O)	(A few days)	(NA)	(NA)	(NA)	1,000 to 3,000	1,000 to 3,000	(NA)
Carbon dioxide (CO_2)	About 1,000*	1	1	1	280,000	387,000	Fossil fuel, cement production, land use change
Methane (CH_4)	11	67	23	6.9	250	1,750	Fossil fuel, rice paddies, waste dumps, livestock
Nitrous oxide (N_2O)	114	291	298	153	270	315	Fertilizers, combustion, industrial processes

Source: Adapted from [8] and [15]

The concentration of emitted carbon dioxide drops rapidly at first, but significant carbon dioxide remains in the atmosphere even after 1,000 years. This is why carbon dioxide we put into the atmosphere will be trouble for a very long time. It is common to refer to the global-warming potential (also known as GWP) of an airborne compound over a 100-year time frame. Clearly, time does remove some of the greenhouse punch of methane or nitrous oxide, but time has much less effect in reducing the effect of our most pernicious gas, carbon dioxide.

The True Cost of Carbon (Or Who Owns the Sky?)

In the United States, land of the free, the carbon dioxide we all produce by turning on cars or by burning coal to produce electricity has been essentially free. Why? Because the atmosphere is free. Greenhouse gas emission comes without a direct price tag to those who use the atmosphere as a dumping ground and are not paying for the right to do so. Essentially, it is as if we were treating the atmosphere as a dumping ground of infinite capacity. But we know that is not the case. And we know that putting too much stuff in the atmosphere will have and is already having steep, disruptive consequences on health, safety, and the welfare of humans. We also know that the atmosphere knows no

boundaries. Unlike land, which can be yours or mine, air belongs to all of us equally.

When no property rights are assigned to a resource, it is often overused by those able to do so. They suffer no immediate penalty for using more. We are realizing that placing a property-right value on a resource like the atmosphere, or more precisely on the right to use the atmosphere, makes the user pay more attention to how he or she is using that resource. In practice, if we place a monetary value on a ton of carbon dioxide emitted, we begin to use market signals about the value of avoiding that cost (by emitting less). How we decide what price to use could take an entire book to explore. In short, economists tell us that whether the price for using the air (that is, emitting the greenhouse gas) is a tax on the gas emitted by volume or an allowance permit with a cost by volume (a cap and trade scheme), " the price under an efficient cap and trade policy will be exactly equal to the efficient tax." [9]

While climate scientists have been urging that we take action on climate change before it is too late, many economists have been suggesting that we wait until we are more prosperous to be able to afford the cost of taking action. Their argument against taking action now is that retooling to decarbonize our energy infrastructure would be very expensive. In late 2006 the chief economist of the United Kingdom, Sir Nicholas Stern, released a lengthy, detailed report, the *Stern Review on the Economics of Climate Change*. It was commissioned, not by the environment ministry, but by the treasury.

Sir Nicholas reached a profound conclusion: Doing nothing about climate change is far more expensive and risky than taking strong proactive and immediate measures. One reason past economic analysis has underestimated the true economic costs of business as usual is that "climate change is a result of the externality associated with greenhouse-gas emissions—it entails costs that are not paid for by those who create the emissions." [13, ch. 2]) Others, namely our children or residents in distant lands, will pay for these costs. Furthermore, Stern reported that the monetary cost of climate change would be much higher than previously expected, because earlier estimates had not included some of most uncertain, but highest, impact consequences. And the cost of doing nothing, the "business as usual" approach, would mean, at a minimum, a 5% average reduction in global per capita consumption "now and forever."

Taking into account other risk factors such as direct impact on human health of climate disruption, Stern suggested, could make doing nothing about climate change today mean a permanent per capita consumption reduction of 20%. That means a US economy up to one-fifth smaller than today. [13, ch. 6] The good news is that actually paying for mitigation of greenhouse gases and adaptation to climate disruption is far less expensive, if we begin now. Stern and his colleagues estimate that stabilizing atmospheric concentrations of greenhouse gases would require deeply reducing the output to three-quarters of today's levels by 2050, and ultimately to one-fifth of today's levels by the century's end. This effort would stabilize atmospheric carbon dioxide equivalent levels between 500 and 550 parts per million. This global effort would require an investment of around 1% of gross domestic product, which is dramatically less expensive than the 5% to 20% costs of doing nothing. [13, ch. 9] In the *Stern Review* Sir Nicholas concludes simply, "Tackling climate change is the pro-growth strategy." Since the report was published, many others in business, government, and the nonprofit sector have come to agree. Many of the same actions that combat climate change, such as technological research and development, education, and infrastructure improvements, also strengthen the long-term economy. [8, 13] We will examine those in chapters that lie ahead. But before we move on, a bit of background about carbon and the other greenhouse gases is available in the online content for this chapter.

To sum up, we have examined some elements of the problem (emissions) and learned we lack a price signal strong enough to reward positive behavior or punish negative behavior. And we have learned that good old-fashioned conservation—"using less"—is a potentially powerful part of the solution.

CONNECT THE DOTS

- American households and transportation they rely on produce somewhere between 32% and 41% of all US greenhouse gas emissions each year. That is 2.1 billion tons, or 8% of the world's total emissions

- No-cost and low-cost behavior changes are at hand that bring real carbon dioxide emission reductions.

- If enough Americans take these actions, we can reduce our emission by roughly 150 million tons of carbon dioxide per year—while saving money.

- The essential question—Do we fit on our planet?—has one overwhelming answer: No!

- The next question is tougher to answer—How do we change behaviors so that we do fit in a way that is fair and just for all humans, regardless of where we were born?

Online Resources

www.eoearth.org/article/Consumption_and_well-being

www.eoearth.org/article/Carbon_footprint

www.eoearth.org/article/Economics_of_climate_change

www.eoearth.org/article/Toward_an_ecological_economy

Global Footprint Network, www.globalfootprint.org

Human Dimensions of Global Change Project, www7.nationalacademies.org/hdgc/

NOAA National Climate Data Center, www.ncdc.noaa.gov

United Kingdom Office of Climate Change, www.occ.gov.uk/

United Nations Environment Programme, www.unep.org

UNEP/GRID-Arendal, www.grida.no/climate

United States DOE Energy Savers, www.energysavers.gov

See also extra content for Chapter 6 online at http://ncseonline.org/climatesolutions

Climate Solution Actions

Action 1: Green Buildings and Building Design

Action 2: Moving Forward—Transportation and Emissions Reduction

Action 5: Mitigating Greenhouse Gases Other Than CO_2

Action 6: Energy Efficiency and Conservation

Action 11: Economics—Setting the Price for Carbon

Works Cited and Consulted

[1] Benfield K (2008) Smart Growth Program. *Natural Resouces Defense Council*, Washington, DC (read December 28, 2008). http://switchboard.nrdc.org/blogs/kbenfield/

[2] Dietz T (2008) Human Action and Climate Change. *Climate Change: Science and Solutions: 8th National Conference on Science, Policy and the Environment.* http://ncseonline.org/2008conference/cms.cfm?id=1716

[3] Dietz T, Stern PC (2008) *Public Participation in Environmental Assessment and Decision Making.* (National Research Council, Washington, DC). www.nap.edu/catalog/12434.html

[4] Ehrlich PR, Pringle RM (2008) *Where does biodiversity go from here? A grim business-as-usual forecast and a hopeful portfolio of partial solutions.* Proceedings of the National Academy of Sciences. http://www.pnas.org/cgi/content/abstract/105/Supplement_1/11579

[5] Ewing R, Kreutzer R (2006) Understanding the Relationship between Public Health and the Built Environment. *US Green Building Council.* New York. www.usgbc.org

[6] FAO (2006) Livestock's Long Shadow. www.fao.org/docrep/010/a0701e/a0701e00.htm

[7] Gardner GT, Stern PC (2008) "The Short List: The Most Effective Actions US Households Can Take to Curb Climate Change." *Environment*, September–October:12–24. www.environmentmagazine.org

[8] Jones V (2008) *The Green Collar Economy.* (HarperCollins Publishers, New York). www.greenforall.org

[9] Keohane NO, Olmstead SM (2007) *Markets and the Environment*, 141–142. (Island Press, Washington, DC). http://islandpress.org

[10] Lovins A (1989) The Negawatt Revolution: Solving the CO_2 Problem. *Green Energy Conference.* CCNR Montreal (read August 21, 2008). www.ccnr.org/amory.html

[11] NOAA (2005) Greenhouse Gases: Frequently Asked Questions. *National Climatic Data Center (NCDC).* National Oceanic and Atmospheric Administration, Washington, DC (read November 1, 2008). www.ncdc.noaa.gov

[12] NRDC (2008) Smart Growth. *Natural Resources Defense Council.* New York (read December 28, 2008). www.nrdc.org/smartgrowth/

[13] Socolow R, Pacala S (2004) *Stabilization wedges: solving the climate problem for the next 50 years with current technologies.* Science 304(5686):968–972. www.sciencemag.org

[14] Stern N (2007) The Stern Review Report on the Economics of Climate Change. HM Treasury London. www.hm-treasury.gov.uk/sternreview_index.htm

[15] Stern PC (2000) *New environmental theories: toward a coherent theory of environmentally significant behavior.* Promoting Environmentalism 56(3):407–424. http://www.blackwell-synergy.com/doi/abs/10.1111/0022-4537.00175

[16] UNEP/GRID-Arendal (2005) National Carbon Dioxide (CO_2) Emissions per Capita. *Vital Climate Change Graphics Update.* Oslo. http://maps.grida.no/go/graphic/national_carbon_dioxide_co2_

e missions_per_capita as published in www
.vitalgraphics.net/climate2.cfm?pageID=8

[17] Vandenbergh M, Barkenbus J, Gilligan J (2008)
Individual carbon emissions: low hanging fruit.
UCLA Law Review. http://ssrn.com/abstract
=1161143

[18] Wackernagel M (2008) Ecological Footprint. *Global
Footprint Network* (read August 20, 2008). www
.footprintstandards.org

[19] Walser M, et al. (2008) "Carbon Footprint," (topic
ed) Nodvin S, et al., in *Encyclopedia of Earth*,
(ed) Cleveland CJ. (Environmental Information
Coalition, National Council for Science and the
Environment, Washington, DC). www.eoearth
.org/article/Carbon_footprint

[20] Wolfson R (2008) *Energy, Environment, and
Climate*, chap 13 (W. W. Norton & Company, New
York). www.wwnorton.com/college/physics

CHAPTER 7

No Silver Bullet, Many Silver Wedges

There are three major ways to reduce greenhouse emissions: reducing energy use, replacing fossil fuels with renewables and increasing energy efficiency. Policy instruments are available for all of them. [11]

SONJA KOEPPEL AND DIANA ÜRGE-VORSATZ, 2007

According to the National Electrical Manufacturers Association, American households use roughly 3.1 billion light-bulbs today. The incandescent lightbulb is a microcosm of the waste built into our current energy habits—not to take anything away from the brilliance of its inventor, Thomas Edison. But with the common incandescent bulb, only 5% of the electricity we pay for becomes light. The rest becomes waste heat. [9] That heat is fine if you want to use incandescent bulbs to keep chicken hatchlings warm, but not when you light your bedroom while running the air conditioner to cool off the same space.*

Japan, Australia, Ireland, Cuba, and Venezuela have pending bans on incandescent lightbulbs, precisely because the technology is so wasteful. The US Congress is also deliberating a ban on incandescent bulbs, to take effect by 2014 or 2015. [7]

In Chapter 6 we learned about the low-hanging fruit for reducing household emissions significantly. Households can and should reduce their carbon dioxide emissions. But household reductions alone will not suffice. Most of the other, more powerful greenhouse gases—and a great deal of carbon dioxide—are not directly produced by household consumption. For example, farming is a significant source of nitrous oxide and methane emissions. Municipal solid waste dumps, as well as agricultural livestock practices, are significant sources of methane. Each new molecule of methane is itself 20 times more powerful as a greenhouse gas than one of carbon dioxide. Cement manufacture alone emits 5% of global human-caused carbon dioxide emissions. Half is from the chemical process (heating calcium carbonate produces lime for cement and carbon dioxide as emission), and 40% is from burning fuel to provide the heat. Yet, no construction anywhere can take place without cement. China has surged ahead of the rest of the world in cement manufacture

---

*See, for example, http://www.18seconds.org for more on switching to compact fluorescent bulbs.

II: How to Think About Climate Solutions

in the past decade. China now manufactures 10 times more cement than the United States and 40% of all the cement worldwide. [23]

No Single Path to Capping Greenhouse Gases

Economic journalist Thomas Friedman describes the historic convergence of global warming, global flattening (the leveling effect of technology diffusion), and global crowding as the arrival today of the "Energy-Climate Era." [6] Stabilizing climate and mitigating the effects of greenhouse gas emissions are complex and formidable issues, but they are critically important. Rising to this challenge will require major undertakings in the energy and technology sectors worldwide, and particularly here in the United States. Indeed, the Energy-Climate Era requires an energy technology revolution for energy sources, production, transmission, distribution, and end user efficiency and recapture. Exploring these energy topics will require that we use some numbers in this and the following two chapters.[†]

> Utilities made their money by building
> stuff . . . because they were rewarded by their
> regulators with increased rates on the basis
> of those capital expenses. The more capital
> they deployed, the more they made. . . . We
> are not going to regulate our way out of the
> problems of the Energy-Climate Era. We can
> only innovate our way. [6: pp. 222, 243]
>
> *Thomas Friedman, 2008*

When we consider all the different human activities that produce greenhouse gases, it becomes crystal clear that we need many reduction strategies in place all at once. Worldwide, greenhouse gas emissions are composed of carbon dioxide (77%), methane (14%), nitrous oxide (8%), and fluorinated gases (1%). The worldwide economic sectors that produce these emissions include energy for transportation, electricity, and heat (61.4%); land use changes such as sprawl (18.2%); agriculture (13.5%); waste (3.6%); and industrial processes (3.4%). [1]

Interestingly, sources for greenhouses gas emissions in the US economy are skewed much more heavily toward household consumption, transportation, and manufacturing processes. Changes in land use, such as deforestation, play a much less significant role in emitting greenhouse gases in the United States than in the developing nations. The greenhouse gas impacts of suburbanized sprawl in the United States come much more directly from the fossil fuel addiction such sprawl induces than from the impact of replacing forest with buildings and roads.

Altogether, the US economy produces about 17% of the entire world's greenhouse gas emissions. Of this total picture, American consumers can directly affect the residential buildings (21.6% of all emissions) and related road transportation (15.3% of all emissions). Those wedges of the emission pie are significant. But almost two-thirds of all US emissions come from non-household sources. These include big sectors of the American economy, such as commercial buildings (12% of all emissions); chemical processing (8.5%); other industrial processes (5.9%); air, rail, and ship transport (5.6%); agricultural soils and fertilizer (3.6%); oil and gas extraction, refining, and processing (3%); cement manufacture (2.3%); livestock and manure (2.5%); iron and steel manufacturing (2.2%); landfills (1.9%); and other waste processes (0.8%). [1]

The American challenge on energy consumption is that our nation is the world's largest energy consumer (see Figure 7.1). The

---

[†]But we will make every effort to keep our geek instincts in check—somewhat! If you want to skip all our fun with the numbers, jump ahead to the "no regrets" section of this chapter for solutions that are ready to be put to work.

FIGURE 7.1 online at ncse.org/climate solutions

US energy flows from source (coal, oil, nuclear) to end use in 2007

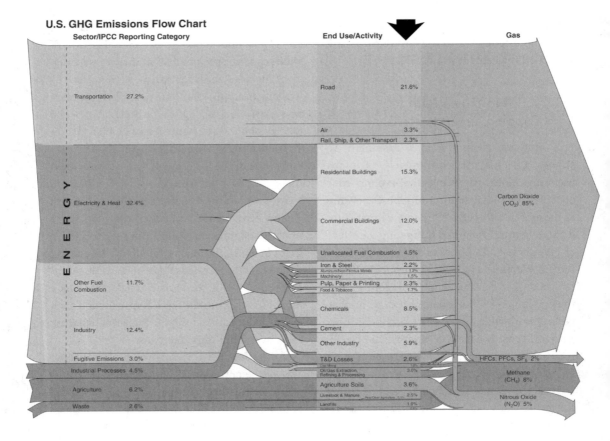

U.S. GHG Emissions Flow Chart

| Sector/IPCC Reporting Category | End Use/Activity | Gas |

FIGURE 7.2 US energy flow from use to greenhouse gas emission

This diagram shows the flow of US energy from its use to the resulting greenhouse gas emissions. This flow reflects the complexity of our diverse society and the energy pathways it uses. All the end user activities located in the center under the dark arrow represent opportunities for higher efficiency, shift to cleaner power sources, or sequestration of the currently resulting emissions. The unit of measure is quadrillion Btus. Source: [1]

slightly good news is that we do use a wide mix of energy sources. This means that at any given moment, someone somewhere is working on improving almost any imaginable energy production and delivery system. The bad news is that we rely very heavily upon dirty energy sources. Currently, liquid petroleum used for transportation is the top energy source (40%), followed by coal (23%), and natural gas (23%), both used for electricity generation. Nuclear power for electricity generation supplies a small but significant slice of our energy (8%). Hydropower for electricity from dams (just over 3%) currently supplies less than half what nuclear

power does. All other renewables together (just under 3%) supply even less than hydropower does. [22]

According to US Geological Survey Director Mark Myers, stabilizing climate requires capturing not only the "low-hanging fruit"—as improvements in energy efficiency and other, relatively simple and economical solutions are often referred to—but also "a lot of different fruit from a lot of different trees." [15] Take a look at all the different opportunities for conservation or for rethinking our energy consumption in the End Use/Activity section in the center of Figure 7.2.

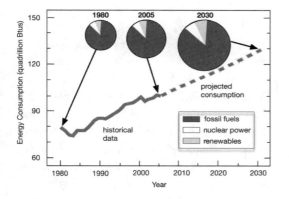

FIGURE 7.3. **US energy consumption: Recent growth and future projections 1980–2030**

The United States relies very heavily on fossil fuels. The proportion of energy that will come from fossil fuels is not expected to fall, according to recent government projections. Source: [15]

According to the US Energy Information Administration (EIA), in 2005 the United States consumed more than 7.5 billion barrels of oil,‡ almost 622.9 trillion liters (22 trillion cubic feet) of natural gas, and 997 million metric tons (1.1 billion short tons) of coal. The EIA projects a 30% growth in energy consumption by 2030. The EIA also projects renewable and non–greenhouse gas–emitting sources of energy will increase over the coming decades. However, under current scenarios, the relative percentages of these cleaner sources—currently 15% for renewables and nuclear combined—would remain largely the same. In other words, in 2030 about 85% of US energy would still come from fossil fuels. Therefore, future fossil fuel use and its greenhouse gas emissions will actually increase significantly, due to overall rise in energy demand and use (see Figure 7.3). We should note that many energy specialists find EIA's assumptions too limiting. The EIA "reference case" cited here assumes no changes in the nation's energy policy, demographics trends, or technology progress (an example of the "business as usual" scenario). Leaving out such factors can seriously underestimate the future potential for conservation or renewable energy sources.

The latest US government forecasts project a modest but steady increase per year in energy consumption from 2005 through 2030, as shown in Figure 7.3. Global energy demand will increase more steeply than the US demand, for the simple reason that the United States is already such a high per capita energy user, especially compared with the world's other more-populous nations, such as India and China. We will discuss more international comparisons in Chapters 8 and 9.

Total primary energy consumption in the United States, including energy for electricity generation, is projected to grow by 0.7% per year from 2006 to 2030 in the business-as-usual case. If we do not radically expand the nation's non–fossil energy capacity, over half of that energy growth will come from burning fossil fuels. Coal use will increase rapidly in the electric power sector. Why? Today's growth in electricity demand and current environmental policies favor coal-fired capacity additions. The latest US government forecast explains the likely growth of coal as follows:

About 54 percent of the projected increase in coal consumption occurs after 2020, when higher natural gas prices make coal the fuel of choice for most new power plants under current laws and regulations, which do not limit greenhouse gas emissions. Increasing demand for natural gas in the buildings and industrial sectors offsets the decline in natural gas use in the electricity sector after 2016, resulting in a net increase of 5 percent from 2006 to 2030. [21]

‡A barrel is about 42 US gallons (159 liters). Many more useful energy definitions await you at www.eia.doe.gov/glossary.

The Long Term Starts Now

Stabilizing climate requires us to replace fossil fuels with sources of energy that do not emit greenhouse gases. Thus, we must make dramatic advances in energy efficiency, clean energy sources, and new technologies to reverse the consumption trajectory of carbon-based energy. Yet each of these energy sources—other than efficiency—has its own unique challenges. Nuclear power raises concerns about plant safety, spent-fuel waste management, cost, and public apprehension. Wind power has issues of intermittency distance of site locations from the urban centers where power is needed, and bird and bat kills from collisions with wind turbine blades. Solar energy is presently relatively expensive per unit of electricity produced. Geothermal, hydropower, tidal energy, biofuels, and each potential new source of energy brings with it challenges that must be carefully considered when developing a plan to transition from a fossil-fuel-based society to one on track for achieving climate stabilization.

The scale and magnitude of the Energy-Climate Era transition make apparent that there will be no single "silver bullet" solutions—many paths are necessary. And each path begins with research and development (as we will explore further in Chapter 9). USGS Director Myers asserts that several key components will be integral to any transition plan: (1) A common knowledge framework—for assessing and understanding the capacity of new energy sources and technologies to be effective—must be developed nationwide and worldwide. (2) New and enhanced energy sources will be part of the solution. (3) Improved technology must be developed quickly, and significant technologies must be invested in right away. (4) More robust and long-term strategies for conservation and efficiency, including incentives and societal commitments, must be developed. [15] We must rethink how cities, suburbs, transportation systems, and buildings work and must redesign them to maximize efficiency. So where do we start?

Taking such measures early is essential to meeting the long-term challenge of mitigating greenhouse gas emissions. Economist Leon Clarke of Pacific Northwest National Laboratory reminds us that, although dialogue regarding energy and technology often deals with the long term, the choices we make in the near term set the basis for future reductions in emissions. Beyond the Department of Energy's own forecasts, the US Climate Change Science Program has assessed a series of climate stabilization scenarios to determine how emission reduction might play out between now and AD 2100. This program integrates federal research on global change and climate change across all the myriad agencies with relevant information. [3]

In the reference scenarios that the Climate Change Science Program examined, US primary energy consumption would grow between one and two and one-half times today's levels over the rest of the 21st century, while global total primary energy consumption grows to between three and four times today's levels over the century. Even when baseline reference scenarios incorporate advancements in energy efficiency and slightly more than business-as-usual technological advancements, the scenarios still show a dramatic expansion in carbon dioxide emissions that become increasingly large as the 21st century bears on. In fact, three different models predict about a 300% increase in annual carbon dioxide emissions under scenarios that assume some efficiency and new technology. [3, 17] Bad news, indeed!

The lesson is that failure to undertake steep emission reductions starting immediately will produce atmospheric conditions that in turn will push the planet toward a warming of 5 degrees Celsius or more by AD 2100. Clarke concludes, "The trajectory we set today will play out in the second half of the century. We are setting the infrastructure for our future ability to mitigate carbon emissions now." [3]

TABLE 7.1 Six Technologies That Could Help Reduce Emissions

Carbon dioxide capture and storage

Carbon capture and storage (CCS) systems offer the potential for continuing to use the Earth's abundant fossil fuel resources, especially coal, while reducing their CO_2 emissions release to the atmosphere. CCS technologies would only be widely deployed as part of a global commitment to reduce greenhouse gas emissions, but their deployment under such a commitment would lower the cost of achieving the necessary emissions reductions. (CCS involves capturing gaseous carbon dioxide emissions at the point of combustion and injecting the gases, in most scenarios, into the ground for long-term storage.)

Bioenergy

Biotechnology includes increasing the quality and quantity of biomass energy supply, the use of bio-based fuels, and the enhancement of carbon sequestration in soils and forests. Biomass fuels, whose combustion-related CO_2 emissions are roughly nullified by the CO_2 removed during plant growth, have both foundations as the oldest energy sources used by people and new promise as engineered fuels that can be utilized in many different economic sectors.

Hydrogen systems

Hydrogen is appealing in the context of climate change because it is a portable energy carrier that does not emit any CO_2 as it is consumed. (However, the greenhouse gas implications of hydrogen depend entirely on the source being used to produce the hydrogen gas.) Hydrogen is also appealing in terms of conventional pollutants because water vapor is the only by-product of its use. Hydrogen can be used to serve transportation energy demands—to operate automobiles, trucks, and other commercial carriers—that are now almost completely met by fossil fuel–based liquids that emit CO_2. Hydrogen can also displace fossil fuel–based end use applications in buildings and industry. (New hydrogen gas storage and transport systems would need to be built.)

Nuclear energy

Nuclear power production has no direct CO_2 emissions and is already a significant component of the global energy system. In 2006, there were 435 operational nuclear power stations around the world generating approximately 16% of global electricity production. Improved economic competitiveness and safety of nuclear power along with concern for energy security and climate change are leading to a steady increase in worldwide nuclear power capacity. Waste disposal and proliferation concerns, including high costs, associated with expanding nuclear energy use remain important and unresolved issues.

Wind and solar power

Wind and solar technologies have enormous potential to meet a significant portion of the world's future energy demands with little impact on the atmosphere. However, large-scale deployment of wind and solar power raises unique research and systems analysis issues. Wind and sunlight are intermittent resources in that their availability, while predictable, cannot be completely controlled. In addition, wind and solar power generators often require an investment in transmission capacity to deliver power to populated load areas. Current wind and solar technologies require large up-front capital investment, although they offer low recurring costs. Technological developments can lower their capital cost. Wind and solar generators are typically much smaller than fossil fuel and nuclear plants, requiring multiple units over a wide area to build up to a large scale, a challenge for land use and environmental aesthetics, but an advantage in pursuing more widely distributed, more diverse, and more stable energy supplies.

End use energy technologies

Energy services, also called energy end uses, include demands such as cooling, heating, and lighting homes; transporting people and freight; and heating and powering a range of industrial processes. Efficiency gains in end use technologies reduce the demand for energy to provide specific energy services such as lighting, allow the use of carbon-free energy sources, and reduce the losses of energy in the process of converting primary fuels to electricity and delivered fuels. More efficient end use technologies also help to conserve natural resources, reduce the impact of energy production on the environment (e.g., air quality, other pollution), and enhance energy security.

Source: Adapted from [4]

TABLE 7.2 Predicted US Primary Energy Use by Fuel: 2006–2030

	Consumption projections by 5-year increments (quadrillion Btus)					
	2005	2010	2015	2020	2025	2030
Liquid petroleum fuels	40.47	40.46	41.80	42.24	42.78	43.99
Natural gas	22.65	23.93	24.35	24.01	23.66	23.39
Coal	22.78	23.03	24.19	25.87	27.75	29.90
Nuclear power	8.16	8.31	8.41	9.05	9.50	9.57
Hydropower	2.70	2.92	2.99	3.00	3.00	3.00
Other renewables (see detail below)	3.33	4.70	5.52	6.67	7.85	8.16
Total	**100.09**	**2,113.35**	**2,122.26**	**2,130.84**	**2,139.54**	**2,148.01**
Biomass	2.45	3.01	3.60	4.50	5.42	5.51
All other renewables (wind, solar, etc.)	0.88	1.69	1.92	2.17	2.43	2.65

Source: Adapted from [21: table 1]

Techniques to Stabilize Climate

The good news is that at least six technologies have the potential to play a major role in a climate-constrained world. Most of these technologies are already at least in the research and development stage, including carbon (dioxide) capture and storage (CCS) from coal-fired power plants. Many are further along, in the application-testing and early deployment stages, including solar, wind, and biomass. Some involve adopting behavior changes as well as technology, such as combining a boost in end user efficiency with conservation, for example, using a more efficient appliance for fewer hours. These emission reduction technologies are presented in a snapshot in Table 7.1.

But, let's not kid ourselves: The challenges are stark. For example, in order to cap atmospheric carbon dioxide by 2100 to 550 parts per million by volume (ppmv, or ppm), which is still a dangerously high level, we would need to improve carbon capture and storage from our actual capacity today about 70-fold by 2020, about 500-fold by 2050, and 6,000-fold by 2095. [4] More sobering, the projections for the amount of energy that society will need show that the US and the world will remain heavily reliant on fossil fuels for decades to come. (See Table 7.2.)

Technology's role is to lower the cost of stabilization, not only in strictly financial terms, but in amenities as well. Good technology should improve human health and other difficult-to-quantify goods and services that determine one's quality of life. Carbon emissions could be mitigated with the technologies of today or with the technologies of the 18th century—by going back to the preindustrial modes of living. But the cost to society of the latter would be tremendously high. Thus, we need to develop the technologies of the 21st century that have the largest benefits or smallest costs to our quality of life. Before we look into these technologies in the next few chapters, let's examine a clever way to think about the many diverse tools that could stabilize future atmospheric emissions. It is a "wedges" concept (and simulation game) devised by researchers at Princeton University.

The Wedges-Within-the-Stabilization-Triangle Game

Atmospheric carbon dioxide concentrations from fossil fuel burning are on track to double in the next 50 years, as we saw in Figure 3.11, from a preindustrial level of 280 ppm to a future level by 2050 that quite possibly approaches 600 ppm. Specifically, the business-as-usual scenario of changing next to nothing will cause annual global carbon emissions to rise from 8

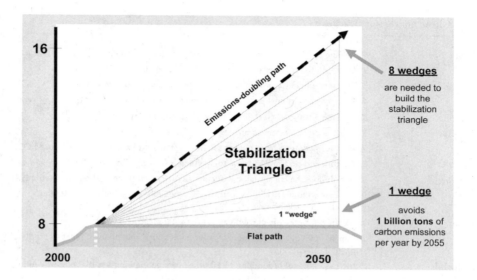

FIGURE 7.4 Eight wedges within the carbon stabilization triangle
The emissions-doubling path is the one we are on now, moving toward tripling carbon dioxide concentrations above preindustrial levels. The desired flat path will require taking 16 billions tons of carbon emissions out of the mix by 2055. Researchers Socolow and Pacala call each potential one-eighth of this triangle a "wedge." A wedge starts in present time with small impact that steadily grows. Examples of wedges are invention of viable carbon storage techniques that we can afford, the scale-up of the conservation of energy program, and implementation of end user efficiency on a society-wide basis. Source: [8]

billion tons today to 16 billion tons in 2050. Only if the future carbon emissions over the next 50 years can be kept flat at today's already high levels can we steer a course toward avoiding a doubling of carbon dioxide concentration and therefore avoid some of the nastier effects of climate disruption. We can lower emissions— starting today—by deploying low-carbon energy technologies and enhancing natural carbon sinks. The amount of carbon emission savings we need to attain is the "stabilization triangle" we see in Figure 7.4.

In 2004 Princeton physicist Robert Socolow and biologist Stephen Pacala devised a simple "wedge" analogy to help us visualize the climate stabilization options we have. To get on track to avoiding dramatic climate change, the world must avoid emitting about 200 billion tons of carbon over the next five decades. Breaking that volume down into eight wedges allows us to

think of what individual strategies might complete one wedge. Socolow and Pacala define a wedge as a reduction strategy that has the potential to grow from zero today to avoiding 1 billion tons of carbon emissions per year by 2050. The total amount of avoided carbon emissions in each wedge over the period between today and 2050 would accumulate to about 25 billion tons. A combination of strategies will be needed to build the eight wedges of a climate stabilization triangle.

The good news is that Socolow and Pacala at the Carbon Mitigation Initiative—as well as others—have identified at least 15 different wedge strategies that have the potential to reduce global carbon emissions by at least 1 billion tons per year by 2054. [19] This is the heart of the Carbon Mitigation Initiative's "climate stabilization wedges" concept, a framework for understanding both the carbon emissions cuts

TABLE 7.3 Fifteen Stabilization Wedges

Efficiency	1. Double fuel efficiency of 2 billion cars from 30 to 60 mpg.
	2. Decrease the number of car miles traveled by half.
	3. Use best efficiency practices in all residential and commercial buildings.
	4. Produce current coal-based electricity with twice today's efficiency.
Fuel switching	5. Replace 1,400 coal electric plants with natural gas facilities.
Carbon capture and storage	6. Capture and store emissions from 800 coal electric plants.
	7. Produce hydrogen from coal at six times today's rate and store the captured CO_2.
	8. Capture carbon from 180 coal-to-synfuels plants and store the CO_2.
Nuclear power	9. Add double the current global nuclear capacity to replace coal-based electricity.
Wind power	10. Increase wind electricity capacity by 50 times relative to today, for a total of 2 million large windmills.
Solar power	11. Install 700 times the current capacity of solar electricity.
	12. Use 40,000 square kilometers of solar panels (or 4 million windmills) to produce hydrogen for fuel cell cars.
Biomass fuels	13. Increase ethanol production by 50 times by creating biomass plantations with area equal to one-sixth of the world's cropland.
Natural sinks	14. Eliminate tropical deforestation and double the current rate of new forest planting.
	15. Adopt conservation tillage in all agricultural soils worldwide.

Source: [19]

needed to avoid dramatic climate change and the tools already available to do so. [8] Many strategies available today can be scaled up to reduce emissions by at least 1 billion tons of carbon per year by five decades from now. As Rebecca Hotinski of the Initiative writes, "We call this reduction a 'wedge' of the triangle. By embarking on several of these wedge strategies now, the world can take a big bite out of the carbon problem instead of passing the whole job on to future generations." [8]

Here is the dark irony. In Figure 3.10 we noted that emissions had accelerated in recent years. Due to that acceleration of emissions since 2004, when the Princeton team first devised the concept, the future emission reductions Socolow and Pacala had identified as a cumulative 175 billions tons in 2004 swelled to 200 billion tons in just 3 years. So the original seven wedges of 1 billion tons of carbon savings has been replaced by the now-required eight wedges of 1 billion tons. This is a sobering example of how delaying action on climate makes future solutions that much harder.

The good news is we have at least 15 different strategies that could be employed as stabiliza-tion wedges, starting with existing off-the-shelf technologies (see Table 7.3). The wedges-within-the-stabilization-triangle game allows students and citizens to discuss what combination of options they would favor. For example, a wedge of emissions savings would be achieved if the fuel efficiency of all the cars projected for 2055 were doubled (from 30 to 60 miles per gallon). Adding new nuclear electric plants to triple the world's current nuclear capacity would cut emissions by one wedge, if the new nuclear fission plants displace coal-fired plants. [8]

No-Regrets Climate Wedges

A rich package of well-developed technologies is ready to be scaled up as societal solutions to climate disruption. So the question becomes How do we choose among potential mitigation technologies, both current and future? Paul Epstein describes how we can organize the potential stabilization wedges into two distinctive categories of solutions to climate disruption: [5]

Certain technologies and strategies—such as energy efficiency and forest retention—have benefits beyond their contribution to a

TABLE 7.4 No-Regret Solutions versus Those Requiring More Study

No-regret solutions to scale up rapidly now	"More study" solutions that need life cycle analysis before wide-scale adoption
1. Energy efficiency and conservation	1. Oil sands and shale oil
2. Smart technologies for intelligent grids	2. Ethanol and biodiesel
3. Green buildings and rooftop gardens	3. Coal with CO_2 capture and storage
4. Efficient appliances	4. Geoengineering
5. Distributed generation with renewable sources	5. Nuclear fission
6. Passive solar heating and daylighting	6. Nanotechnology
7. Ground source heat pumps	7. Wave, current, and tidal energy
8. Cogeneration	
9. Solar thermal arrays	
10. Photovoltaic arrays	
11. Wind farms	
12. Geothermal energy	
13. Industrial efficiency	
14. Green chemistry	
15. Smart urban growth	
16. Healthy cities programs	
17. Public transport and light-rails	
18. Plug-in hybrid electric vehicles	
19. Sustainable forestry	
20. Conservation tillage	
21. Locally grown organic agriculture	
22. Less-intensive livestock practices	
23. Municipal solid waste management	
24. Low-technology/human-powered devices	

Source: [5]

climate stabilization strategy and should be implemented immediately. Energy efficiency saves operating costs, and forest retention preserves biodiversity and the water cycle. Energy efficiency and consumption reduction can be undertaken immediately with "no regrets" and considerable financial savings. Improving public transit, expanding plug-in hybrid vehicles, and building "smart" electric grids (that learn who needs how much power when) all rely on existing, safe, and well-proven technologies. Minimizing agricultural tillage (to increase natural carbon sinks), boosting vehicle economy standards, and reducing red meat consumption are also solutions ready to be adopted now that require behavior changes but no new technology.

A large set of additional no-regrets solutions are nearly ready to be brought to scale. Wind energy, solar power, geothermal power, and hydropower fit into this category. They work well now but need more deployment in the market to reach economies of scale that make them cost-effective. Of these, solar and wind power in particular are sound technologies with huge potential to produce electricity in North America. Yet, these alternative energy technologies have lacked consistent support and sufficient investment, such as the nationwide effort that launched the civilian nuclear power program or put a human on the Moon a generation ago.

Several other energy technologies require more study, because they have not yet been studied fully, are potentially dangerous, or may nullify their carbon mitigation purpose. Biofuels, fossil fuel–based strategies (such as switching from coal to natural gas, carbon capture and storage, hydrogen fuel cells, and coal to liquid fuel), nanotechnology, and nuclear fission each fall into this more-study category. Each must be researched thoroughly and considered carefully before any decisions are made regarding further implementation.

These less-developed solutions may be laden with unintended or unexplored side effects so great that they should not be ramped to the wedge scale. (See Table 7.4 for a full list.) For example, nuclear power does generate electrical power without significant greenhouse gas emissions. But nuclear fission also comes with the significant responsibility of storing the radioactive waste it produces safely for potentially many tens of thousands of years. Such costs only become apparent if we examine the full life cycle input and output for such sources, something we discuss in Chapter 8.

Specifically, Harvard Medical School's Center for Health and the Global Environment documents that many no-regrets solutions are ready now for scaling up. These interconnected climate solutions mutually reinforce each other and collectively have a high payoff in reducing emissions as well as providing a healthier population with new economic-growth opportunities. These "scale up" solutions include smart, cleanly powered grids; healthy city programs; and measures to minimize liquid fossil fuels. For example, the same effort to minimize liquid fuels would boost health in cities and take advantage of a smarter electricity grid by enhanced public transport, promoting walking and biking, plug-in hybrid electric vehicles, and smart urban growth in general. "Healthy city" programs would combat the urban heat-island effect and the concentration of pollution that typifies cities now, with green buildings, white roofs or rooftop gardens, walking paths, biking lanes, tree-lined streets, open space, congestion control, and improved public transport to decrease vehicular miles traveled, promote exercise, save money, and create jobs.

The Healthy Cities program of the World Health Organization (WHO) is a global movement with networks established in all six WHO

TABLE 7.5 The Qualities of a Healthy City

A city should strive to provide the following:
1. A clean, safe physical environment of high quality (including housing quality)
2. An ecosystem that is stable now and sustainable in the long term
3. A strong, mutually supportive and nonexploitative community
4. A high degree of participation and control by the public over the decisions affecting their lives, health, and well-being
5. The meeting of basic needs (for food, water, shelter, income, safety, and work) for all the city's people
6. Access to a wide variety of experiences and resources, with the chance for a wide variety of contact, interactions, and communication
7. A diverse, vital, and innovative city economy
8. The encouragement of connectedness with the past, with the cultural and biological heritage of city dwellers, and with other groups and individuals
9. A form that is compatible with and enhances the preceding characteristics
10. An optimum level of appropriate public health and sick care services accessible to all
11. High health status (high levels of positive health and low levels of disease)

Source: [24]

regions but particularly widespread in Europe from Amsterdam to Zagreb. (See Table 7.5.) The current focus of cities participating in the network is on healthy aging, healthy urban planning, and health impact assessment, which is an additional complementary theme of physical activity and active living. [25]

Naturally, smart growth efforts require long-term, integrated planning and adequate investment. Infrastructure may seem expensive to replace, but we may have little choice. For example, the American electricity grid reflects an antiquated approach in which each state-based utility strung spoke-and-hub transmission lines to serve its major markets. Little thought was given to moving power across states lines or how these networks would connect in larger regional settings. The condition of today's electricity grid in the United States is not unlike the condition of our state-based road system before the federal government designated the linking routes

INSIGHT 7: THE SMART GRID ADVANTAGES OF THE ENERGY INTERNET

Imagine if your electric company told your dishwasher to run itself at 3 a.m. because that would be far cheaper than running at 7 p.m. Or, on the verge of a citywide brownout on a hot summer afternoon, the electric company could dial down everyone's air conditioner by 1 degree for 1 hour to shave the peak of the demand and avert any brownout that day.

If smart grid technologies made the United States grid just 5% more efficient, it would be equal to eliminating the fuel and greenhouse gas emissions from 53 million cars. [14] One US Department of Energy study in 2003 calculated that internal modernization of US grids with smart grid capabilities would save between $46 billion and $117 billion over the next 20 years. [10] Those savings would be even higher today. Smart grid technology exists today yet has not been widely implemented. A smart grid is a system that monitors both consumption demand and electricity supply instantaneously and reroutes supplies as needed. It would also be able to send instructions to end users about powering down unneeded appliances or information about when power would be less expensive.

Efficiency sensors in buildings would help maximize the comfort while minimizing the energy required. Finally, in the event of electricity grid disruptions due to overloads or storms, a smart grid would reroute electricity to keep critical facilities such as hospitals and emergency service offices online during power losses. Did we mention that all this smart grid technology exists already?

Smart grids could additionally rely on distributed generation capacity, which simply means power produced near the point of use. [14] For example, small natural gas–fired plants that burn cleaner than coal are smaller, are easier to locate closer to high-demand urban areas, and can turn on when needed more quickly than coal. Using a wider array of lower-carbon energy sources (e.g., solar, wind, geothermal, natural gas and other renewables, and cogeneration of steam and power) lowers the risks of shortage of any particular one. Improving storage capacities for electricity generated by wind on windy days and by sunlight during daylight hours would enhance the value of these two renewable sources to the grid as a whole. But we have not really tried to innovate in this direction yet in the United States.

of the interstate highway system. The massive Northeast blackout of August 2003 showed how vulnerable the grid is to simple failure. That blackout may have cost $6 billion in direct losses as it affected 10 million Canadians and 40 million Americans. [2]

We generate most of our electricity in large centralized facilities (e.g., coal-fired plants), which are often located great distances from the end users. Utility grids start at power generating facilities and include transmission, distribu-tion, storage, and delivery to the end user. Over time, power plants have become more reliable at generating electricity than the patchwork grid has been at delivering it. The unreliability of the system is one reason that most power plants are built to have excess capacities to make up for weaknesses in the supply grid. Yet, the empha-sis among utilities has remained on massive power plants quite distant from their markets. Placing more, smaller power plants closer to the customers and using more-reliable grid technology would improve the reliability of the system. In addition, switching from the current use of highly inefficient alternating current in our transmission lines to digital direct current transmission would have double benefit: reduce loss of energy during transmission (line loss) and enable real-time sensor reports (to better balance supply against demand). [14]

In addition, cogeneration is an increasingly attractive option. Cogeneration, also known as combined heat and power (CHP), means generating both heat and power simultaneously and capturing and using them both. Facilities that already burn fuel on a large scale, for example, municipal waste incinerators, create tremendous amounts of heat. Rather than being lost up a chimney, that heat could be used to turn a steam turbine to generate electricity for the grid. Or the heat could be piped to nearby facilities for an industrial process or for driving the heat pumps of nearby buildings.

Decentralizing the grid, making it smarter, and increasing the number of combined heat and power plants and the number of renewable power plants sounds like a dream, right? But other countries are already doing this. If Shakespeare wrote *Hamlet* today, the famous line might become "there's something to be gotten in the state of Denmark."

"Denmark transitioned from being 99 percent dependent on foreign energy sources such as oil and coal in 1970 to becoming a net exporter of natural gas, oil and electricity today," writes Benjamin Sovacool. "The country is the unchallenged world leader in terms of wind technology, exporting US$7.45 billion in energy technology and equipment in 2005." [20] Danish energy consumption has grown only 4% from 1980 to 2004, even though the economy grew more than 64% in fixed prices in that period.

How? Through a consistent three-decade effort following the oil price shocks of the early 1970s, Denmark has completely overhauled its once heavily centralized, fossil fuel–driven, import-dependent energy economy. Denmark set renewable energy policies into motion (with

a stiff energy tax and other regulations) that focused on funding and supporting decentralized, small-scale, bottom-up innovation and research. In the mid-1980s, Denmark had 15 large centralized CHP plants but no smaller-scale regional plants. By 2008, Denmark had added over 415 wind power plants in all regions of the country and a similar number of small CHP plants. The Danes are now net energy exporters and have the fifth largest installed wind power capacity of any nation, remarkable for a country that ranks 134th in size and has a population of about 5.5 million (including 50,000 residents of Greenland).

Energy Waste and Efficiency

Let's come full circle. In Chapter 6, we mentioned "negawatts" and the role efficiency can play in reducing growth in future energy demand. So how efficient is the overall energy flow in the United States today? Not very! How much more efficiency can we squeeze out of our energy system? A lot! The plain truth is we have enormous waste built into our current energy flows.§

Imagine the following: In order to feed just yourself each day, you have to make enough food for three people but are forced to throw away the first two servings before each meal. That is the situation in our current electric power grid. Take a look at the top right-hand corner of Figure 7.5. The data in the figure are assembled on a regular basis by US Department of Energy's Lawrence Livermore National Laboratory and depict what happens in our electric energy system. What major energy sources are used to generate electricity? Coal, nuclear, natural gas, and hydro dams, in descending order. How much of the electricity that we generate is lost along the way before reaching its final goal of your

---

§In chapter 9 we will examine the energy efficiency in a cross-national economic context, but here we will stick to the US case.

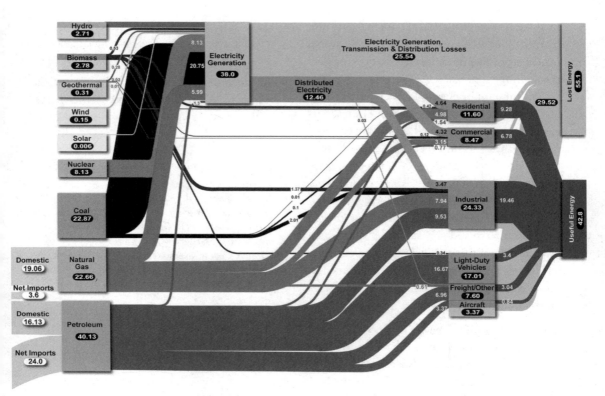

FIGURE 7.5 Energy generation, transmission, and distribution losses

The left side of this chart shows, by percentage, the "input" sources of energy in the United States in 2004. The boxes in the middle of the chart show the percentages of "throughput," that is, what happens to the energy as it is converted for end use. Note that 38% of all our incoming energy is used for electricity generation. More startling, of the electricity we do generate, only one-third is actually distributed to the end user, with the rest wasted in electrical generation, transmission, and distribution losses. The right side of this chart shows the total resulting lost energy from all sectors as more than 55% versus the actual useful delivered energy (about 43%). Source: [13]

computer or light fixture? Two-thirds! (Figure 7.5 shows the complete picture.)

Where are the two biggest opportunities for major efficiency gains? Reducing electrical transmission losses and boosting vehicle efficiency would have a big impact on getting the most kilowatt-hours of electricity or the most distance traveled per unit of energy consumed.

Why is so much energy "lost" from the oil well to the car or the coal mine to the light-bulb? The major reason is the energy conversion process itself. A natural resource, such as a lump of coal, has energy embodied in it prior to undergoing any human-made conversions or

transformations. Because the machinery we use to convert an energy source into electricity is never 100% efficient, some fraction in the initial conversion is "lost" as heat energy. Heat is inevitable when fossil fuels are burned. But the total accumulated energy conversion loss is astounding, as we see in Figure 7.5. Capturing some of that heat would have a double payback—first,

FIGURE 7.6 online at ncse.org/climate solutions

Flow chart shows US electricity sources, uses, and conversion loss in 2007.

by getting something useful out of what is now discarded, and second, by using that recovered heat to replace some of the electricity being generated in the first place.

The amount of energy lost in a year to heat and friction in converting fuels to electricity is almost as much as the total amount of coal and natural gas burned in a year to produce that electricity (as shown in Figure 7.6, with even more recent data than in Figure 7.5). Overhauling the continent's electrical grid has enormous potential to both reduce the amount of energy needed in the first place and to reach those sites where new energy sources will be creating electricity. But, we also need more professionals trained in power engineering, as this field once more rises in importance.

> The shortage of the power engineering workforce is a national security issue. . . . 46% of all engineering jobs could become vacant by 2012, due to retirements by the aging workforce and other forms of attrition. [12]
> *National Science Foundation Workshop on the Future Power Engineering Workforce, 2008*

In sum, while the challenge of reducing greenhouses gases to stabilize the future climate is steep, we have a great many points in the energy system where specific long-term improvements can and must be made. But we have to start now—in investment in energy research, development, and demonstration initiatives. Once a coal-fired power plant or a cement factory is built, it will emit carbon dioxide for decades. So any sharp improvements now have an effect that lasts decades.

CONNECT THE DOTS

- Climate stabilization and greenhouse gas emissions mitigation are complex and formidable issues.
- An energy technology revolution is needed for sources, production, transmission, distribution, and end user efficiency and recapture.

- Many different no-regrets greenhouse gas reduction strategies are already available for development and implementation.
- The biggest potential lies in reducing energy demand and in increasing the efficiencies in the electrical system and transportation sector.

Online Resources

www.eoearth.org/article/
    Consumption_and_well-being
www.eoearth.org/article/Carbon_capture_and_storage
www.eoearth.org/article/Greenhouse_gas
www.eoearth.org/article/Primary_energy
US Climate Change Science Program, www.climate
    science.gov
US Energy Information Agency (Annual Energy
    Report), www.eia.doe.gov
US Energy Information Agency Glossary, www.eia.doe
    .gov/glossary
International Energy Agency, www.iea.org
Smart Grid, www.oe.energy.gov/smartgrid.htm
Grid Wise, www.gridwise.org
See also extra content for Chapter 7 online at http://
    ncseonline.org/climatesolutions

Climate Solution Actions

Action 1: Green Buildings and Building Design
Action 2: Moving Forward—Transportation and Emissions Reduction
Action 7: Biofuel Industry and CO_2 Emissions—Implications for Policy Development
Action 9: How to Ensure Wind Energy Is Green Energy
Action 10: Nuclear Energy—Using Science to Make Hard Choices

Works Cited and Consulted

[1] Baumert K, Herzog T, Pershing J (2005) *Navigating the Numbers: Greenhouse Gas Data and International Climate Policy,* 122. (World Resources Institute). www.wri.org
[2] CBC News (2003) Power Outage Background November 13, 2003 (read September 4, 2008). www.cbc.ca/news/background/poweroutage/
[3] Clarke L, Edmonds J, Jacoby H, Pitcher H, Reilly J, Richels R (2007) Scenarios of Greenhouse Gas Emissions and Atmospheric Concentrations. (Subreport 2.1A of Synthesis and Assessment Product 2.1). www.climatescience.gov/Library/sap/sap2-1/finalreport/default.htm
[4] Edmonds JA, Wise MA, Dooley JJ, Kim SH, Smith SJ, Runci PJ, Clarke LE, Malone EL, Stokes GM (2007) Global Energy Technology Strategy:

Addressing Climate Change. Global Energy Technology Strategy Program. www.pnl.gov/gtsp/publications/

[5] Epstein P (2008) Healthy Solutions for the Low Carbon Economy: Guidelines for Investors, Insurers and Policy Makers. *The Center for Health and the Global Environment*. Harvard Medical School. http://chge.med.harvard.edu/programs/ccf/

[6] Friedman T (2008) *Hot, Flat and Crowded: Why We Need a New Green Revolution and How It Can Renew America*, 410. (Farrar, Straus, and Giroux, New York). www.thomaslfriedman.com

[7] Govtrack.us (2007) Energy Efficient Lighting for a Brighter Tomorrow Act of 2007. *GovTrack.us* (Database of federal legislation) (read October 25, 2008). http://www.govtrack.us/congress/bill.xpd?bill=s110-2017

[8] Hotinski R (2007) Stablization Wedges: A Concept & Game. *Climate Mitigation Initiative*. Princeton University. www.princeton.edu/wedges

[9] Johnson J (2007) *The end of the light bulb*. Chemical & Engineering News 85 (December 3). http://pubs.acs.org/cen/

[10] Kannberg LD, Kintner-Meyer MC, Chassin DP, Pratt RG, DeSteese JG, Schienbein LA, Hauser SG, Warwick WM (2003) GridWise: The Benefits of a Transformed Energy System. Pacific Northwest National Laboratory under Contract with the United States Department of Energy. *US Department of Energy* (read December 2, 2008). www.pnl.gov/energy/eed/etd/pdfs/pnnl-14396.pdf

[11] Koeppel S, ürge-Vorsatz D (2007) *Assessment of Policy Instruments for Reducing Greenhouse Gas Emissions from Buildings*. (United Nations Environment Programme, Central European University, Budapest). www.unepsbci.org

[12] Lauby M, et al. (2008) in Executive Summary. *National Science Foundation Workshop on the Future Power Engineering Workforce*. November 29–30, 2007. http://ecpe.ece.iastate.edu/nsfws/

[13] Lawrence Livermore National Laboratory (LLNL) (2006) Energy and Environment Division. Livermore, CA (read October 1, 2008). https://eed.llnl.gov/

[14] Litos Strategic Communication (2008) *The Smart Grid: An Introduction*, 7. (Litos Strategic Communication (under contract for US Department of Energy) (read December 22, 2008). www.oe.energy.gov/DocumentsandMedia/DOE_SG_Book_Single_Pages.pdf

[15] Myers M (2008) An Enlightened Energy Future: Understanding the Challenge. *National Conference on Science, Policy, and the Environment*. http://ncseonline.org/2008conference/

[16] Ohanian HC, Markert JT (2007) *Physics for Scientists and Engineers*, 3rd ed, p 674. (W. W. Norton & Company, New York). www.wwnortion.com/physics

[17] Shindell DT, Levy II H, Gilliland A, Schwarzkopf MD, Horowitz LW (2008) Climate Change from Short-Lived Emissions Due to Human Activities in Climate Projections Based on Emissions Scenarios for Long-Lived and Short-Lived Radiatively Active Gases and Aerosols. In A Report by the US Climate Change Science Program and the Subcommittee on Global Change Research, (eds) Levy II H, Shindell DT, Gilliland A, Schwarzkopf MD, Horowitz LW. (Department of Commerce, NOAA's National Climatic Data Center). www.climatescience.gov

[18] Singh G (2007) FY 2007 Progress Report: Advanced Combustion Engine Technologies. DOE EERE OVT 2007. www1.eere.energy.gov/vehiclesandfuels/

[19] Socolow R, Pacala S (2004) *Stabilization wedges: solving the climate problem for the next 50 years with current technologies*. Science 304(5686):968–972. www.sciencemag.org

[20] Sovacool BK (2008) Is the Danish Renewable Energy Model Replicable? *Technology: Future Energies*. Scitizen.com (read December 22, 2008). www.scitizen.com/stories/Future-Energies/2008/03/Is-the-Danish-Renewable-Energy-Model-Replicable/

[21] US EIA (2008) Annual Energy Outlook 2008 with Projections to 2030. Department of Energy. DOE/EIA-0383(2008)(read October 1, 2008). www.eia.doe.gov/oiaf/aeo/demand.html

[22] US EIA (2008) Annual Energy Review 2007. DOE/EIA-0384 (2007). www.eia.doe.gov/emeu/aer/

[23] USGS (2008) Mineral Commodity Summaries. http://minerals.usgs.gov/minerals/pubs/mcs/

[24] WHO (1997) Twenty Steps for Developing a Healthy Cities Project. www.euro.who.int

[25] WHO (2003) National Healthy Cities Networks: A Powerful Force for Health and Sustainable Development in Europe. www.euro.who.int

CHAPTER 8

Energy in the Cycle of Material Life

A life cycle approach is a way of thinking which helps us
recognize how our selections—such as buying electricity or
a new t-shirt—are one part of a whole system of events. [8]

JIM FAVA AND JENNIFER HALL, 2008

Researchers at Carnegie Mellon University compared the impact of "foodmiles" (a measure of how far food travels between its production and the final consumer) with the impact of what kind of food we eat, regardless of where it is grown. The surprising conclusion is that just one day during which we replace red meat and dairy with chicken, fish, eggs, or a vegetable-based diet has a much greater emission reduction impact than a week of eating strictly local food. The researchers found that "on average, red meat is around 150% more [greenhouse gas]-intensive than chicken or fish. Thus, we suggest that dietary shift can be a more effective means of lowering an average household's food-related climate footprint than 'buying local.'" [41]

If we analyze the flow of materials throughout the entire cycle of their use, from what goes into creating them and what happens to the materials once we are through, we get a much more complete view. For food, most of the climate impact occurs in growing it and shipping it to the grocery store. The life cycle supply chain is 6,760 kilometers (km), or 4,200 miles, on average for all foods, and it is highest for red

meats (20,400 km or 12,700 mi) and lowest for beverages (1,200 km or 750 mi). [41] Red meat food miles include the distance that feed travels to reach the livestock and other inputs. The distance that ground beef travels from the feedlot to your butcher is actually only about one-tenth the total food miles needed to produce that pound of future hamburger. Even though food is transported long distances, more than fourth-fifths of the greenhouse gas emissions associated with food comes from what happens before we ship it. The production phase of our food is a massive web of energy-hungry activities. We produce the fertilizer that we ship to the fields where we grow the grain that we mill and ship to where we feed the cattle. On average, red meat is around 150% more greenhouse gas intensive than chicken or fish. [41]

Life Cycle Thinking: Analyzing the Flow of Materials

What about those "more study" solutions we briefly explored in Chapter 7? How should we study them? Physician Paul Epstein and his col-

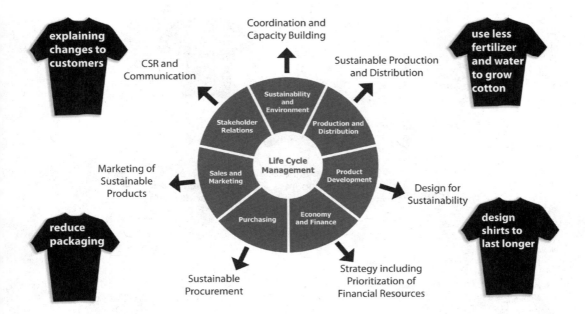

FIGURE 8.1 Life cycle thinking applied to a t-shirt

Meet the Connect the Dots t-shirt company. A product system, or life cycle, can begin with extracting raw materials from natural resources in the ground and generating energy. Materials and energy are then part of production, packaging, distribution, use, maintenance, and eventually recycling, reuse, recovery, or final disposal. In the case of a simple t-shirt, the stages involve a wide variety of impacts due to fertilizer used to grow the cotton, dyes and water used to manufacture the shirt, and bleaches and detergents used by the owner to wash the shirt. (CSR stands for corporate social responsibility.) Source: Adapted from [27] and [35]

leagues comment, "Comparing life-cycle costs— the health, ecologic, and economic dimension— of proposed solutions can help differentiate safe solutions from those warranting further study, and from those with risks prohibiting wide-scale adoption. Solutions meeting multiple goals merit high ratings." [5]

Life cycle assessment—often abbreviated LCA and also termed "life cycle analysis," "cradle-to-grave analysis," or "material-flow analysis"—is the study and valuation of the environmental impacts of a specific product or service made necessary by its existence. In other words, we should study and quantify both the process of how a product is created and the impact of using and disposing of the product itself. Environmental management standards used by many industry sectors are now beginning to require such life cycle assessment, or to

stress the benefits of voluntary compliance. The main goals of life cycle thinking are to reduce a product's resource use and emissions to the environment as well as improve its socioeconomic performance throughout its life cycle. Such thinking may create links between the economic, social, and environmental dimensions within an organization and throughout its entire value chain. [35] What would a life cycle analysis for the humble t-shirt look like? Figure 8.1 offers one hypothetical example.

Cradle to grave designs dominate modern manufacturing. [28: p. 27]
Architect Bill McDonough and chemist Michael Braungart

Some economists estimate that 90% of the raw materials, such as iron ore, extracted to make durable goods, such as refrigerators, in

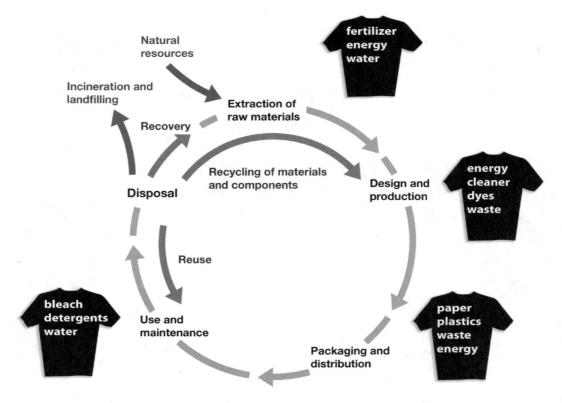

FIGURE 8.2 Life cycle management roles in a t-shirt organization

All functions in an organization play an important role in life cycle management. The figure shows examples of how different departments in an organization can contribute to the overall life cycle management program. The t-shirt ideas here are merely illustrative and only the tip of the iceberg of all the life cycle opportunities that could be examined and connected to each other. Source: Adapted from [35]

the United States become waste destined for a grave almost immediately. But it does not have to be that way. McDonough and Braungart's "waste equals food" thesis informs their "cradle-to-cradle" belief that almost all manufacturing processes can be redesigned if we grasp that waste does not exist. Waste has to go somewhere. In a manufacturing or extraction process, waste may go to a landfill or an incinerator. In an energy production process, waste by-products (e.g., heat, particulates, and greenhouse gases) often go into the atmosphere or biosphere at no immediate financial cost to the producer.

Now that we have examined life cycle implications for t-shirts, let's tackle some tougher subjects.

The Life Cycle of Fossil Fuels in Brief

Each fossil fuel has distinct benefits. High-energy liquid fuels such gasoline or diesel are a good fit for vehicles as fuel that travels along with the mobile user. Natural gas is the easiest of the three fossil fuels to deliver by underground pipe, and therefore it is widely used for home heating and cooking. Rocklike coal delivered by the railcar requires industrial handling equipment only found at the massive scale of centralized electric generation plants.

Each fossil fuel also has a significant downside for human and ecosystem health. Consider the life cycle of oil, our nation's current primary transportation energy source. The United States uses a quarter of the world's petroleum—22 million barrels a day. Half is imported. [38] And liquid fuel is needed for vehicles. Oil exploration disturbs wildlife breeding grounds and coastal and arctic habitat. Oil extraction is a messy process that degrades the land, especially river deltas and forest habitats. Oil transport leads to over 30 million gallons (700,000 barrels) in spills and leaks that contaminate coastlines. Oil refineries emit carcinogens, such as benzene. Oil combustion creates air pollutants, acid rain, and greenhouse gases that accelerate climate disruption. From this broader perspective of health and safety, the true costs of this fuel to human and ecosystem health must be factored into expanding its use. [6]

Consider coal's life cycle. Coal represents a quarter of the world's energy consumption but generates 30% of the global carbon dioxide emissions. Half of the electricity in the U.S. is generated by burning coal. [38] Coal consumption worldwide has grown twice as quickly as any other energy source, up 30% since 2002. [38] In 2008, as worldwide demand rose, the United States became a major global exporter of coal for the first time since the early 1990s. [24] Underground coal mining leads to injuries, chronic illness, and mortality. Each year, approximately 1,000 miners in the United States alone die from coal workers' pneumoconiosis, also known as black lung disease, a preventable illness caused by exposure to coal mine dust. [30]

Surface or strip mining leads to mountaintop removal, toxic releases, deforestation, water contamination, and cancer clusters in downstream communities. When coal surfaces are exposed, pyrite (iron sulfide) is also exposed to water and air and forms sulfuric acid. As long as rain falls on the mine's slag heaps, the sulfuric acid production continues and leaches into waterways, whether the mine is still operating or not. Coal and coal waste products, including fly ash, bottom ash, and boiler slag, contain many heavy metals, including arsenic, lead, mercury, nickel, vanadium, beryllium, cadmium, barium, chromium, copper, molybdenum, zinc,

selenium, and radium, all of which are dangerous if released into the environment. Coal also contains low levels of uranium, thorium, and other naturally occurring radioactive isotopes whose release into the environment may lead to radioactive contamination. [12, 39] While these substances are trace impurities, enough coal is burned that significant amounts of these substances are released, resulting in more radioactive waste than from nuclear power plants. [12] Coal combustion, even with carbon captured through the chimney, still yields nitrates, sulfates, neurotoxin mercury, and particulate emissions, with their link to asthma.

A stunning 70% of all US rail freight is devoted to supplying coal-fired plants with coal to burn. While "clean coal" sounds great, the "cleaning process" requires 40% more energy to capture and store coal emissions beyond what is needed for traditional coal-fired plants. [29b] Therefore, every unit of energy generated from "clean coal" requires even more railcars supplying even more coal to fuel the "cleaning" process itself. That means higher extraction costs, land degradation, more release of chemicals into the environment, and increased potential health hazards to mine workers. You can see where a fuller life cycle analysis of burning coal might be headed.

The Life Cycle of Nuclear Fuel in Brief

Nuclear power is used exclusively for electricity generation. In essence, a nuclear power plant is a heat engine fueled by fission reaction of its radioactive fuel (rather than fossil fuel combustion) that produces steam that turns a turbine that generates electricity. The radioactive decay of its fuel indicates the rate of the reactions. Currently 104 commercial nuclear power plants provide about one-fifth of the total electricity produced in the United States, making the United States the largest nuclear power user among the 31 countries that operate about 439 plants today. [38] France generates four-fifths of its electricity from nuclear power. The United

States, France, and Japan account for over half of all the nuclear power produced each year. As many as 30 new nuclear plants are on the drawing boards in the United States alone, pending approvals in anticipation of the nation's growing electricity demands. Other nations, such as China and India, are actively planning nuclear plants using newer reactor technologies. Currently, the International Atomic Energy Agency projects growth in nuclear generation through 2030 at the high end that will match the 3.2% per year growth in overall electricity generation. [16]

Consider the life cycle of nuclear power. Uranium ore is mined, enriched into higher-grade material, processed, and manufactured into the fuel shape and size needed for reactors. After delivery to a nuclear power plant and use there, the fuel is eventually spent, meaning it is still radioactive but not at sufficient levels to generate the tremendous heat used by plants to turn steam turbines. Spent fuel is either placed in a permanent repository or is reprocessed for reuse to capture the remaining radioactivity. Current technologies allow up to 95% of the fissionable material to be reprocessed. [29a]

The life cycle of nuclear fission raises the issues of "storage, security, and safety"—the "three S's" as Dr. Epstein calls them. About 70,000 metric tons (150 million pounds) of spent nuclear fuel and high-level radioactive waste are currently stored at 121 sites around the United States. Each year a typical reactor produces about 20 tons of radioactive waste. [40] So, if we did create a long-term repository at a place like Yucca Mountain, it would take many decades to clear up this backlog alone. Yucca Mountain, located in an isolated desert site on federal land, would have provided long-term, deep underground, and centralized storage for hazardous nuclear waste. Yucca Mountain is adjacent to the US Department of Energy's Nevada Test Site about 80 miles from Las Vegas. Doubts regarding the safety of such disposal methods were amplified by the 2007 discovery by the US Geological Survey of 10 seismic fault lines within a 20-mile range of Yucca Mountain.

In his budget request for 2010, President Obama stripped federal funding for the Yucca Mountain repository, effectively killing this proposal after two decades of controversy.

Even if we do solve the issue of storing spent nuclear fuel rods, nuclear reactors require enriched uranium as fuel. Uranium is a rare mineral and usually extracted as yellow uranium oxide. Its life cycle, known as the "yellow cake road," is littered with well-documented health hazards, from increased lung cancer among uranium miners (from radon exposure), to increased death rates for uranium processors (from leukemia), to increased mortality for workers at nuclear power and weapons facilities (from all cancers—lung, multiple myeloma, and others). [6]

Beyond the risk of obtaining enriched uranium, the new advanced "pebble" technologies remove the risk of runaway fission reactions. But external hazards remain. The largest plant in the world, in Kashiwazaki in Japan, released radiation into the nearby sea after suffering a 6.8-magnitude earthquake. [31] Transport of radioactive material exposes the public to potential accidents. Nuclear power plants are often located directly at the water's edge to have access to large quantities of water used for cooling the reactors. Sea level rise may threaten nuclear plant facilities built on coastal plains; key US facilities and all 13 in the United Kingdom have shoreline locations.

If we were to create one wedge (of 1 billion tons of carbon dioxide avoided) out of nuclear energy, we would need one new Yucca Mountain every 5 to 10 years to store the future buildup of associated radioactive waste. The federal regulatory criteria for Yucca Mountain required that the groundwater under the mountain be protected for at least 10,000 years, the period of the waste's active radioactive decay. As Epstein and his colleague note, "Ensuring the safe storage of radioactive waste for tens-to-hundreds of thousands of years remains a serious obstacle to expanding nuclear energy." [6]

But nuclear storage and safety are not the

TABLE 8.1 Costs of Power Plant Construction

Nuclear fission plant	$6 to $12 billion
Coal-fired plant	around $2 billion
Natural gas–fired plant	around $1.6 billion (less steel, concrete, and labor than for a coal plant)
Wind (offshore)	around $1.8 billion
Wind (on land)	less than $1 billion

Source: [6]

only obstacles. In an uncertain political world with rising conflicts over resources, the reprocessing of spent fuel may be more susceptible to abuse than the processing of the original uranium. Only a few nations—France, the United Kingdom, Russia, Japan, India, and the United States—have nuclear reprocessing technology to reduce the radioactivity levels of spent fuel. But many more nations have the enrichment technology needed to increase and concentrate the radioactivity level of potential fuel. In 2006, in a controversial attempt to expand nuclear power production worldwide, the United States inaugurated an effort to organize the world into "fuel supplier nations," who would supply fuel and reprocess spent fuel, and a larger group of "user nations," who operated nuclear power plants. The specter of nuclear fuel landing in unreliable hands is a legitimate security concern, given the high negative consequences of radioactivity, regardless of the likelihood.

Finally, nuclear plant costs and the long time that planning and development take remain significant roadblocks to bringing a nuclear power wedge online soon enough. Nuclear power plants take about 10 years to construct. No new plants have been built in the United States for a quarter century, largely due to safety and political concerns. The projected costs of constructing a new 1-gigawatt-generation nuclear power plant recently rose from under $4,000 per kilowatt of generating power to $5,000 to $7,000 per kilowatt of generating power. [25] Major insurers and investors such as Swiss Re and ING are grappling with the basic question: Is this insurable? [6]

Analyzing Return on Investment for an Energy Source

Alternatives to petroleum-based fuels should be evaluated over the course of their life cycles as well. In order to create fuels, energy is required. Therefore, part of the life cycle analysis of energy sources is the simple question, Does creating this fuel require more energy than it produces? We should do this, of course, for each greenhouse gas emission reduction strategy.

Biofuels—solid, liquid, or gas fuels extracted from recently dead biological material—are gaining public attention. (In contrast, fossil fuels are derived from long-dead biological matter.) Also called biomass, biofuels include dung, wood, and crop residues, as well as crops grown specifically for such use, such as corn and switchgrass. Biofuels can be burned directly or converted into liquid fuels such as ethanol. The energy-in-to-energy-out balance of biofuels should be considered in what is known as net energy analysis. Net energy analysis includes the calculation of the energy return on the energy investment (EROI, or EOI in some literature), the fuel equivalent of the financial return on investment that is common in the business world. EROI is the ratio of the energy delivered by a process to the energy used directly and indirectly in that process: Ideally, we want the EROI to be greater than 1, meaning the energy produced is greater than the energy used to produce that energy (see Equation 1).

Equation 1

Energy return on energy investment (EROI) =

$$\frac{\text{Energy returned to society}}{\text{Energy required to get that energy}} =$$

$$\frac{\text{Energy output}}{\text{Energy input}}$$

Charles Hall and his colleagues at SUNY-ESF write, "We believe . . . net energy analysis offers the possibility of a very useful approach for looking at the advantages and disadvantages of a given fuel and offers the possibility of looking into the future in a way that markets seem

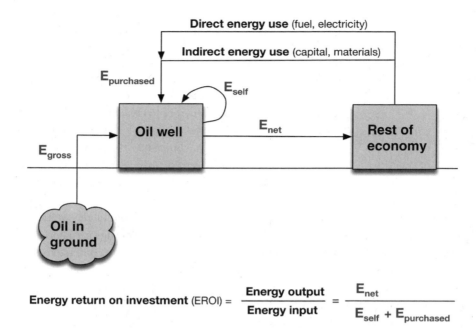

$$\text{Energy return on investment (EROI)} = \frac{\text{Energy output}}{\text{Energy input}} = \frac{E_{net}}{E_{self} + E_{purchased}}$$

$$\text{Energy surplus (Net Energy Value)} = \text{Energy output} - \text{Energy input}$$
$$= E_{net} - (E_{self} + E_{purchased})$$

FIGURE 8.3 Simple EROI of an oil well

An energy return on energy investment (EROI) analysis for an oil well would include the cost of the capital and materials needs to create and maintain the well, as well as the more obvious energy costs for transporting the crude, refining it, transporting the finished product, and mitigating the emissions upon its combustion. Source: [4]

unable to do." [15] Figure 8.3 depicts the energy return on investment for a typical oil well.

In addition to considering the energy-in and energy-out at the oil well, mine mouth, or farm gate, we can move further down the energy "food chain" to consider the EROI as it enters the point of use, such as factory or city cogeneration plant. Hall defines this EROI at the point of use as follows:

Equation 2

$$EROI_{point\ of\ use} =$$

$$\frac{\text{Energy returned to society}}{\text{Energy required to get and } deliver \text{ that energy}}$$

Consider a home furnace as the point of use. The fuel oil that is extracted and shipped con-

siderable distance now has to be burned in the furnace to heat the home. The furnace requires electricity for its ignitor and thermostat. Inevitably some of the energy in the burned oil is lost to the home as waste heat up the chimney. Using the delivered energy requires additional energy or costs energy along the way. Hall terms this the extended EROI and defines it as follows:

Equation 3

$$EROI_{extended} =$$

$$\frac{\text{Energy returned to society}}{\text{Energy required to get, deliver, and } use \text{ that energy}}$$

In order to deliver one barrel of fuel to the end user and to use it there requires about three barrels to be extracted from the ground, with

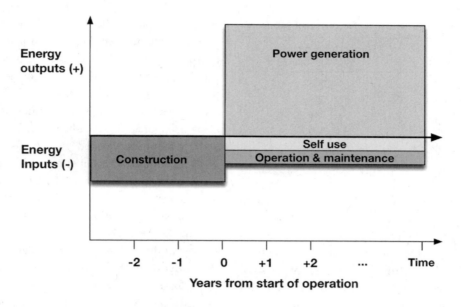

FIGURE 8.4 Energy payback for a gas well

The energy payback period for a source includes subtracting from the energy it generates until it has used up all the energy cost of its construction, operation, and self-use. Source: [4]

two being used indirectly and therefore lost to the end user. Thus three barrels in for one barrel out is the situation we find ourselves in today for most liquid fossil fuels. To paraphrase Professor Hall and his colleagues, twice as much energy is consumed in the process of extracting and delivering and using liquid fuel than is in the fuel itself at its point of use. [15]

A common related concept is the energy payback period. Every energy system has initial investments of energy in the construction of a facility. The facility then produces energy for a number of years until it reaches the end of its effective lifetime. Along the way, additional energy costs are incurred in the operation and maintenance of the facility, including any self-use of energy. An example of the latter is the natural gas produced by a gas well that is then used to pump more gas out of the ground, or the electricity from a power plant that is used to run the computers and lights in the plant. The energy payback period is the time it takes a facility to "pay back," or produce an amount of energy equivalent to, that invested in its start-

up. Figure 8.4 depicts the energy payback for a typical natural gas well in simple chart form. (See the online content for this chapter for more on this concept.)

Old versus New Sunlight

The nation relies on fossil fuels that are subject to supply shocks and price volatility as well as foreign entanglements. A recent burst of enthusiasm for biological fuels has gripped the nation. The "biofuel" category is extremely broad, from wood pellets to gasified ethanol. Biofuel sources can include trees, bamboo, corn, sugarcane, and a host of grasses. For many biofuels now in active production, the energy balance may be negligible or even negative. A proper EROI enegy balance analysis would take the following energy use into account throughout the entire life cycle of the material. The stages of biofuel production that need to be examined for total net resource product include (1) growing the crop—including the impact of clearing land to do so; (2) producing and dispensing fertilizers,

TABLE 8.2 Energy Input in Biofuels—Out of Balance

Biofuel source	% fossil fuel energy input minus % output
Sunflower	118%
Woody biomass	57%
Switchgrass	45%
Corn to ethanol	29%
Soybean to biodiesel	27%

Source: [33]

pesticides, and herbicides (often manufactured from oil or natural gas); (3) operating farm machinery; (4) irrigating cropland; (5) grinding and transporting harvested crops; (6) fermenting and distilling the fuel; (7) processing; (8) packaging; (9) transport to market; and (10) marketing. Many biofuels also require converting forest or peatlands into biofuel farms. When the negative impact of resulting greenhouse gases are considered, converting corn to ethanol emits twice as much greenhouse gas emissions as gasoline per unit of energy produced. Similarly, cellulosic switchgrass-to-ethanol raises net emissions by 50% over gasoline. Soybean biodiesel refineries foul local rivers with glycerin and methanol. Palm oil plantations displace biologically diverse tropical forests and release such large stores of carbon that this activity has shot Indonesia into third place—just behind the United States and China—on the list of top greenhouse gas emitters. [33]

The EROI method takes into consideration all the factors of production in creating the fuel, including the fertilizer used to prepare the fields for the biomass growth (and its associated environmental issues, such as nitrous oxide and smog pollution), the tremendous amount of land needed to bring biofuels to scale, the fermentation of sugar (which requires energy input), and the burning of biofuel, which releases carbon dioxide along with volatile organic compounds such as acid aldehyde, which combines with nitrous oxide to create lung-irritating, heat-trapping tropospheric ozone. In addition, aldehydes such as formaldehyde are toxic to living cells but not regulated in current emission

laws, which predate liquid biofuel use. Adding even a small amount of ethanol to petroleum-based gasoline, as in American E10 gasohol, can increase the tailpipe emission of aldehydes produced by combustion. In short, the biofuel sector needs a great deal more study before a sustainable practice about how best to produce and use them is developed.*

In an admirable recent study, Tad Patzek of the University of California, Berkeley, College of Engineering summarized the energy use in the United States from all sources to compare with the total energy production of all biological sources in the nation. He sought to answer one simple question: "If we wanted to convert all our 'dirty' fossil or nuclear energy sources to 'clean' biological sources, would America have enough biomass sources to do so?" The resounding answer is No! The main reason is very sobering. Our total energy use per year now outstrips the entire biomass energy production that we use for food and livestock feed plus the biomass energy production of industrial wood for shelter, roots of all plants, vegetation in our forests, and lastly, the smallest slice of our biostock that we now already convert to energy (as ethanol, firewood, etc.). [31]

Current proposals to replace a good part of the fossil energy devoured each year by us with the biomass-derived fuels are pure fantasy. The only way to increase the biomass share of primary energy use in the U.S. is to decrease the fossil fuel consumption. To make the U.S. competitive with the rest of the developed world, we should strive to decrease our fossil energy consumption by a factor of two, so that each American uses daily only 50 times more energy than we need as food to live. [31]

Tad Patzek, 2006

---

*Such work across nations is already underway. The Roundtable on Sustainable Biofuels has released proposed standards for sustainable biofuels in draft for wide public comment as the "Principles on Sustainable Biofuels Production." (See www.bioenergywiki.net).

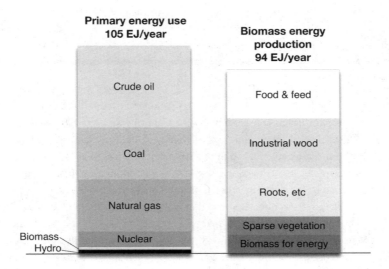

FIGURE 8.5 Primary energy use versus biomass energy production in the United States

Annual fossil and nuclear energy consumption in the United States is now greater than all biomass growth over the entire US territory. The left bar shows the relative quantity that each major energy source contributes in the United States per year, for a total of 105 EJ in 2003. The right bar shows an estimate of annual biomass production over the entire country in a year. Over three-fourths of the biomass is committed already for food, feed, paper, fiber, or lumber or is stored as roots or is otherwise inaccessible. The "food and feed" sector is heavily subsidized by fossil fuel use in agriculture, so expanding production there under our current practices would accelerate fossil fuel consumption before increasing biofuel potential. And one-half of the remaining biomass is locked up in sparse vegetation not practical to harvest. So the "harvestable" biomass for energy that is not already being used for biofuels may only add up to 2% or 3% of US energy consumption today. Source: [32]

Additional emerging technologies are both exciting but largely experimental at this stage. Nanotechnology, in which components measure one-billionth of a meter, holds the promise of greatly increasing efficiencies and reducing costs in many different energy production fields, from solar power to circuitry. But we know too little about nanotechnology durability, safety, and long-term performance. Nanomaterials represent high capital investment opportunities, but research into their health and safety risks needs to be robust as well.

Another promising but young field is energy derived from waves, tides, and currents. We do not yet have adequate research and development on how to capture the potential energy from naturally occurring, zero-carbon, base-load power, such as harnessing the Gulf Stream off the southeast coast of the United States. Tidal and wave turbines are likely to be useful as local resources but currently represent an extremely small installed base. Microbial energy sources also show good potential but need to be studied, developed, and then scaled up.

INSIGHT 8: ZERO ENERGY HOUSING

The Beddington Zero Energy Development, or BedZED, is the United Kingdom's largest eco-village. All the homes built at BedZED are zero–fossil energy galleried apartments. Eco-construction and developing green lifestyles based on integrating work and living areas with transit hubs can be easy, accessible, and affordable and can provide a good quality of life. For example, the heating requirements of BedZED homes are about 10% that of a typical home. The housing development is one of the most coherent examples of sustainable living in the United Kingdom. Initiated by BioRegional, BedZED was developed by the Peabody Trust in partnership with BioRegional Development Group and designed by Bill Dunster Architects. Located in Wallington, South London, BedZED comprises 100 homes, community facilities, and work space for 100 people. Residents have been living at BedZED since March 2002. Dunster and his team designed these homes as a pilot project integrating sustainable technologies. The BedZED monitoring results from the BedZED's first year of occupation show that building performance and transport patterns have been very much as expected. Table 8.3 shows BedZED comparisons with the national average for space heating and hot water, with new homes built to year 2000 building regulations in brackets. [1]

TABLE 8.3 BedZED Home Energy Reduction Through Sustainable Design

BedZED energy categories	Monitored reduction	Target
Space heating	88% (73%)	90%
Hot water	57% (44%)	33%
Electricity	25%	33%
Mains water	50%	33%
Fossil fuel car mileage	65%	50%

Source: [1]

The Benefits of No-Regrets Wedges

Fortunately, several solutions stand up well in a life cycle analysis with ancillary benefits to society—if scaled up to the size of a climate stabilization wedge. Green building features such as increased natural lighting and proximity to green spaces have positive impacts on physical and mental health.

These health and cost-savings benefits apply to residential buildings as well as to work places where a business can readily gain a more productive workforce. When designed into medical facilities, features such as daylighting and improved passive air circulation can even speed a patient's recovery. As we mentioned in Chapter 7, more pedestrian-friendly cities with transportation networks and proximity of residences and workplaces also can lead to better air quality and healthier residents. [7]

Sustainable forestry, a potential wedge for its ability to stop uncontrolled deforestation and protect an important natural carbon sink, can improve social and economic capital, particularly in the developing world. Logging, land-clearing, drought, pests, and demand for wild meat coupled with fisheries decline collaborate to degrade the condition of the world's tropical, temperate, and boreal forests. This decline in the forest carbon sink causes about one-fifth of the greenhouse gas problem. Bark beetle infestations in British Columbia have recently turned

II: How to Think About Climate Solutions

TABLE 8.4 Estimated Savings from Green Buildings in the United States

Health and energy savings (in 1996 $US)
Respiratory disease: $6 to $14 billion
Allergies and asthma: $1 to $4 billion
Sick building syndrome: $10 to $30 billion
Worker performance: $20 to $160 billion
Total energy savings: $70 billion
Schools with natural light
20% speed increase on math tests
26% speed increase on reading tests
Stores with natural light
40% increase in sales
Hospitals with better lighting and ventilation
Improved patient outcomes and reduced hospital stays

Source: Adapted from [6] from, respectively, [9, 20, 21, 11]

vast pine forests from carbon sinks into carbon sources as the trees die. [25]

Green chemistry is transforming industrial processes from linear systems—which require many inputs and generate excessive and often toxic waste—to closed-loop systems in which materials can be recycled and reused with far less waste by-product. Green chemistry has the potential to save tremendous energy and resources in a cradle-to-cradle manner. [28]

These wedges must be integrated together into systems that allow the benefits to work together. For example, a healthy cities program would include green buildings, rooftop gardens, improved public transportation, more bike lanes, more walking paths, and so on. [42] Energy systems must be redesigned for hybrid power—geothermal, wind, and solar generated regionally and distributed in an intelligent, more flexible, and decentralized electricity grid.[†] [2] Clean energy must be made available for developing nations, allowing them to leapfrog over outdated, fossil fuel–reliant technologies as they

[†]For a recent, highly readable report on the challenges and opportunities in transforming the electric power sector, see "The Electric Economy" at the Resource Center at www.globalenvironmentfund.com.

work toward sustainable development and the UN Millennium Development Goals.

From all reasonable accounts, the long-term economic benefits of deploying green technologies that begin to displace existing high-emission technologies surpass the growth available by sticking to business as usual. In other words, as Al Gore says, "Doing what's best for the environment also happens to be what's best for the economy." Van Jones, formerly of the Ella Baker Center for Human Rights in Oakland, California, now green jobs advisor to President Obama, points out that not only can such job expansion efforts provide real and long-term employment in higher numbers than older technology jobs can, but this kind of economic development has particular positive potential for benefitting those workers in blue-collar fields that are shrinking or in communities of color or less economically privileged communities. [16]

Local economic "green collar" jobs will become a driver of growth, because doing what is good for the long-term economy will be the same thing that is good for the environment. European researchers examined energy efficiency measures being tried around the world for the UN's Sustainable Buildings and Construction Initiative. In that 2007 study, Sonja Koeppel and Diana Ürge-Vorsatz found that significant barriers to implementing energy efficiency measures in many sectors of the economy from industry to banking were the fear of losing jobs, fear about sharing trade secrets, and lack of understanding or trust that efficiency provides economic paybacks. Fortunately, they also found many ways in which economic and market-based instruments could be made more transparent to the consumers and decision makers as a whole. The solutions to such barriers include demonstration programs, accreditation systems, and standardization of contract procedures. For Europe, the shift from national currencies to the common market's euro was effectively a dry run for communicating how to adapt to a new system across all sectors and national borders. [23]

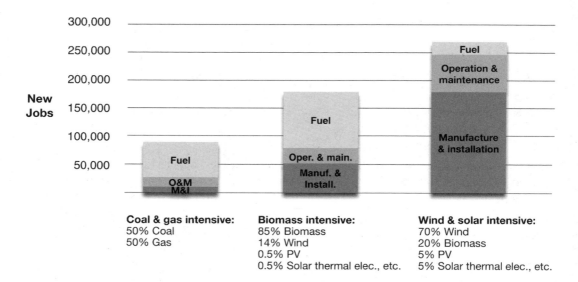

Coal & gas intensive:
50% Coal
50% Gas

Biomass intensive:
85% Biomass
14% Wind
0.5% PV
0.5% Solar thermal elec., etc.

Wind & solar intensive:
70% Wind
20% Biomass
5% PV
5% Solar thermal elec., etc.

FIGURE 8.6 Green collar effect of growing green energy

The bars show the estimated employment created by meeting the equivalent of 20% of current US electricity demand via (left) expansion of fossil fuels or (center and right) two different portfolios of renewables-based electricity generation. Of these three scenarios, the largest job growth would come from expanding wind and solar power so that they provide 80% of the new energy produced. The different fuel mixtures come from current state or federal renewable energy portfolio standards. It is assumed that biofuel would be mixed in solid, liquid, or gas state with existing fossil fuels. Biofuel emits less net greenhouse gas emissions than coal only or gas only because the biomass is regrown in subsequent years. (See also US Department of Energy, The Billion Ton Feedstock Supply, 2004). Source: [18]

Michael Renner of Worldwatch Institute reports the number of green jobs is on the rise: "'Climate-proofing' the global economy will involve large-scale investments in new technologies, equipment, buildings, and infrastructure, which will provide a major stimulus for much-needed new employment and an opportunity for retaining and transforming existing jobs." [36]

In *Hot, Flat and Crowded*, journalist Thomas Friedman writes, "We are living at the hinge of history that is going to determine just which way this Energy-Climate Era will swing. If we are going to manage what is already unavoidable and avoid what will be truly unmanageable, we need to make sure everything we do from here on helps build a real sustainable, scalable solution. The cheap and easy paths are all closed." [10: p. 410]

We will know the green revolution has succeeded on the day we no longer hear about "green cars" or "green buildings" but just "cars" and "buildings" that are energy and resource misers, just because that is the best business practice for us all. Dan Kammen at the University of California, Berkeley, and his colleagues used the government's own numbers to project how many new jobs would be created to increase US energy production by 20%. They looked at three scenarios for adding more energy production: by expanding fossil fuels, biofuels, and wind power. Wind power, as the largest portion of a mix of other renewables, would generate triple the number of new jobs as the same energy volume from coal, oil, or gas, and about 30% more new jobs than adding that much new energy mainly with biofuels. [18] Indeed, worldwide,

II: How to Think About Climate Solutions

wind power is the renewable with the highest growth rate in this decade. Wind power has also become one of the renewable technologies with the broadest base of support, with installations in more than 70 countries. And, wind power is among the least expensive to install per kilowatt of power delivered. [35]

Across a range of scenarios, the renewable energy sector generates more jobs than the fossil fuel–based energy sector per unit of energy delivered (i.e., per average megawatt). In addition, we find that supporting renewables within a comprehensive and coordinated energy policy that also supports energy efficiency and sustainable transportation will yield far greater employment benefits than supporting one or two of these sectors separately. [18]

Dan Kammen, Renewable and Appropriate Energy Laboratory

Achieving such systems will require incentives, both positive and negative, private-and public-sector alignment, and an infrastructure for reaching each wedge designed by the federal government. If we are well informed of the true costs, benefits, and impacts of these stabilization wedges, Paul Epstein and his colleagues conclude, we will be able to make decisions that will benefit the environment, the economy, and our security and help stabilize climate.

CONNECT THE DOTS

- Several climate solutions—scaled up to the size of a stabilization wedge—stand up well in a life cycle analysis and have ancillary benefits to society; these include green buildings, sustainable forestry, and green chemistry, among others.

- Consumers as energy producers with solar roofing and small-scale wind power combined with efficiency measures can make a positive difference, but only if these local technologies are widely installed.

- These wedges must be integrated together into systems that allow the benefits to work together, for example, in healthy city programs.

- Clean energy must be made available for developing nations, allowing them to leapfrog over outdated, fossil fuel–reliant technologies as they work toward sustainable development.

- Several solutions have hidden costs that need to be examined in a life cycle analysis—biomass fuels, nuclear fission, and carbon capture from coal, among others.

- It is important to take the entire return on investment into account for any given energy source.

- Most forms of renewable energy provide co-benefits in public health and job creation.

- These renewable energy benefits are maximized if coupled with energy efficiency programs and reduction of demand programs to reduce the overall energy demand.

Online Resources

www.eoearth.org/article/Carbon_capture_and_storage
www.eoearth.org/article/Energy_return_on_investment_(EROI)
www.eoearth.org/article/Fossil_fuel_power_plant
www.eoearth.org/article/Net_energy_analysis
www.eoearth.org/article/Nuclear_power_reactor
www.eoearth.org/article/Renewable_energy
www.eoearth.org/article/Ten_fundamental_principles_of_net_energy
Renewable Energy Policy Network for the 21st Century, www.ren21.net
Roundtable for Sustainable Biofuels, www.bioenergy wiki.net/index.php/Version_Zero
UNEP Sustainable Building and Construction Initiative, www.unepsbci.org
UNEP Life Cycle Initiative, www.unep.fr/scp/initiative
See also extra content for Chapter 8 online at http://ncseonline.org/climatesolutions

Climate Solution Actions

Action 1: Green Buildings and Building Design
Action 7: Biofuel Industry and CO_2 Emissions—Implications for Policy
Action 9: How to Ensure Wind Energy Is Green Energy
Action 10: Nuclear Energy—Using Science to Make Hard Choices

Works Cited and Consulted

[1] BedZED (2008) Beddington Zero Energy Development. *BioRegional*. South London, UK (read December 13, 2008). www.bioregional.com/programme_ projects/ecohous_prog/bedzed/bedzed_hpg.htm

[2] Berst J, Bane P, Burkhalter M, Zheng A (2008) The Electricity Economy: New Opportunities from the Transformation of the Electric Power Sector. www.globalenvironmentfund.com; www.global smartenergy.com

[3] Cleveland CJ (2005) *Net energy from the extraction of oil and gas in the United States.* Energy 30(5):769–782. http://linkinghub.elsevier.com/retrieve/pii/ S0360544204002890

[4] Cleveland CJ (2008) "Energy Return on Investment," (topic ed) Constanza R, in *Encyclopedia of Earth*, (ed) Cleveland CJ. (Environmental Information Coalition, National Council for Science and the Environment, Washington, DC). www.eoearth.org/article/ Energy_return_on_investment_%28EROI%29

[5] Epstein P (2007) "Climate Change: Healthy Solutions." Guest Editorial. *Environmental Health Perspectives*, April 115(4):A 180. http://chge.med. harvard.edu/programs/ccf/

[6] Epstein P (2008) Healthy Solutions for the Low Carbon Economy: Guidelines for Investors, Insurers and Policy Makers. *The Center for Health and the Global Environment*. Harvard Medical School. http://chge.med.harvard.edu/programs/ccf/

[7] Ewing R, Kreutzer R (2006) Understanding the Relationship between Public Health and the Built Environment. *US Green Building Council*. New York. www.usgbc.org

[8] Fava J, Hall J (2004) Why Take a Life Cycle Approach? *UNEP Life Cycle Initiative*. United Nations Environment Programme Division of Technology, Industry and Economics, Paris. www .unep.fr/scp/lcinitiative/publications

[9] Fisk WJ (2000) *Health and productivity gains from better indoor environments and their relationship with building energy efficiency.* Annual Review of Energy and the Environment 25:537–566. http:// arjournals.annualreviews.org/loi/energy

[10] Friedman T (2008) *Hot, Flat and Crowded: Why We Need a New Green Revolution and How It Can Renew America*, 410. (Farrar, Straus, and Giroux, New York). www.thomaslfriedman.com

[11] Frumkin H, Frank L, Jackson RJ (2004) *Urban Sprawl and Public Health: Designing, Planning, and Building for Healthy Communities.* (Island Press, Washington, DC). http://islandpress.org

[12] Gabbard A (2008) Coal Combustion: Nuclear Resource or Danger. *Oak Ridge National Laboratory*. Oak Ridge, TN (read December 22, 2008). www.ornl.gov/info/ornlreview/rev26-34/text/ colmain.html

[13] Gagnon L, Bélanger C, Uchiyama Y (2002) *Life-cycle assessment of electricity generation options: the status of research in year 2001.* Energy Policy 30(14):1267–1278. http://linkinghub.elsevier.com/ retrieve/pii/S0301421502000885

[14] Graedel TE (1996) *On the concept of industrial ecology.* Annual Review of Energy and the Environment 21:69–98. http://arjournals.annualreviews. org/doi/abs/10.1146/annurev.energy.21.1.69

[15] Hall C, Balog S, Murphy DJR (2009) *What Is the Minimum EROI That a Sustainable Society Must Have?* Energies 2:25–47. http://web.mac.com/ biophysicalecon/iWeb/Site/Downloads.html

[16] IAEA (2008) Nuclear's Great Expectation: Projections Continue to Rise for Nuclear Power, but Relative Generation Share Declines. International Atomic Energy Agency. September 11, 2008 (read October 20, 2008). www.iaea.org/NewsCenter/ News/2008/np2008.html

[17] Jones V (2008) *The Green Collar Economy*. (HarperCollins Publishers, New York). www .greenforall.org

[18] Kammen D (2007) The Low-Carbon Imperative and Economic Opportunity. US House of Representatives Committee on Oversight and Government Reform Testimony for the November 8, 2007, Hearing on: Opportunities for Greenhouse Gas Emissions Reductions. http://socrates.berkeley .edu/~kammen/

[19] Kammen D, Farrell AE, Plevin RJ, Jones AD, Delucchi MA, Nemet GF (2007) *Energy and Greenhouse Impacts of Biofuels: A Framework for Analysis*, in *OECD Research Round Table: Biofuels: Linking Support to Performance*, (ed) Perkins S. (Organization for Economic Cooperation and Development, Paris). www.internationaltransportforum.org/jtrc/ RoundTables/BiofuelsOutline.pdf

[20] Kats G (2006) Greening America's Schools: Costs and Benefits. http://dsforgau.ozstaging.com/ downloads/greening_americas_schools.pdf

[21] Kellert SR, Heerwagen J, Mador M (2008) *Biophilic Design: The Theory, Science and Practice of Bringing Buildings to Life.* www.wiley.com/WileyCDA/ WileyTitle/productCd-0470163348.html

[22] Knapp K, Jester T (2001) *Empirical investigation of the energy payback time for photovoltaic modules.* Solar Energy 71(3):165–172. http://linkinghub .elsevier.com/retrieve/pii/S0038092X01000330

[23] Koeppel S, Ürge-Vorsatz D (2007) *Assessment of Policy Instruments for Reducing Greenhouse Gas Emissions from Buildings.* (United Nations Environment Programme, Central European University, Budapest). www.unepsbci.org

[24] Krauss C (2008) "An Export in Solid Supply." *New York Times*, March 19, 2008. www.nytimes.com

[25] Kurz W (2008) *Making the paper: mountain pine beetles contribute to carbon release and climate change.* Nature 452:987–990. www.nature.com/nature/

[26] Lovins A, Sheikh I (2008) *The nuclear illusion.* Ambio, in November 2008 preprint. www.rmi.org/sitepages/pid257.php#E08-01

[27] Mastny L (2003) *Purchasing Power: Harnessing Institutional Procurement for People and the Planet.* Worldwatch Paper #166. www.worldwatch.org

[28] McDonough W, Braungart M (2002) *Cradle to Cradle: Remaking the Way We Make Things.* (North Point Press/FSG, New York). www.mbdc.com

[29a] MIT (2003) the Future of Nuclear Power. http://web.mit.edu/nuclearpower

[29b] MIT (2007) MIT Study on the Future of Coal: Options for a Carbon-Constrained World. http://web.mit.edu/coal/

[30] NIOSH (2008) Faces of Black Lung. *National Institute for Occupational Safety and Health (NIOSH).* Centers for Disease Control and Prevention. August 18, 2008 (read September 4, 2008). www.cdc.gov/niosh/blog/

[31] NIRS (2007) Report on Earthquake Damage to Japan's Kashiwazaki-Kariwa Nuclear Power Facility. *Nuclear Information and Resource Service* (read September 12, 2008). www.nirs.org

[32] Patzek TW (2006) The Earth, Energy, and Agriculture. Paper presented at *Climate Change and the Future of the American West: Exploring the Legal and Policy Dimensions.* http://petroleum.berkeley.edu/papers/Biofuels/BiofuelsTop.htm

[33] Pimentel D, Patzek TW (2005) *Ethanol production using corn, switchgrass and wood; biodiesel production using soybean and sunflower.* Natural Resources Research 14:65–76. www.springerlink.com

[34] Randolph J, Masters GM (2008) *Energy for Sustainability: Technology, Planning, and Policy.* (Island Press, Washington, DC). www.energyforsustainability.org

[35] Remmen A, Jensen AA, Frydendal J (2007) *Life Cycle Management: A Business Guide to Sustainability.* (Life Cycle Initiative, United Nations Environment Programme). www.unep.fr/scp/lcinitiative/publications/

[36] REN21 (2008) Renewables 2007 Global Status Report. (Deutsche Gesellschaft für Technische Zusammenarbeit, Worldwatch Institute, Paris: REN21 Secretariat and Washington, DC:Worldwatch Institute). http://www.ren21.net

[37] Renner M (2008) *Working for People and the Environment.* Worldwatch Report #177. www.worldwatch.org

[38] US EIA (2008) Annual Energy Review 2007. DOE/EIA-0384(2007). www.eia.doe.gov/emeu/aer/

[39] USGS (1997) Radioactive Elements in Coal and Fly Ash: Abundance, Forms, and Environmental Significance. Fact Sheet FS-163-97. *US Geological Survey.* Denver, CO. http://pubs.usgs.gov/fs/1997/fs163-97/FS-163-97.html

[40] Wald ML (2008) "As Nuclear Waste Languishes, Expense to US Rises." *New York Times*, February 19, 2008. www.nytimes.com

[41] Weber CL, Matthews HS (2008) *Food-miles and the relative climate impacts of food choices in the United States.* Environmental Science and Technology. http://pubs.acs.org/cgi-bin/abstract.cgi/esthag/asap/abs/es702969f.html

[42] WHO (2003) National Healthy Cities Networks: A Powerful Force for Health and Sustainable Development in Europe. www.euro.who.int

CHAPTER 9

Multiple Intensity Disorder

> We have not put in place the basic requirement for trying: a
> coordinated set of policies, tax incentives and disincentives,
> and regulations that would stimulate the marketplace to
> produce an Energy Internet, to move the clean power technolo-
> gies we already have—like wind and solar—down the learning
> curve much faster, and to spur the massive, no-holds-barred-
> everybody-in-their-garage-or-laboratory innovation we need for
> new sources of clean electrons. [7: p. 243]
>
> THOMAS FRIEDMAN, 2008

Before 2001, the world's scientists lacked international standards for collating all the climate data they were gathering in their individual research projects. The scale of the global change each laboratory was observing demanded a common pool into which all the research results could be collected for comparison and context. How else would we learn from each other's work and, just as important, learn what we collectively do not yet know? The Global Carbon Project (GCP) is exactly the kind of cross-national and cross-disciplinary organization we need to grapple with climate disruption.[*] The GCP mission is both simple and complex. The simple goal is to help scientists worldwide build a common repository about the greenhouse gases in the Earth's atmosphere. The complex part of this mission is that

climate researchers are scattered all over the globe and did not previously have standards for measuring or methods for presenting their data. But since its founding in 2001, GCP has begun reporting, in plain and direct terms, the greenhouse gas emissions story. Its scientific steering committee includes scientists in 13 countries, who in turn represent many different national research initiatives. Thanks to the collation work of the GCP, we have, at last, some reliable data on atmospheric trends. That is the good news.

Worldwide in 2007, we added 30 billion tons of carbon dioxide to the atmosphere. As we learned in prior chapters, it is already too late to avoid significant warming on the scale of 2 to 2.5 degrees Celsius (°C) in this century. Global emissions of carbon dioxide had been accelerat-

---

[*]Original data to complete the global carbon budget are generated by multiple agencies and research groups around the world and are collated annually by the Global Carbon Project (www.globalcarbonproject.org). Data are available for the period of 1850–2006 and can be downloaded from www.globalcarbonproject.org/carbontrends/.

ing at a rate of about 1.3% per year in the 1990s. However, most recently from 2000 to 2006, the global emission growth rate more than doubled to the current annual rate of 3.3%, as reported in a major paper led by the GCP team. [1] That is profoundly bad news. Under this acceleration of emissions, limiting overall warming below 2.5°C will be a monumental challenge. Why are emissions growing more quickly today than a few years ago?

Six Economies and Four Factors

Let's take a brief look at a handful of key economies that each dominate a region of the globe. We will examine "economies" rather than "nations" because the former term simply reflects the reality of several large regional trading networks. The European Union has a common carbon emissions regime and common currency shared among its 25 nations. The states of the former Soviet Union (FSU) are centered around Russia, and Russia's energy policy dictates that of the other states in the Russian Federation, such as Belarus and Ukraine.

The six largest carbon dioxide–emitting economies represent very different levels of emissions per capita and per million dollars of economic product. The United States produces the most total tons (Figure 9.1, Chart A) and tons per person (Chart B). China is leading the pack in annual emissions growth rate, as China continues to bring more major coal-fired power plants online (Chart C). The former Soviet Union states produced the most tons of carbon dioxide in 2004 per dollar of economic product (Chart D). Meanwhile, the European Union, including Germany, is trying to do the right thing by cutting emissions.

But we know the world economy is always changing. The industrialized nations, the United States, European Union, and Japan, have economies that exhibit higher current levels of economic output per person with lower net population growth. Directly parallel to that, they also emit more greenhouse gases per person.

Overall, global population growth is slowing down but still increasing, especially in the developing economies. Eighty percent of the world's population lives in developing economies, represented here by China and India. Today, China and India together represent 35 out of every 100 people on Earth, and 45 of out every 100 who live in a developing nation. These two nations are industrializing rapidly, as evidenced by India's emerging automobile industry and China's rapid growth in coal-fired power plants. So why has the global emission rate of greenhouse gases accelerated sharply since 2000?

If we examine these same six economies and four factors over a 4-year time span, from 2000 to 2004, we see sharp differences emerge. Frank Princiotta, a senior scientist at the US Environmental Protection Agency (EPA) did just that and compiled the analysis we see in Figure 9.2. Developing economies accounted for 73% of the global growth in CO_2 emissions in 2004. However, these economies accounted for only 41% of emissions themselves and only 23% of emissions since 1800. Figure 9.2 shows some of the same data as in Figure 9.1. For example, the growth rate in emissions in Figure 9.1 Chart C is represented in Figure 9.2 by the small triangles for each economy. But Dr. Princiotta added the economic growth (gross domestic product, or GDP) per person as well for each of the six economies. Historically, GDP per capita (person) has risen by about 1.6% per year over the past century, when adjusted to 1990 US dollars. (We use 1990 dollars here because many of the climate indicators are calibrated to 1990 as a base-year reference point).

We see that high growth in GDP per population parallels high growth in annual carbon dioxide emissions in China, India, and the republic of the former Soviet Union (comparable for our purposes to the Russian Federation in Figure 9.1). You might ask why energy use per GDP shrank between 2000 and 2004 for each economy except China. The hidden reason is that in 2004 these economies squeezed more economic product out of each unit of energy con-

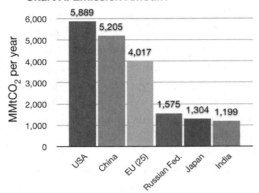

Chart A: Emission Amount

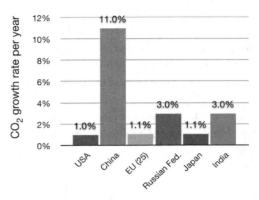

Chart C: Emission Growth Rate

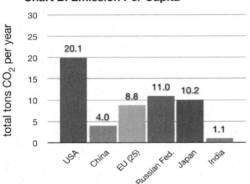

Chart B: Emission Per Capita

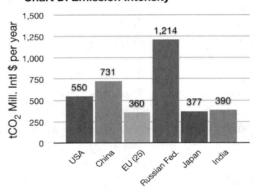

Chart D: Emission Intensity

FIGURE 9.1 Six economies and four emission factors: A 2004 snapshot

These four charts represents carbon dioxide emissions in 2004 for six major economies. The United States leads in total CO_2 emitted (Chart A) and total per person (Chart B). China leads in emissions growth at 11% (Chart C). And the Russian Federation leads in tons of emission per dollar of economic product (Chart D). The data exclude the effect of land use changes (deforestation or reforestation) and use the following units: total emissions in carbon dioxide equivalent (CO_2) in million metric tons ($MMtCO_2$) in Chart A; tons carbon dioxide equivalent per person (tCO_2) in Chart B; percent annual growth rate from prior year of carbon dioxide equivalent in Chart C; metric tons of carbon dioxide per million international dollars (tCO_2/Mill. Intl. $) in Chart D. For fun, search the European Environment Agency (www.eea.europa.eu) for "energy intensity graph." Source: Adapted from [32]

sumed than they did in 2000, except for China. In other words, the economies became more efficient in converting energy into product.

What Drives Emissions?

What lessons can we draw from this set of economic and emissions data? Can we express emissions as a combination of factors? First, we can see that both population size and growth of population matter. We can also see that income (GDP per person) is a factor. Of these six economies, those with higher GDP growth also had higher emissions growth. Finally, we see that the efficiency of the economy in turning energy into product is a factor. Let's look at this last factor slightly differently.

China and Russia (as the FSU) show the

II: How to Think About Climate Solutions

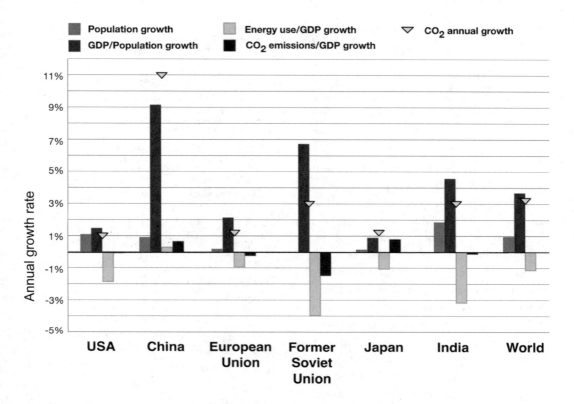

| Population growth | Energy use/GDP growth | ▽ CO₂ annual growth |
| GDP/Population growth | CO₂ emissions/GDP growth | |

FIGURE 9.2 Six economies and four emission factors over time: 2000–2004

Solid fuels, such as coal, produced a larger and more rapidly growing share of emissions in developing regions (the sum of China, India, and other developing nations) than in developed regions (USA, European Union, Japan), and the former Soviet Union region had a much stronger reliance on gas than the world average. Source: [19] as adapted from [1]

highest growth rate in GDP per person for the 4 years covered in Figure 9.2, at over 9% and almost 7%, respectively. Yet the energy sources they use to power their economies are very different. China relies much more heavily on domestic coal to fire its electricity stations and its industrial plants. Russia relies more heavily on tapping its significant natural gas reserves. We know that burning a ton of coal emits more carbon dioxide than burning the equivalent mass of natural gas. So on this basis, a natural gas–fired economy would be less intensive in its carbon dioxide emissions than the same economy fired by coal plants. We can call this fuel mix the carbon intensity of the energy use, or how many tons of carbon is emitted per watt

of energy used. In short, China's economy is more carbon intensive than that of Russia per watt of energy consumed.

A watt (W) is a common unit of measure for power and is equal to the rate of work done by 1 joule of energy per second. A common electrical power measure is the kilowatt (kW), 1,000 (10^3) watts. A kilowatt-hour (kWh) is the amount of work 1,000 watts do in 1 hour and a common unit of measure for our electric bills. Burning a 100 W lightbulb for 1 hour uses 0.1 kW and represents 0.1 kWh. A megawatt (MW) is equal to 1 million (10^6) watts. A gigawatt (GW) is 1 billion (10^9) watts. A terawatt (TW) is equal to 1 trillion (10^{12}) watts. The current annual average power usage by all humans is about 15 TW. One

TABLE 9.1 Four Quantifiable Factors That Drive Emissions

Population:	The number of people	(Population growth rate projections)
Income:	The growth of the economy*	(GDP/population)
Energy intensity:	The amount of energy used in the economy*	(Energy use/GDP)
Emission intensity:	The amount of carbon based energy used*	(CO_2e emissions/energy use)

*These quantities are often calculated on a per capita basis to make comparisons between nations easier.

terawatt of power is the equivalent to the annual output of about 1,000 nuclear power plants.

Something else is going on that we hinted about earlier. Over time, an economy can become more efficient at turning that ton of fuel into a product than it was the year before. Newer product designs may use less raw material. Manufacturing processes improve, so factories become more efficient. The amount of energy (e.g., watts) required to produce one unit of economic output (e.g., a dollar) may go down—a good thing. We can describe this as lowering the energy intensity of the economy. Energy intensity comprises, in turn, economic efficiency, energy conservation, and the overall economic structure of industry and society (e.g., light manufacturing versus heavy manufacturing, mass transit use versus personal vehicles, size of service versus manufacturing sector). Historically, the global energy intensity had been declining at a rate of about 1% per year, which means as a whole we have been becoming 1% better each year at extracting dollars out of watts.

These intensity relationships can be expressed as formulas for annual carbon dioxide emissions.[†]

Equation 1

$$CO_2 \text{ emissions} = \text{Population} \times \text{Income} \times \text{Energy intensity} \times \text{Emissions intensity}$$

or

[†]This expression is also known as the Kaya identity, a concept that the IPCC now uses in projecting its emissions scenarios. See also www.realclimate.org/index.php?p=164. Source: [6 and 21] originally using values published in [33]

Equation 2

$$CO_2 \text{ emissions} = \text{Population} \times \frac{GDP}{\text{Population}} \times \frac{\text{Energy use}}{GDP} \times \frac{CO_2 \text{ emissions}}{\text{Energy use}}$$

These four factors—population, income, energy intensity, and emissions intensity—constitute an economic analysis of what drives greenhouse gas emissions. If we measure each of the major factors that drive emissions against their 1990 levels, we can assign the 1990 level each a value of 1, which yields $1 \times 1 \times 1 \times 1 = 1$. Any gain in one factor in a different year would be a plus, and any drop would be a minus. For example, a doubling of population would be a 2, and a drop by one-tenth would be −0.1. In this simple Kaya identity equation, we can lay out all four factors—population, income wealth (GDP), energy intensity, and carbon emission intensity—to see where differences between them lie. If we take a political science perspective of what drives emissions, we realize that the following social factors play determining roles: knowledge and values (what do we know and what do we think is most important?), social organizations (how are we organized for action?), policy (what specific actions do we want to undertake?), and institutions (what mechanisms do we have for reaching consensus and undertaking collective action over long periods of time?). We will discuss these political questions in later chapters. But these factors are useful to keep in the back of one's mind, because all of the work ahead requires systems-level thinking and action.

Relatively small increases in income, population, and fuel mix can result in large increases in total emissions. Conversely,

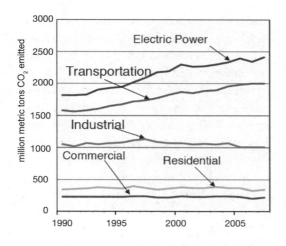

FIGURE 9.4 Increased carbon emissions from US fossil fuel power plants in 2007

The electric power sector represents the single largest source of US energy-related carbon dioxide emissions. American transportation emissions are lower and have grown at about the same rate as electric power emissions since 1990. Direct-use emissions in the residential, commercial, and industrial sectors, which do not include the emissions associated with the generation of electric power used by those sectors, have remained relatively flat since 1990. Source: [28]

large increases in income growth can be at least partially offset by improved energy intensity or fuel mix. [8]

World Resources Institute

Multiple Intensity Disorder: Coal = Electricity = Emissions = Problem

In short, lowering the carbon-producing component of a nation's energy infrastructure can have long-term positive results for emissions. For example, the switch in Russia from many old coal-fired power plants to newer natural gas–fired plants has lowered the carbon intensity of the overall Russian economy. Worldwide today, fossil sources account for four-fifths of energy supply in use: coal (25%), natural gas (21%), petroleum (34%). These are followed by non–fossil fuel sources such as nuclear (6.5%), hydro (2.2%), and burning biomass and waste (11%). Finally, only 0.4% of global energy demand is currently met by renewables such as geothermal, solar, and wind. [18]

Let's take a look closer to home. In the United States we still rely quite heavily on carbon-intensive coal-fired power plants for electricity.

FIGURE 9.3 online at ncse.org/climate solutions

Distribution and capacity of US coal-burning power plants in 2002

Half of the electricity generated in the United States is from coal. The map in Figure 9.3 shows the size and location of the coal-fired generating plants in the continental United States. The United States has over 1,500 coal-fired power plants, with each averaging about 220 MW of power supply. [28] The United States produces about 1.5 billion tons of carbon dioxide per year from these coal-burning power plants.

The worldwide scale of burning coal to make electricity is enormous. China is currently constructing the equivalent of two 500 MW coal-fired power plants per week and a capacity comparable to the entire UK power grid each year. One 500 MW coal-fired power plant produces approximately 3 million tons per year of carbon dioxide. [15] In the United States as elsewhere, coal equals electricity that in turn equals carbon-intensive emissions.

In the United States, our electric power generation with coal and natural gas and, to a lesser extent, our petroleum-fired transportation are the key factors driving up the volume of greenhouse gas emissions. Since 2004, the news on electric power carbon emissions has gotten worse. As we see in Figure 9.4, electric power demands in the United States are projected to rise. Electricty demand is expected to grow at about 1.1% per year through 2030, largely because of increases from commercial and residential building demands. Interestingly, industrial electricity needs are projected to level off and not rise beyond what they are today. Slow growth in industrial production, particularly in the energy-intensive industries, will likely limit growth of electricity demand in the industrial sector. The largest increase in electricity demand will come in the commercial sector, projected to grow at 49% from 2006 to 2030, as service industries continue to drive growth. Residential demand will grow by 27%. As the Energy Information Administration (EIA) wryly notes about the increase projected for residential uses, "Population shifts to warmer regions also increase the need for cooling." [26] If we kept the same electricity supply fuel mix we have today, we would meet half that demand from carbon-intensive coal plants. The EIA also notes, "Continuing efficiency gains in electric heat pumps, air conditioners, refrigerators, lighting (notably LED lighting), cooking appliances, and computer screens slow the growth of electricity demand." But steady modest population growth alone will cause the overall electricity demand to rise, even after such efficiency gains.

Why Conservation Is Better and Efficiency Is Not Enough

Let's look at a best-case efficiency scenario for the United States. The Department of Energy (DOE) has recently estimated that energy efficiency could have the technical potential to level off energy demand growth in the country through 2030. See Figure 9.5 for the major efficiency technologies that would be required. Many of these technologies exist now. But we have yet to actually fund the deployment on a nationwide scale of most of these technologies. For example, the DOE projection for the building sector includes a widespread adoption of zero-emission buildings, yet the nation has virtually no such buildings today. For the industrial sectors, the efficiency technologies from potential nanomanufacturing are in their infancy. For transportation, the DOE assumes massive expansion of hybrid vehicle and advanced diesel engine technology. Yet the research and development (R&D) funding for such innovations is quite modest to date. Hence, few such high-efficiency vehicles are on the road today.

One reason that relying solely on improving efficiency will not work is that we cannot impose efficiency quickly enough, given the vast size of the existing inefficient infrastructure from buildings to transportation networks and power grids. Conservation, on the other hand, will work and has a huge capacity to reduce emissions. As we saw in Chapter 6, the cheapest carbon is the emissions that we do not produce. We could all drive cars that get double the gas mileage of our current cars as an efficiency gain. Or we could just cut our driving in half as a conservation measure. But combining both efficiency and conservation is much more powerful. If we get 40 instead of 20 miles per gallon and cut the total miles driven by half, we reduce emissions by 75%.

Even if the United States does all it can (and it should) on efficiency, the growth potential among developing nations for efficiency gains is enormous too. Three billion more people by 2050 will be using too much carbon-based energy unless we transform the energy supply system and maximize both efficiency and conservation. These go hand in hand. For example, shaping new urban centers to avoid energy-hungry sprawl can reduce commuting distances (increase conservation) for millions of workers while taking advantage of highly efficient next-generation building materials and designs. "Energy for buildings is the most important sector of the energy demand in the United States. To provide heating, cooling, lighting, water heating, as well as all of the other things that we use energy for in our residential and commercial buildings accounts for 39% of US primary energy demand," write John Randolph and Gilbert Masters in their authorita-

FIGURE 9.5 online at ncse.org/climate solutions

Potential gains in energy efficiency could reduce energy demand.

II: How to Think About Climate Solutions

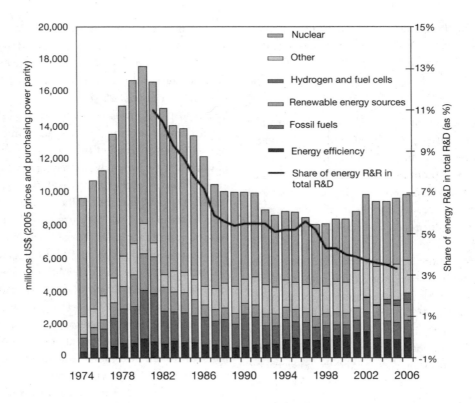

FIGURE 9.6 Public energy research, development, and demonstration investment trends: 1970–2006

Investment in new energy technologies has shrunk considerably since the energy crises of the 1970s. Source: [25]

tive (790-page!) book *Energy for Sustainability.* "Transportation accounts for 29% and . . . industry uses 32%." [20: p. 215] If we add to the energy burden of manufacturing building materials, transporting them to the site, and constructing the building, this "embodied" energy raises the total energy burden of America's buildings to nearly half (48%) of the nation's entire current annual energy consumption. To get to new system-wide and society-wide levels of efficiency, we have to make radical increases in research, design, development, and deployment, all of which are the prerequisites for commercialization of products and widespread adoption in the marketplace. But the investments in basic and applied science R&D for efficiency technologies have not kept pace with the needs for R&D. As Thomas Friedman has been saying, we need

10,000 innovators innovating, because we don't yet know which 100 or even 10 of those ideas will have a big payoff. [7]

Federal energy technology budgets are now half what they were 30 years ago, both as overall dollars allocated and as a share of all federal R&D dollars. The single largest benefactor of federal attention is nuclear technology. Investments in energy efficiency R&D have fallen steadily since 2002, just as we most need the insights from basic and applied research on efficient energy use (see Figure 9.6).

Investing in energy efficiency is a win-win solution that could save consumers billions, promote America's energy independence, and generate stable, low-risk returns for investors. But investors need a strong, national commitment to energy efficiency to overcome persistent barriers and to realize this opportunity. [5]

Mindy Lubber, president of Ceres and director of Investor Network on Climate Risk

Activities that emit carbon dioxide and other greenhouse gases expand with every home and every road built. Population growth itself will drive up emissions, even if we are as efficient as possible. Even if we become even more efficient in producing electricity, the cement we require for roads and buildings is a major carbon dioxide producer. Cement production generates more carbon dioxide emissions than any other industrial process; it is responsible for more than 5% of human-caused carbon dioxide emissions. Half the carbon dioxide from cement manufacture comes directly from the necessary calcification process; the rest comes from fuels needed to heat the process and from electricity to power the factories themselves. The point is that even more of the best energy-efficient technology for the fuel used in making cement will leave untouched the considerable emissions from baking the lime to make cement. One irony of any effort to renew the world's infrastructure with better transit networks and more efficient buildings is that producing the cement to do so will drive up emissions.

Concrete is second only to water as the most consumed substance on earth, with nearly three tons used annually for each person on the planet. Cement is the critical ingredient in concrete, locking together the sand and gravel constituents in an inert matrix. [31]

Cement Sustainability Initiative

In online Figure 9.7 we can see the impact of both the burgeoning cement production and the stalling of our prior gains in energy needed per dollar of economic value produced. We know that carbon intensity of the economy fell each year from 1980 until 2002—a good thing for the environment and the economy. Then, for the first time in almost 25 years, carbon intensity began to climb; more carbon dioxide is now required for the same relative economic output—bad news for all of us. As a direct consequence, not surprisingly, we have seen carbon emissions rates accelerate since 2003. [21]

What does all this mean? Rather than a best-case scenario, that is, 350 parts per million by volume (ppmv) for 2 degrees of warming, or a worst-case scenario (600 ppmv for 6 degrees of warming), let's use a middle-of-the-pack emissions scenario for 2040: 450 CO_2 ppmv (verus 380 today) that unleashes 3°C warming. How much could energy efficiency reduce the global business-as-usual emissions growth over the next 30 years to limit the peak to 450? About 43%, according to US EPA scientist Frank Princiotta. [19] We should note that Dr. Princiotta's examination of efficiency's potential includes supply-side (more efficient combustion in power production) as well as demand-side (more efficient end use) factors. And how much emissions growth does that leave for other tactics to address? About 57%, including the cement calcification emissions. [19] Put simply, the total carbon emissions per year are projected to double by 2040. So reducing emissions with maximum-efficiency technology (most of which is woefully underfunded today) will still leave emissions higher in 2040 than they are today (see online Figure 9.8). [25]

In other words, we still have to remove about 6 out of 10 new tons of potential atmospheric carbon dioxide by some means beyond efficiency. [19] If reaching 9 billion people by mid-century with adequate standards of living means that efficiency alone does not adequately curb total emissions, then what does?

FIGURE 9.7 online at ncse.org/climate solutions

Worldwide cement production in 2007

FIGURE 9.8 online at ncse.org/climate solutions

Carbon is on the rise since 2003.

II: How to Think About Climate Solutions

Technology Wedges: The Potential of Carbon Capture and Storage

If we cannot avoid making carbon dioxide, can we capture the carbon dioxide at the source, before it goes up the exhaust stack into the atmosphere? The International Energy Agency (IEA) has devoted a great deal of time and talent to this question. They surmise that a number of emission stabilization strategies (or wedges) show promise. But these wedges vary quite widely in how much emission reduction they will likely achieve by 2050.

For example, the IEA judges that all the producer and end user efficiency gains (better distribution efficiencies for electricity, higher-mileage vehicles, more-efficient appliances, etc.) represent a combined emissions saving wedge of 43% (as we saw above). A wide-scale nuclear power program will likely reduce emissions from the baseline business-as-usual projection by a 6% wedge. Renewables are almost three and one-half times more effective, at a 21% wedge. End users switching to lower-carbon fuels (such as natural gas) will make almost twice as big a wedge, at 11%, as nuclear power. And then the IEA projects that carbon capture and storage (CCS) techniques could remove a wedge of 20%, or one-fifth of all the carbon produced. That is a big savings!

> In a greenhouse-gas-constrained world, carbon dioxide capture and storage technologies offer the potential for continuing to use the Earth's resources of fossil fuels while preventing their CO_2 emissions from being released to the atmosphere. [4: p. 13]
>
> *Jae Edmonds*

At the end of the cycle for any fossil fuel, and especially for coal-fired plants, capturing

---

FIGURE 9.9 online at ncse.org/climate solutions

Potential carbon mitigation from different technology wedges shows some room for optimism.

---

the carbon dioxide before it leaves the smoke-stack is technically feasible. This process of CCS (alternately called carbon capture and storage or carbon capture and sequestration) is expensive, experimental, and controversial. CCS refers to a set of technologies designed to reduce carbon dioxide emissions from large point sources, including coal-fired power plants, to mitigate climate change. CCS technology involves capturing carbon dioxide and then storing the carbon in a reservoir other than the atmosphere, rather than allowing it to be released into the atmosphere where its accumulation would contribute to climate change. [16]

Carbon capture involves trapping the carbon dioxide emission gas at the source and storing it, usually by injecting it underground as a gas. Several different ways to store carbon gas are possible and have been studied, including storing carbon in terrestrial ecosystems, the oceans, and underground in geologic formations. Terrestrial carbon storage refers primarily to biological carbon sequestration in plants, relying on the photosynthetic process of capturing atmospheric carbon dioxide and converting it into organic carbon held within plants and soils. Ocean storage generally refers to the injection of captured carbon dioxide directly into the oceans, but it also includes other mechanisms of enhancing oceanic uptake of carbon (such as iron fertilization to increase phytoplankton growth for more carbon uptake). Geologic carbon storage refers to the injection of captured carbon dioxide into underground, naturally occurring geologic reservoirs that will trap the gas to prevent it from reentering the atmosphere. Among these different carbon storage approaches, geologic storage has emerged as the method with the greatest potential for large-scale CO_2 emissions reductions in the near term. [16]

CCS is not risk free and certainly would require significant investment. Pumping carbon dioxide directly into an aquifer—a deep geologic layer of porous rock—could lead to acidification of the groundwater, much as excess carbon dioxide is acidifying the ocean. When a saline aquifer becomes more acid, heavy metals such as lead and arsenic dissolve from surrounding minerals into the groundwater. Large amounts of concentrated carbon dioxide are toxic to plants and animals. Increased underground pressures may lead to releases or leaks from prior drill holes. Pumping in carbon dioxide could also lead to disruption of underground microbial communities, leading to releases of other gases. [14] Acidifying aquifers enhances calcite dissolution, which can lead to fractures in calcium carbonate (limestone), allowing the trapped carbon dioxide or other gases to leak out. [22]

Because coal-fired power plants are such a major source of carbon emissions, let's examine how carbon capture might actually work with them—trapping the carbon dioxide emission gas at the source and storing it, usually by injecting it underground as gas. In each case below, the electricity output is 500 MW. A power output of that size is roughly the demand that a half million households would create (if the average home's instantaneous demand is 1,000 kW). [15]

Removing carbon dioxide from flue gas requires energy, primarily in the form of low-pressure steam both for maintaining the chemical conditions needed to capture the carbon solution and for compressing the resulting gas into liquid form for storage. The good news is that CCS technology can substantially reduce the carbon emissions from the plant—by 90%. The bad news is that doing so is quite expensive in at least two ways: (1) The power plant's size itself needs to be over one-third larger, and (2) the coal feed increases substantially, in this case by 40%,

because of the extra energy needed to harvest, compress, and store the carbon from the flue gas. More-efficient CCS technologies are being developed that would divert less energy from the output back into the storage function. But these higher-efficiency capture methods are even more expensive and experimental at this point. [15] Absent a coherent federal policy on carbon capture and storage, the public utility commissions of individual states will be left to wrestle with the options and policy choices. While that may offer a range of solutions in our state "laboratories of democracy," it may also fail to create the critical mass needed to complete the expensive final stages of research, development, and demonstration at sufficient scale to be cost effective. [2]

So where are we in developing better, cleaner energy sources?

RD3: Research, Development, Demonstration, and Deployment

In order to have an adequate carbon storage technology to apply to fossil fuel–fired plants or to cement production, we have a lot of work to do. Any ideas that have merit need more research to develop into prototypes. Prototypes need to be tested under real-life conditions as demonstration projects. Demonstrations that succeed can lead to deployment strategies that allow early adopters access to the technology. And finally, once deployable, a technology will need to be commercialized so it may be shipped to customers and sites all over the world. Hence, we need not just R&D, but RDD&D, and even RDD&D, or RD3.

> My belief is, and I am going to say this in a careful way, the best way to fund R&D might simply be a charge on every kilowatt hour delivered to every customer in America, so that money is earmarked and focused on technologies.
>
> *James Rogers, chairman, president, and CEO, Duke Energy, 2008*

If all of the carbon dioxide that US coal-fired plants produced were transported for sequestra-

---

FIGURE 9.10 online at ncse.org/climate solutions

A 500 MW pulverized coal plant without carbon capture

FIGURE 9.11 online at ncse.org/climate solutions

A 500 MW pulverized coal plant with carbon capture

II: How to Think About Climate Solutions

TABLE 9.2 IEA's Blue Scenario Technology Time Line

Carbon capture and storage time line from R&D to wide-scale commercial use					
Year	2005	2010	2020	2030	2040	2050
RD3 stage	Research and Development		Demonstration	Deployment	Commercialization	
Technology advances	10 demo plants* (cost $US 15 billion)	20 full-scale plants built (cost $US 30 billion)		12% of power generation with carbon capture by 2030	30% of power generation with carbon capture by 2050	
	Major DSF† storage validated: 2008–2012	Development of transport infrastructure: 2010–2020				

*Several CCS demonstration plants exist today, of which Norway's Sleipner plant is the earliest (1996). As of mid-2008, two or three CCS trials were underway in the USA, compared with 1,500 existing coal-fired power plants in the country.

†DSF is a desulfurization technology for further cleaning the emissions from coal-fired plants.

Source: [17]

tion, the quantity would be equivalent to three times the weight and, under typical operating conditions, one-third the annual volume of natural gas transported by the US gas pipeline system. If 60% of the CO_2 produced from US coal-based power generation were to be captured and compressed to a liquid for geologic sequestration, its volume would about equal the total US oil consumption of 20 million barrels per day. At present, the largest sequestration project is injecting 1 million tons/year of CO_2 from the Sleipner gas field into a saline aquifer under the North Sea. One significant component is the transportation infrastructure that would deliver the liquid carbon dioxide to permanent underground storage sites. [15]

If we examine other technologies in energy, we see that a great many possible tools are presently somewhere between the initial R&D stage and the deployment stage. Many of the efficiency technologies are quite well along in this pipeline to market. Note how many differ-

FIGURE 9.12 online at ncse.org/climate solutions

Technology RDD&D includes basic science, applied R&D, demonstration, deployment, and commercial-scale production and installation.

ent CCS technologies are in the early stages of RD3 (see Figure 9.13).

Coda: Americans and Efficiency

"American taxpayers are already making a strong investment in the solutions we need," says David Rodgers, the DOE's deputy assistant secretary for Energy Efficiency and Renewable Energy during the Bush administration. The question is whether the investments are strong enough.

The Office of Energy Efficiency and Renewable Energy (EERE) manages America's investment in research, development, and deployment of the DOE's diverse energy efficiency and renewable energy applied science portfolio. It had been starved for funds under the Bush years. Congress appropriated 25% more in fiscal 2009 than it had in 2008 for EERE programs, in anticipation of a sea change in the political support for such work, to invest in advancing clean energy technology, renewable energies, best practices in energy efficiency, and improving energy regulations. By the time the Obama administration arrived in Washington, the EERE budget request for fiscal 2010, $2.3 billion, reflected a doubling of EERE's request for fiscal 2008 to match the new president's commitment to meaningful energy R&D. And just to help EERE learn to drink from a fire hose, Congress

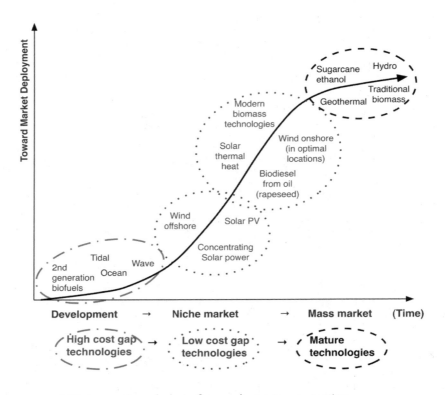

FIGURE 9.13 Innovation chain in future electricity generation

Over time, technologies develop from early innovation to niche markets and eventually to mature status in mature mass markets. Source: [17] via [3]

and the White House designated a stunning total of $16.8 billion for EERE programs in passing the American Recovery and Reinvestment Act of 2009. This infusion meant a tenfold increase overnight in resources for energy efficiency and renewable energy R&D. At least one third of this money was targeted at efficiency work, such as the highly effective weatherization and energy assistance programs, that rely on mature technologies with immediate conservation results.

Technology innovations will be a large part of the solution to climate change as we replace current carbon-emitting energy sources with those that reduce or completely eliminate carbon dioxide from energy production.

The US economy is twice as efficient in energy use per dollar output than China, and Japan is twice as efficient as the United States. With efficiency, nations can make a big step to where we need to go. But energy experts, such as

Amory Lovins, point out that the gains in American energy intensity (total energy used divided by GDP) of the past 30 years happened for largely accidental reasons, and not because of the kind of focused effort long the norm in Japan. [13]

By just expanding use of biofuels, improving fuel economy, and tightening appliance standards in lighting, we could reduce emissions in 2030 by 625 million metric tons. Furthermore, if we were to replace projected growth in demand with renewable energy sources (solar, wind, and geothermal), we could reduce carbon dioxide emissions by over 300 million metric tons in 10 years. [23]

As a new technology enters a market, early adopters may be willing to pay more. But to penetrate a market deeply enough to make a dent in greenhouse gases, new technologies will need to be cost-effective as replacements for old technologies. Government policies can

II: How to Think About Climate Solutions

While the energy efficiency potential is large, most experts agree that using end-user efficiency to flatten energy demand is not the same thing as reducing emissions, for two reasons. First, the continued economic advantage to reducing energy consumption provides incentives for firms to reduce their energy intensity. Improvements in energy intensity would likely have occurred in the absence of such government efficiency programs, as was the case in the 1980s and 1990s. Second, unless the mix of energy production technologies changes radically, the emissions growth may not slow down enough to matter. However, collaborations between government and industry on voluntary programs, and the expertise gained in promoting enhanced end use efficiency technologies, could provide a solid foundation for a more aggressive program. Such programs are particularly significant in the power generation and transportation sectors.

"But the pace isn't fast enough," says David Rodgers of the US Department of Energy's Office of Energy Efficiency and Renewable Energy (EERE). "We are trying to shorten the length of the pipeline, to get technologies to the marketplace sooner." Shortening the pipeline between ideas and marketplace requires the nation to (1) improve technology through funding cutting-edge research and development, (2) develop and implement durable policies to foster technology use through new tax and regulation policies that reduce barriers and create proper incentives, and (3) facilitate access to capital through programs such as the Department of Energy's Technology Commercialization Fund.*

TABLE 9.3 Priorities for Energy Efficiency at EERE

Technology
- Continue fundamental and applied R&D for enabling technologies to reduce the energy consumption and transform the carbon footprint of the built environment (homes, offices, and manufacturing).

Regulation, codes, standards
- Accelerate, modernize, and elevate appliance standards with greater-consensus rule making.
- Promote superior model building codes with executable plans of coordinated implementation by the states.
- Provide utilities with returns on energy efficiency comparable or superior to investments in generation; provide industry with pathways for best practices.

Voluntary and market-based deployment
- Establish the National Action Plan for Energy Efficiency.
- Expand and modernize the Energy Star program concurrent with technology.
- Expand advocacy for energy-efficient, solid-state lighting (e.g., CFLs, LEDs).
- Target civic infrastructure (e.g., Energy Smart schools, hospitals, libraries, municipal facilities) to be energy-efficient, secure sites for distributed generation.

Education and outreach
- Expand the capacity for educating all age groups, reach targeted populations with superior communications, and develop effective behavioral modification tools.

Source: Adapted from [23]

---

*www1.eere.energy.gov/commercialization/abouttheprogram.html

(continued)

Production tax credits for renewable sources are an example of a policy that would help get technologies in the marketplace sooner—if the credits remain in place on a timescale that matches how long it takes to build and to break even on such facilities. It will be critical to accelerate access to capital for clean energy technologies and to secure the money needed to take promising technologies—such as solar photovoltaic technology and geothermal power—from the demonstration lab to many thousands of homes and businesses. Many of the barriers to energy efficiency are not technological, says Rodgers. "We need to better understand human behavior. People make less informed economic decisions when it comes to energy efficiency." [23]

accelerate market penetration by outlawing old technology, mandating new technology, or providing market-based incentives for switching. We create the necessary preconditions for the technological breakthroughs that will supply the energy, abundant and clean, of the mid-21st century. We must also balance this longer time line with our crucial need to begin to dramatically reduce emissions now, ensuring that we are placing ourselves on a trajectory that realistically leads to where we need to go. A comprehensive and collaborative energy and technology plan that is flexible and robust must be developed to ensure our energy future will safeguard our own future. As we think ahead, we should recall that electric power generation emits only two-fifths of US and world carbon dioxide and that transportation emissions are almost as large as those from electricity. So improving energy efficiency for end users in both electrical and transportation sectors would address two and a half times as much carbon dioxide emission as an electricity-only strategy. [26]

In sum, we need efficiency tools, carbon capture tools, and renewable energy tools deployed on a very large scale. Ultimately, technologies that succeed follow a life cycle of their own in

the market. Lower-cost technologies ready now, such as wind, photovoltaic solar, and concentrated solar, need less time to market. Higher-cost technologies that may show great potential will take longer to develop. But the market overall for a wide variety of low-carbon energy technologies will continue to mature over time.

As Amory Lovins points out, "Carbon displacements should be both fast in collective deployment and effective [in carbon displaced per dollar]." [13] Fortunately, many options that should be economic priorities (as more effective per dollar invested) are also environmental priorities because they exhibit lower emissions impacts over their life cycles than more-costly options. So what would that plan look like? We'll tackle some of these better options in the next two chapters.

CONNECT THE DOTS

- The energy intensity of our economies must be reduced so that we use less energy to produce each dollar of economic output.

- The carbon intensity of our energy supply must be reduced so that each unit of energy consumed by the economy produces lower carbon and other greenhouse gas emissions than at present.

- Two key economic sectors—power generation and transportation—continue to grow in both size and emissions. Both need a strategy to

---

FIGURE 9.14 online at ncse.org/climate solutions

Costs of saving or delivering 1 kWh of new electricity

change the trajectory of their coinciding contributions to global carbon dioxide emissions.

- Programs to displace carbon emissions should include the following components.

- In the near term, we must actualize the low-hanging fruit of increasing energy efficiency (e.g., switching to higher-mpg vehicles) and conservation opportunities (e.g., driving less to begin with) and preservation of natural carbon sinks such as forests.

- Carbon displacement that comes from end use efficiency is both profitable—cheaper than the energy it saves—and quick to deploy as many good efficiency technologies are ready to commercialize.

- We need to accelerate the new technology development phase. Lowering the greenhouse gas impact of coal-fired energy and mobile fuel sources must be a RDD&D priority. Current incentives should be modified to encourage energy reduction.

- A massive RDD&D program to mitigate carbon should be initiated. Key technologies such as carbon capture and storage, next-generation mobile fuels (for partial zero-emissions and hybrid vehicles, etc.), nuclear energy, and renewables should be subject to collaborative research to develop the next round of these technologies.

- We should explore research on geoengineering issues. Radical solutions, such as seeding the atmosphere with sulfur dioxide and other albedo-changing and carbon-sequestering concepts, need to be seriously considered to buy us some time.

- Fundamental research is needed to assess the efficacy and impact of these geoengineering approaches and to prepare the technologies, if they are desirable to deploy.

Online Resources

www.eere1.energy.gov
www.eoearth.org/article/Carbon_capture_and_storage
www.eoearth.org/article/Energy_Information_
Administration_(EIA),_United_States

www.eoearth.org/article/Energy_transitions
www.eoearth.org/article/Nuclear_power_reactor
www.eoearth.org/article/Rebound_effect
www.eoearth.org/article/Ten_most_distortionary_
energy_subsidies
www.globalcarbonproject.org
www.iea.org
www.pewclimate.org/white_papers/coal_initiative
www.ren21.int
http://unitconversion.org
See also extra content for Chapter 9 online at http://
ncseonline.org/climatesolutions

Climate Solution Actions

Action 6: Energy Efficiency and Conservation

Action 7: Biofuel Industry and CO_2 Emissions—Implications for Policy Development

Action 8: Solar Energy:Scaling Up—Science and Policy Needs

Action 9: How to Ensure Wind Energy Is Green Energy

Action 10: Nuclear Energy—Using Science to Make Hard Choices

Action 21: The US Global Change Research Program (USGCRP)—What Do We Want from the Next Administration?

Works Cited and Consulted

[1] Canadell JG, Le Quéré C, Raupach MR, Field CB, Buitenhuis ET, Ciais P, Conway TJ, Gillett NP, Houghton RA, Marland G (2007) *Contributions to accelerating atmospheric CO_2 growth from economic activity, carbon intensity, and efficiency of natural sinks.* Proceedings of the National Academy of Sciences 104(47):18866. http://www.pnas.org/cgi/content/abstract/104/47/18866

[2] Cowart R, Vale S, Bushinsky J, Hogan P (2008) State Options for Low-Carbon Coal Policy. www.pewclimate.org/white_papers/coal_initiative/

[3] Doornbosch R, Gielen G, Koutstaal P (2008) Mobilising Investments in Low-Emission Energy Technologies on the Scale Needed to Reduce the Risks of Climate Change. Round Table on Sustainable Development. www.iea.org

[4] Edmonds JA, Wise MA, Dooley JJ, Kim SH, Smith SJ, Runci PJ, Clarke LE, Malone EL, Stokes GM (2007) Global Energy Technology Strategy: Addressing Climate Change. Global Energy Technology Strategy Program. www.pnl.gov/gtsp/publications/

[5] Energy Future Coalition (2008) Diverse Coalition of Organizations Calls upon the Congress and Next President to Make Energy Efficiency a National Priority. *Press Release* (September 16, 2008). www.energyfuturecoalition.org

[6] Field CB, Raupach MR, eds (2004) *The Global*

Carbon Cycle: Integrating Humans, Climate and the Natural World (Island Press, Washington, DC). http://islandpress.org

[7] Friedman T (2008) Hot, Flat and Crowded: Why We Need a New Green Revolution and How It Can Renew America, 410. (Farrar, Straus, and Giroux, New York). www.thomaslfriedman.com

[8] Herzog T, Baumert K, Pershing J (2006) Target: Intensity: An Analysis of Greenhouse Gas Intensity Targets. http://wri.org

[9] Hoffert MI, Caldeira K, Benford G, Criswell DR, Green C, Herzog H, Jain AK, Kheshgi HS, Lackner KS, Lewis JS (2002) Advanced Technology paths to global climate stability: energy for a greenhouse planet. Science 298 (5595): 981. www.sciencemag.org

[10] IAEA (2008) Nuclear's Great Expectation: Projections Continue to Rise for Nuclear Power, but Relative Generation Share Declines. International Atomic Energy Agency. September 11, 2008 (read October 20, 2008). www.iaea.org/NewsCenter/News/2008/np2008.html

[11] Leinen M (2008) Oceans: A Carbon Sink or Sinking Ecosystems? National Conference on Science, Policy, and the Environment: Climate Science and Solutions. http://ncseonline.org/2008conference/

[12] Lovins A, Sheikh I (2008) The nuclear illusion. Ambio, in November 2008 preprint. www.rmi.org/sitepages/pid257.php#E08-01

[13] Lovins A (2008) RMI Stanford Energy Lectures: Lecture 5: Implications (March 2008). Rocky Mountain Institute (read September 4, 2008). www.rmi.org/stanford

[14] Metz B, Davidson O, de Coninck H, Loos M, Meyer L (2005) Special Report: Carbon Dioxide Capture and Storage. IPCC. www.ipcc.ch

[15] MIT (2007) MIT Study on the Future of Coal: Options for a Carbon-Constrained World. http://web.mit.edu/coal/

[16] NETL (2001) Coal and Power Systems Strategic and Multi-year Program Plans. www.doe.gov/sciencetech/carbonsequestration.htm

[17] OECD IEA (2008) Energy Technology Perspectives 2008: Scenarios & Strategies to 2050: Executive Summary. www.iea.org

[18] OECD IEA (2008) Key World Energy Statistics from the IEA 2008. www.iea.org

[19] Princiotta F (2009) Global climate change and the mitigation challenge. Journal of the American Waste and Air Management Association (forthcoming, 2009). www.epa.gov/appcdwww

[20] Randolph J, Masters GM (2008) Energy for Sustainability: Technology, Planning, and Policy. (Island Press, Washington, DC). www.energyforsustainability.org

[21] Raupach MR, Marland G, Ciais P, Le Quéré C, Canadell JG, Klepper G, Field CB (2007) Global and regional drivers of accelerating CO_2 emissions. Proceedings of the National Academy of Sciences 104(24):10288. http://www.pnas.org/cgi/content/long/104/24/10288

[22] Renard F, Gundersen E, Hellmann R, Collombet M, Le Guen Y. (2005) Numerical modeling of the effect of carbon dioxide sequestration on the rate of pressure solution. Oil & Gas Science and Technology Review 60:381–399. http://ogst.ifp.fr/

[23] Rodgers D (2008) Commentary on Energy and Technological Challenges. Climate Change Science and Solutions: Eighth National Conference on Science, Policy, and the Environment. http://ncseonline.org/2008conference/

[24] Smith R (2008) "New Wave of Nuclear Plants Faces High Costs." Wall Street Journal, May 12, 2008. http://online.wsj.com/

[25] Taylor P (2008) Energy Technology Perspectives 2008: Scenarios and Strategies to 2050. IEEJ Workshop, July 7, 2008. Energy Technology Policy Division, Tokyo (read September 3, 2008). www.iea.org/Textbase/publications/free_new_Desc.asp?PUBS_ID=2012

[26] US EIA (2008) Annual Energy Outlook 2008 with Projections to 2030. Department of Energy. DOE/EIA-0383(2008)(read October 1, 2008). www.eia.doe.gov/oiaf/aeo/demand.html

[27] US EIA (2008) Scenario Projections. Department of Energy Energy Information Agency (read September 3, 2008). www.eia.doe.gov

[28] US EIA (2008) US Carbon Dioxide Emissions from Energy Sources 2007 Flash Estimate. May 2008 (read September 5, 2008). www.eia.doe.gov/oiaf/1605/flash/flash.html

[29] US EPA (2007) Emissions & Generation Resource Integrated Database (read September 2, 2008). www.epa.gov/egrid/

[30] USGS (2008) Mineral Commodity Summaries: Cement. http://usgs.gov

[31] WBCSD (2005) Cement Sustainability Initiative Progress Report. www.wbcsdcement.org

[32] WRI CAIT (2008) Climate Analysis Indicators Tool (CAIT) 5.0. http://cait.wri.org

[33] Yamaji K, Matsuhashi R, Nagata Y, Kaya Y (1991) An integrated systems for CO_2/energy/GNP Analysis: Case studies on economic measures for CO_2 reduction in Japan. Workshop on CO_2 reduction and removal. www.iiasa.ac.at

PART III

How We Work Together Now

CHAPTER 10

Carbon Meets Wall Street

The first element of policy is carbon pricing. Greenhouse gases are, in economic terms, an externality: those who produce greenhouse-gas emissions are bringing about climate change, thereby imposing costs on the world and on future genera- tions, but they do not face the full consequences of their actions themselves. [21]

SIR NICHOLAS STERN, 2006

A s a financial services provider, Swiss Re's direct business activities do not have a major environmental impact. Swiss Re is a major reinsurance provider, essentially insuring the insurers. Global reinsurers, like Swiss Re, are in business to enable risk taking essential to enterprise and progress. From the point of view of Swiss Re, unsustainable social or environmental trends may reinforce known risks or create new ones. To raise awareness in the global insurance community, the United Nations Environment Programme (UNEP) in 1995 launched the Statement of Environmental Commitment by the Insurance Industry. Swiss Re was one of the first insurers to sign. Today with almost 12,000 employees worldwide and US$30 billion of shareholder equity at stake, Swiss Re wants to stay in business by paying closer attention to the sustainability of its own practices and those of its clients. [22] And the firm has good company.

In the past 15 years, the businesses that lend the money and underwrite the insurance that all major industries require have begun to wake up to the implications of climate disruption at a planetary scale. These days, if your industry is a major emitter or polluter, you may run into tougher questions from your bank or board of directors than you would have in the past. But the awareness was not always as high as today, and it has waxed and waned in the past.

Emerging Awareness

By the time the United States was racing to put a man on the Moon, Americans began to widely recognize that the manufacturing and indus- trial processes, which produced enormous economic growth, also contributed to declines in quality of air, water, land, and species. In response, after decades of inactivity, in just 6 years from 1963 to 1969, the US Congress wrote a dozen truly landmark environmental laws. These new laws sought to clear smog from the air and toxic pollutants from lakes and rivers as the economy boomed with widespread industrialization and suburbanization. The National Environmental Policy Act of 1969 paved the way for President Nixon to create the Environmental Protection Agency in 1970. Nixon gave the new agency broad regulatory powers for setting and enforcing national environmental standards.

In 1972, the United Nations established UNEP as the environmental agency of the United Nations system. The international effort confirmed both the ubiquitous nature of these environmental issues and their intimate link to the economics of global development. Major national environmental agencies in other industrialized nations, such as the departments of the environment in the UK (1970), Canada (1971), Australia (1971), South Africa (1973), and France (1974) among others, also date from this initial wave of governmental action. These new federal agencies drove compliance up among polluters, initially in cleaning the nations' air and water from local pollution sources. Cleanup became a new norm. Ecological systems benefited, and the health of citizens improved.

Beginning with the oil crises of 1973 and 1979–1980, concern of political leaders led to expansion of renewable energy as support for research and development of new technologies increased. The United States made initial, limited efforts in clean energy technology development, especially solar power, under the Carter administration. The highway speed limit of 55 miles per hour (88 kilometers per hour) was widely adopted to reduce gasoline consumption. Many homes in the 1970s sprouted solar panels of all kinds on their roofs, partly spurred by favorable government incentives. Wind, wave, and solar energy technologies all benefited during this period as their range of applications expanded under federal policies friendly to renewable energy sources.

Throughout the 1970s, the US Congress and its counterparts in other nations wrote a great many environmental laws. Reacting to regulatory pressure, many business sectors, such as heavy industry and manufacturing,

eventually took the lead on complying with the new policies. However, the financial services sector—investment banks and brokerages, diversified commercial banks, custodial banks, private equity firms, and insurance companies—remained largely absent from participating in or commenting on environmental issues. The finance sector continued to play a traditionally active role in financing the technologies and processes to produce clean air and water and to improve waste management. But Wall Street did not actively define how issues of sustainability could affect its own bottom line. Indeed, it was not entirely apparent that sustainable business practices should have an impact on investment returns, either positively or negatively.

But these efforts and government incentive policies were not sustained in the 1980s. With the fundamental shift in political philosophy during the Reagan administration, most of the efficiency and conservation measures from Jimmy Carter's presidency were abandoned and environmental protection was rolled back. Governmental funding for cleaner-energy research and development was slashed, just as many of these emerging technologies reached the brink of technical breakthroughs.

Yet, the close relationship between economic development and the potential for debilitating environmental degradation persisted, especially in economically poor but natural-resource-rich nations struggling to lift themselves into full participation in global markets. The problems of sustaining economic development in all the world's nations remained frustratingly real. The financial sector broke its silence in May 1992 in New York City. A small group of major international banks—including Deutsche Bank, HSBC Holdings, Natwest, Royal Bank of Canada, and Westpac—joined forces with UNEP to spark the banking industry's awareness of and involvement in the environmental agenda. The UN Conference on Environment and Development—the "Earth Summit" in Rio de Janeiro in 1992—had placed strong emphasis on promoting sustainable development, specifically "development that meets the needs of the present without compromising the ability of future generations to meet their own needs." [25, 30] The Rio summit considerably enhanced the UNEP role to encourage economic development compatible with the protection of the environment. As a run-up to the 1992 Rio Earth Summit, this group of financial companies issued the UNEP Statement by Banks on the Environment and Sustainable Development. [26]

Three years later in 1995, leading global insurance and reinsurance firms—General Accident, Gerling Global Re, National Provident, Storebrand, Sumitomo Marine & Fire, and Swiss Re—and major pension funds also grasped the enormity of the financial implications of climate change and joined this inaugural group. In 1999, the UNEP Finance Initiative established a working group on climate change. [27] Today the UNEP Finance Initiative has over 160 signatory institutions from over 44 countries, including AIG, Bank of America, Calvert, Citigroup, Innovest Strategic Value Advisors, and JP MorganChase & Co. Simultaneous with the release of the Intergovernmental Panel on Climate Change's Fourth Assessment Report in mid-2007, Goldman Sachs and McKinsey & Company, two members of the UN's voluntary Global Compact, issued reports that strongly emphasized the role of business in addressing climate disruption and provided well-grounded opportunities and challenges.

The "Greatest Market Failure"

We are well beyond articulating the risks and opportunities. Climate impact management is becoming a corporate governance issue. Sir Nicholas Stern, former head of the UK's Economic Service and its Office of Climate Change, rates his own highly regarded citations as grossly conservative. Stern has famously called climate change "the greatest market failure the world has ever seen." [21]

Why would Sir Nicholas think of market failure? A market fails when it does not reward participants who create fair value to exchange or rewards participants who remove value. Specifically, economists consider a market to fail when it is not efficient in allocating goods and services. At least two reasons explain why rapidly rising greenhouse gas concentrations represent a massive market failure. First, the participants (emitters) have been relying on external goods (the air for emissions) from which they benefited without incurring any transaction cost. In essence, the emitters have been relying on a public good (the air) without obtaining the right to use that good. Second, markets are supposed to organize the exchange of control over commodities in which the property right attached to the commodity would define the nature and duration of the control. We have never previously assigned a property right to the Earth's atmosphere. We are beginning to do so now! The European Climate Exchange, the Chicago Climate Exchange, and the Regional Greenhouse Gas Initiative of the northeast states are all examples of the formation of market institutions based on assigned rights and values to goods previously deemed "free" of charge or ownership.

"Putting an appropriate price on carbon—explicitly through tax or trading, or implicitly through regulation—means that people are faced with the full social cost of their actions," writes Sir Nicholas. "This will lead individuals and businesses to switch away from high-carbon goods and services, and to invest in low-carbon alternatives. Economic efficiency points to the advantages of a common global carbon price: emissions reductions will then take place wherever they are cheapest." [21: Executive Summary]

Climate change action—such as putting a price on heretofore-free carbon emissions—will have a massive effect on the asset values and credit ratings of corporations. These shifts change the way banks deal with them. The sectors most affected by climate change and

Financial leaders are in near universal agreement that right now, without a framework from government for a price on carbon, capital markets will continue to play a limited role and investors will remain in the dark on the potential for returns. All major financial houses have produced reports that examine how climate change will transform the financial transactions of key business sectors, both globally and nationally. Even though major firms such as Citigroup are taking important steps in the area of energy efficiency and even though the advice of firms like Global Environment Fund and Goldman Sachs is heading government in the right direction, these efforts remain ad hoc and opportunistic. More than 90% of chief executive officers claim they are doing more than 5 years ago to incorporate environmental, social, and governance issues into corporate strategy and operations, according to the McKinsey report. [16]

In a 1934 report to Congress, the GDP's chief architect, Simon Kuznets, cautioned, "The welfare of a nation can scarcely be inferred from a measurement of national income." [12]

FIGURE 10.1 Triple-Win for people, profit, and planet: Defining the business case for sustainability

We could start a veritable alphabet zoo with the monikers that public-relations specialists have invented to describe corporate sustainability. "Triple-Win for people, profit, and planet" is just one such concept. To define the business case for sustainability, many corporations have undertaken efforts to publish corporate sustainability reports. The challenge comes in translating such concepts into verifiable measures. A good quantifiable measure exists for the economic-profit domain but has been more elusive for the quality-of-life and health-of-planet domains. Source: [19]

The economists at Redefining Progress have a much more ambitious goal: replacing the traditional gross domestic product measure with the "genuine progress indicator," or GPI. They argue the GDP fails to measure the actual, full environmental costs or benefits inherent in our economic activities. Yet, the GDP serves as a basis for setting economic and tax policy. As the GPI project director, John Talberth writes, "For decades, many economists have acknowledged that the GDP has fundamental shortcomings. . . . The GDP is simply a gross tally of everything produced in the U.S.—products and services, good things and bad." [24] The GPI is one of the first alternatives to GDP vetted by the scientific community and used regularly by government and nongovernmental organizations worldwide. The GPI versus GDP comparison (shown in Figure 10.2) "implies that since 1980 or so the marginal benefits associated with growth in personal consumption expenditures, non-market time, and capital services have been offset by the marginal costs associated with income inequality, natural capital depletion, consumer durable expenditures, defensive expenditures, undesirable side effects of growth, and net foreign borrowing." [24]

(continued)

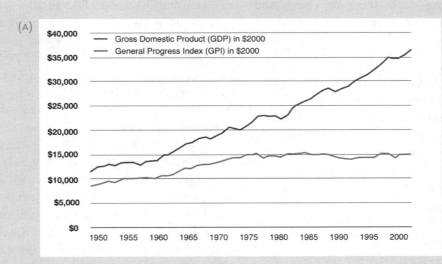

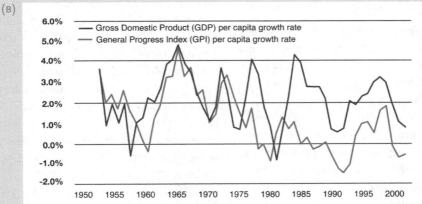

FIGURE 10.2 Genuine progress indicator versus gross domestic product: 1950–2004

These two figures show how changes in the gross domestic product (GDP) track compare with the genuine progress indicator (GPI). (a) Per capita GDP is plotted along with GPI in billions of constant 2000 US dollars. GPI per capita has barely moved since 1978, remaining near $15,000 since that time. Over the period 1950 – 2004, GPI grew at an extremely sluggish rate of just 1.33%. In contrast, GDP per capita rose precipitously from $11,672 in 1950 to $36,596 in 2004—an annual growth rate of 3.81%. (b) Annual GDP and GPI per capita growth rates are compared using a rolling 3-year average to smooth out year-to-year fluctuations. Here, we find a rather striking trend: While GDP growth rates have more or less fluctuated within a positive range, GPI growth rates fall into two distinct periods. In the first period, spanning 1950 to 1980, GPI per capita growth rates more or less match those of the GDP and are generally positive, ranging as high as 4%. Beginning in 1980, GPI growth rates are more frequently negative, bottoming out at −1.64% in 1992. GPI per capita has more or less stagnated since 1978, when it surpassed $15,000 for the first time. Source: [24]

carbon pricing will be transport, energy, and infrastructure (cement production is a big carbon emitter, as we learned in the prior chapter).

> GDP gives no indication of sustainability because it fails to account for depletion of either human or natural capital. It is oblivious to the extinction of local economic systems and knowledge; to disappearing forests, wetlands, or farmland; to the depletion of oil, minerals, or groundwater; to the deaths, displacements, and destruction caused by war and natural disasters. [23]
>
> *John Talberth, 2008*

Energy Market Opportunities

A number of organizations state that government should provide a mandatory national energy policy that creates a cap for carbon emissions, rules for carbon trading, and clear guidance for carbon offsets. A clear price signal results in market incentives that stimulate investment and technological innovation. Financial markets could then step in to respond to supply and demand and help fund the research, development, and deployment, just as had started to happen in the 1970s.

> The higher relative prices of energy will create incentives for businesses to create new, energy-saving technologies and for energy consumers to adopt them. The market for alternative fuels is growing rapidly and will help to shift consumption away from petroleum-based fuels.[†]
>
> *Ben Bernanke, chairman,*
> *US Federal Reserve Bank, 2006*

In fact, we already have a good working example of buying and selling the right to pollute and emit. It is called the Chicago Climate Exchange (whose trading symbol is CCX), "North Amer-

ica's only and the world's first global marketplace for integrating voluntary legally binding emissions reductions with emissions trading and offsets for all six greenhouse gases." [5] The key word here is *voluntary*, as its member clients participate to meet annual greenhouse gas emission reduction targets. Those who reduce below the targets have surplus allowances to sell or bank. Those who emit above the targets comply by purchasing contracts. The price of the contract right to emit fluctuates according to supply and demand. The Regional Greenhouse Gas Initiative (RGGI) permits for 10 northeastern states are traded at the Chicago Climate Futures Exchange, a subsidiary of the CCX. More on these a bit later. [18]

While the CCX trades emission rights on a voluntary basis for six greenhouses gases for several hundred firms from many different sectors, the RGGI market will trade only carbon dioxide emission rights on a mandatory basis for electric power generators in 10 states. On the heels of the launch of RGGI in the Northeast, governors of seven western states and premiers of four Canadian provinces are developing the Western Climate Initiative (WCI) to reduce greenhouse gases by means of a market-based cap and trade system. The WCI states plainly that the benefits it projects include "reducing air pollutants, diversifying energy sources, and advancing economic, environmental, and public health objectives while avoiding localized or disproportionate environmental or economic impacts." [31] See Table 10.1 for a roundup of the emission marketplace.

We see action on the governmental front at the level of states and provinces in seeking to establish market mechanisms to control harmful emissions. That is good news and may be the precursor to scaling up a uniform set of federal initiatives or regulations. Cap and trade has some fans at the federal level already. For example, Congressman Jay Inslee attributes the lack of wide support at present for cap and trade to the fact that it is largely misunderstood here, even as it has strong support in Europe

---

[†]See Chairman Bernanke's speech, "Energy and the Economy," of June 15, 2006 before the Economic Club of Chicago, available at www.federalreserve.gov.

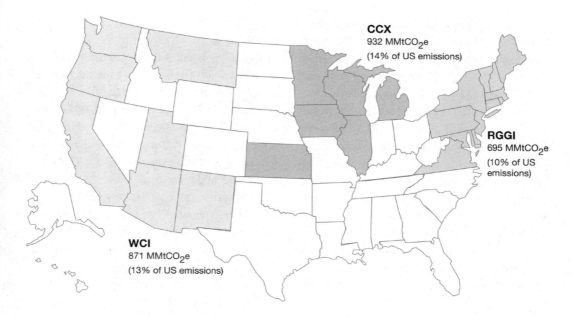

FIGURE 10.3 Regional map on greenhouse gas abatement: 2008

This map indicates the confluence of regional greenhouse gas abatement work by the three major inter-state initiatives. The Western Climate Initiative (WCI), the Chicago Climate Exchange (CCX), and the Northeast's Regional Greenhouse Gas Initiative (RGGI) involve states representing nearly half the US population, and several provinces representing a sizable portion of Canada's population. Together the 10 states represent more than one-third of US carbon dioxide emissions. The emission unit in this figure is in million metric tons of carbon dioxide equivalent (MMtCO$_2$e). Source: [13]

TABLE 10.1 Some Emission Market Examples

	Emissions cap and trade participation	GHGs included	Industries involved	Geographic scope	Trade commencement date
European Union Emission Trading Scheme (EU ETS)*	Mandatory emissions cap and trade market	Carbon dioxide	Multisector	15 EU nations	2005
Chicago Climate Futures Exchange†	Voluntary emissions cap and trade market	6 gases‡	Multisector	USA, national in scope	2003
Regional Greenhouse Gas Initiative (RGGI)§	Mandatory cap and trade market	Carbon dioxide	One sector (electric power industry)	10 states in the US Northeast	2008
Western Climate Initiative (WCI)**	Mandatory with expanding phases	6 gases‡	Multisector	7 western US states and 4 Canadian provinces	2012

*http://ec.europa.eu/environment/index_en.htm

†www.chicagoclimatex.com

‡carbon dioxide, methane, nitrous oxide, hydrofluorocarbons, perfluorocarbons, and sulfur hexafluoride

§www.rggi.org

**www.westernclimateinitiative.org

with a wide range of political groups. Inslee is a member of the House of Representatives from Washington State and coauthor with Bracken Hendricks of *Apollo's Fire: Igniting America's Clean Energy Economy*. [11] Inslee and other proponents argue that a cap and trade system does two things. First, it puts a cap on, that is, sets a limit on, the total amount of carbon dioxide going into the atmosphere—something that we already have successfully done with sulfur dioxide. Second, it creates a market where none previously existed. The government would use the market to allocate the amount of greenhouse gas that any particular industry could emit. Inslee predicts that people in the United States will support such a system once they understand that the cap and trade system is a binding commitment that is enforceable and uses market forces to change behavior. [10]

Green Business Principles

Speakers at the January 2008 National Conference on Science, Policy, and the Environment represented a strong cross-section of firms and organizations leading the way in introducing green principles to the market. We examine a few examples of thought and action leaders here.*

Ceres is a national network of investors, environmental organizations, and other public interest groups dedicated to integrating sustainability into capital markets. In 1989, Ceres announced the creation of the Ceres Principles, a 10-point code of corporate environmental conduct to be publicly endorsed by companies as an environmental mission statement or ethic. Embedded in that code of conduct was the mandate to report periodically on environmental management structures and results. In 1993, Sunoco became the first Fortune 500 company to endorse the Ceres Principles. Today, over 100 companies have endorsed the Ceres Principles (see Table 10.2).

Ceres president Mindy Lubber notes that today nearly all companies have put a dollar value on sustainability. In 2006 for the first time, the annual World Economic Forum in Davos, Switzerland, offered its 2,000 attendees a choice of 18 sessions dedicated to climate change. Banks alone published 97 climate reports in 2007. The voluntary sustainability measure known as the Global Reporting Initiative now has over 1,500 users.

Lubber explains, "If we're going to achieve an 80% reduction of carbon dioxide by 2050, we will need a massive reformation of the world's energy system. With carbon remaining unpriced, while China and India build carbon-based economies mostly fueled by coal, it is impossible to assure that target."†

Bruce Schlein of the Citigroup Foundation emphasizes the importance of energy efficiency as a crucial way to start now to mitigate the effects of climate change. A recent McKinsey study has indicated cost curves for energy efficiency that show that more than half of the actions that could be taken would lead to net cost reduction. [16] These are productive actions that should be taken today. Energy efficiency savings can help forego building new capacity, putting saving on a par with new capital investments. Individual businesses do not need to wait for a marketplace that trades and a government that regulates to mitigate greenhouse gas emissions. For example, Citigroup's current energy efficiency program incorporates the stabilization wedges concept created by Princeton researchers Socolow and Pacala (described in Chapter 7). Citigroup is also a participant in the Clinton Climate Initiative. Former President Bill Clinton launched the Clinton Foundation's Climate Initiative in August 2006, with the mission of applying the foundation's business-oriented approach to fight against climate change in

---

*The case examples featured here include those used by speakers at the 2008 NCSE conference instrumental in the efforts they describe. For a video presentation, visit http://ncseonline.org/climatesolutions.

†For more from the NCSE conference participants, see their complete talks at http://ncseonline.org/climatesolutions.

TABLE 10.2 The Ceres Principles for Sustainable Business Practices

Protection of the biosphere

We will reduce and make continual progress toward eliminating the release of any substance that may cause environmental damage to the air, the water, or the earth or its inhabitants. We will safeguard all habitats affected by our operations and will protect open spaces and wilderness, while preserving biodiversity.

Sustainable use of natural resources

We will make sustainable use of renewable natural resources, such as water, soils, and forests. We will conserve nonrenewable natural resources through efficient use and careful planning.

Reduction and disposal of wastes

We will reduce and where possible eliminate waste through source reduction and recycling. All waste will be handled and disposed of through safe and responsible methods.

Energy conservation

We will conserve energy and improve the energy efficiency of our internal operations and of the goods and services we sell. We will make every effort to use environmentally safe and sustainable energy sources.

Risk reduction

We will strive to minimize the environmental, health, and safety risks to our employees and the communities in which we operate through safe technologies, facilities, and operating procedures and by being prepared for emergencies.

Safe products and services

We will reduce and where possible eliminate the use, manufacture, or sale of products and services that cause environmental damage or health or safety hazards. We will inform our customers of the environmental impacts of our products or services and try to correct unsafe use.

Environmental restoration

We will promptly and responsibly correct conditions we have caused that endanger health, safety, or the environment. To the extent feasible, we will redress injuries we have caused to persons or damage we have caused to the environment and will restore the environment.

Informing the public

We will inform in a timely manner everyone who may be affected by conditions caused by our company that might endanger health, safety, or the environment. We will regularly seek advice and counsel through dialogue with persons in communities near our facilities. We will not take any action against employees for reporting dangerous incidents or conditions to management or to appropriate authorities.

Management commitment

We will implement these principles and sustain a process that ensures that the board of directors and chief executive officer are fully informed about pertinent environmental issues and are fully responsible for environmental policy. In selecting our board of directors, we will consider demonstrated environmental commitment as a factor.

Audits and reports

We will conduct an annual self-evaluation of our progress in implementing these principles. We will support the timely creation of generally accepted environmental audit procedures. We will annually complete the Ceres Report, which will be made available to the public.

Source: [4]

practical, measurable, and significant ways. At the 2007 Clinton Global Initiative, President Clinton announced the 1Sky campaign to accelerate bold federal policy on global warming. The 1Sky campaign supports at least an 80% reduction in climate pollution levels by 2050. [1]

In early 2008, Citigroup, JPMorgan Chase, and Morgan Stanley rolled out the Carbon Prin-

ciples to serve as climate change guidelines for advisors and lenders to US power companies. The work is the first instance of a group of leading banks coming together with power companies and environmental groups to develop a process for understanding carbon risk around electric power sector investments. [6]

The first of the three principles is energy efficiency: "An effective way to limit carbon dioxide emissions is to not produce them." The second principle entails the considerable promise that renewable energy and low-carbon distributed energy technologies hold. The third principle addresses investments in conventional or advanced power generation, such as in natural gas, coal, and nuclear technologies. In a nod to the lack of federal leadership for climate change–based energy policy prior to 2009, the banks note that "due to evolving climate policy, investing in carbon dioxide–emitting fossil fuel generation entails uncertain financial, regulatory and certain environmental liability risks." [6]

However, while companies can make concerted and organized efforts to respond to climate change, extending this to individual companies is much more difficult. One of the most promising ways of sparking efficiency of individual businesses is through the energy service company (ESCO) model, which demonstrates to customers how their energy use is related to the business that they conduct and helps them reduce their energy use. An ESCO develops, installs, and finances projects designed to improve the energy efficiency and maintenance costs for facilities over a 7- to 20-year time period. ESCOs generally act as project developers for a wide range of tasks and assume the technical and performance risk associated with the project. [15]

Currently, used by a limited number of public utility companies and government entities, ESCO projects often include contracts for high-efficiency lighting, high-efficiency heating and air conditioning, efficient motors and variable speed drives, and centralized energy management systems. The debt payments on initial capital investments are paid for with the dollar savings generated by the amount of energy that is actually saved. [29] Of the estimated $20 billion of ESCO projects installed to date, approximately $7 billion—or one-third—has gone directly to pay for labor employment. Job creation or retraining is a benefit flowing from many of the ESCO projects.

The ESCO framework is being examined to overcome barriers in applying it to individual energy use. Current stumbling blocks include aggregation issues, that is, many small households with a variety of cost structures (renting, owning, sharing cost, etc.); standardizing industry monitoring technology; developing incentives that ESCOs can flow through to reward people for changing their energy-intense lifestyles; and developing new risk mitigation packages for cities and towns with low audit ratings. Citigroup's Bruce Schlein concludes that ESCOs themselves need prodding to go past business as usual.

The Global Environment Fund has specialized in emerging markets and clean technology investment programs since 1990. Its founder and president, Jeff Leonard, noted that despite today's acute attention to climate, the global impact of fossil fuel use will not change in the next two decades.

According to Dr. Leonard, national energy independence is misguided and may spawn international policy and trade problems. Simple energy independence does not beneficially alter the underlying reliance on fossil fuel. Biofuels will produce few economic benefits and may cause price distortions amidst shortages in agriculture. Few and fleeting lifestyle changes will not contribute to what is needed for mitigation. In short, after 60 years of economic policies and international diplomacy favoring free trade in and high consumption of oil, the country must now reckon with the fact that our high dependence on oil for transportation, in particular, poses serious national security threats, exacerbates the trade imbalance, and has massive negative environmental consequences at the local and global levels.

With a glimmer of hope, Leonard sums up the challenge, as follows:

I have come to the conclusion that America needs a clear, bold energy strategy to guide it through the next four decades. The strategy must prioritize policies, public infrastructure investments and long-term technology development around one central theme. The theme is electrification—the pervasive use of electricity throughout the economy, and particularly the substitution of petroleum-based fuels with electricity as the core energy supply for transportation uses. A national energy strategy to promote greater electrification of the economy is the most practical, expedient and efficient path to achieving energy security for America, and ultimately of addressing global climate change challenges. [2: p. 5]

Without major new investment in electricity sector infrastructure, our country will face energy shortages, power disruptions, and blackouts on a grand scale in the decades to come. Leonard explains, "The good news is that technologies already exist, or are rapidly evolving, to meet all the challenges outlined in this paper necessary to sustain the electricity grid of the future."[†]

As director of Goldman Sachs Center for Environmental Markets, Mark Tercek maintained that the function of financial markets should be to maximize shareholder value, maintain a neutral position, think hard about risks and opportunities, and retain smart people.[‡] Markets perceive climate change as both big risk and big opportunity. Goldman Sachs looks to renewables for investing and puts its own dollars there. It advises responsible long-term energy investment; for example, in 2007 it dissuaded TXU, a Texas utility company, from building new coal-powered plants.

Goldman Sachs considers carbon a potentially huge asset class and is now very active in carbon trading in the European Union. The firm has confidence that the limited US experience with voluntary trading on the Chicago Climate Exchange can be ramped up quickly with eventual good results. However, any genuine

solution involving cap and trade will have to be global—that means recognizing emissions of carbon dioxide are orders of magnitude greater in complexity than those that caused acid rain, the issue for which cap and trade originated.

What is holding back full market participation in mitigating climate change? Primarily, it is the lack of federal climate regulations. Government needs to take the lead. It is no longer possible that carbon emission can be free or that energy usage can be low-cost. When federal climate legislation is enacted, businesses that have anticipated the change will be well positioned with informed people, a good understanding of energy issues in their companies, and the capacity to respond quickly.

A Goldman Sachs report of 2007 for the UN Global Compact effort showed that among six sectors covered—energy, mining, steel, food, beverages, and media—companies that are considered leaders in implementing environmental, social, and governance policies to create sustained competitive advantage have outperformed the general stock market by 25% since August 2005. [8]

Tercek concludes bluntly, "We need to understand and accept that energy must be more expensive. Voters need to accept higher carbon costs with government providing the cushion for short term societal disruptions. Business is not in the driver's seat; right now government leadership is needed. We need to get started now."

From Principles to Partnerships

In early 2007, in order to encourage the passage of legislation and reduce greenhouse gas emissions, 14 major businesses and leading environmental organizations formed the United States Climate Action Partnership (USCAP) to spearhead an action agenda for business and government. They adopted and published *A Call for Action*, a blueprint for a mandatory economy-wide, market-driven approach to climate protection. The 14 founding firms and nongovern-

[‡]Tercek has since become chief executive officer of The Nature Conservancy.

mental organizations and additional members of USCAP include Alcoa, American International Group (AIG), Boston Scientific Corporation, BP America, Caterpillar, ConocoPhillips, the Chrysler Group, Deere & Company, the Dow Chemical Company, Duke Energy, DuPont, Environmental Defense, Exelon Corporation, Ford Motor Company, FPL Group, General Electric, General Motors, Johnson & Johnson, Marsh, National Wildlife Federation, Natural Resources Defense Council, NRG Energy, The Nature Conservancy, PepsiCo, Pew Center on Global Climate Change, PG&E Corporation, PNM Resources, Rio Tinto, Shell, Siemens Corporation, World Resources Institute, and Xerox Corporation.

Specifically, USCAP seeks to address the global dimensions of climate change, to create incentives for technological innovation, to be environmentally effective, to create economic opportunity and advantage, to be fair to sectors disproportionately impacted, and to encourage early action.

According to USCAP, a desirable US climate change policy would include mandatory approaches to reduce greenhouse gas emissions from the major emitting sectors, including from large stationary sources, transportation, and energy use in commercial and residential buildings, that could be phased in over time, with attention to near-, mid- and long-term time horizons. Flexible approaches to establish a price signal for carbon may vary by economic sector and could include market-based incentives; performance standards; cap and trade; tax reform; incentives for technology research, development, and deployment; and other appropriate policy tools. Approaches that create incentives and encourage actions by other countries, including large emitting economies in the developing world, are also needed to implement greenhouse gas emission reduction strategies.

Members of USCAP state that a cap and trade approach ensures efficiency and the overall lowest cost, guarantees the pollution cuts needed to protect the climate, has the flexibility for creativity and innovation, enables banking of emission allowances, and has had past success in controlling acid rain in the United States.

We, the members of the U.S. Climate Action Partnership, pledge to work with the President, the Congress, and all other stakeholders to enact an environmentally effective, economically sustainable, and fair climate change program consistent with our principles at the earliest practicable date. [28: p. 3]

USCAP, A Call For Action

In order to forge ahead, USCAP expanded the size of the group from 14 to 33; has new working groups on transportation, international policy, and others to tackle difficult issues; and has released new consensus policy recommendations on energy efficiency and geologic carbon storage. The USCAP members help to guide legislation by working with key members of Congress and committees in the Capitol. A visual representation of the emission reduction targets that USCAP recommends appears in Figure 10.4.

Eric Haxthausen, the senior policy advisor for climate change at The Nature Conservancy, an original USCAP member, explains that delaying action on climate change will only lead to even greater economic and environmental costs in the future. Tim Julian, who works on markets and business strategy for the Pew Center on Global Climate Change, echoes the sentiment, arguing that key USCAP provisions must be fast-tracked through Congress. There should be the establishment of a national greenhouse gas inventory and registry, credit for early action to reduce greenhouse gas levels, aggressive technology research and development, and policies to accelerate the deployment of zero- and low-emitting technology. To take a stepwise, cost-effective approach will stabilize concentrations of carbon dioxide over a long-term period and achieve a target zone of a 60% to 80% reduction.

Even big businesses like DuPont appear to favor a clearer, long-term signal from Uncle Sam

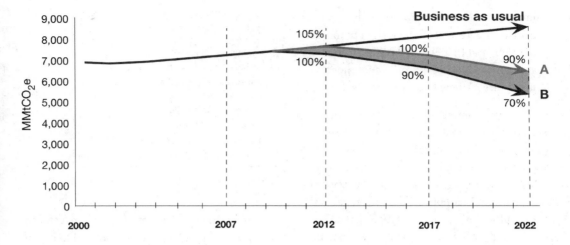

FIGURE 10.4 USCAP's emission reduction scenario

Unless aggressive emission reduction targets are set with a price on carbon as the market, most observers expect the United States to fall short. The emission unit in this figure is in million metric tons of carbon dioxide equivalent (MMtCO2e). Source: [28] using US Energy Information Administration data

on emission policy. Michael Parr, the senior manager of government affairs for DuPont, says, "Striking a balance between environmentally effective and economically sustainable is hard but absolutely critical. . . . At the core of it, we're pressuring the political constituents to get it done." [†]

> The science is pretty bloody compelling. If you don't act and you're wrong, it's a pretty big oops. But if you act and you're wrong, at least those acts are still going towards the greater good of the environment and economy. It's a rational insurance policy.[†]
>
> Michael Parr, DuPont

Lehman Brothers was among the financial firms that have begun to examine the environmental risks inherent in failing to alter a business-as-usual approach for their clients. John Llewellyn, a senior analyst for Lehman, has prepared reports on why climate change

matters to business. It is worth reading a few words from his 2007 report, "The Business of Climate Change":

> Many clients have asked for our view on the argument that, even assuming that scientists' projections of the likely effects of climate change are broadly correct, the effects will be felt only slowly, with little effect on asset prices over most investors' time horizons.
>
> We judge this argument as flawed, for three, linked, reasons. First, markets anticipate even slow-moving variables, such as climate change. Second, policy made in the name of climate change could have an almost immediate, up-front effect on asset prices. And third, markets anticipate policy itself. In this way, expected future effects of climate change become brought right forward to the present.
>
> Fundamentally, the economic case for considering climate change ultimately depends on the science. Our judgment is that the science will increasingly be seen as

[†]For more from the NCSE conference participants, see their complete talks at http://ncseonline.org/climatesoutions.

broadly correct; that this view will be progressively accepted by the weight of market opinion; and that, while the adjustment of asset prices has begun, full adjustment will take years, rather than months. [14]

The Pew Center's Business Environmental Leadership Council comprises the largest US-based association of corporations focused on addressing the challenges of climate change. [17] These companies from all business sectors represent $2.5 trillion in market capitalization and employ over 3.3 million people. In testimony before Congress in early 2007, Pew's president Eileen Claussen summed up the key steps that the Business Environmental Leadership Council states are needed to develop and deploy climate-friendly technologies and to diffuse those technologies on a global scale: "First, we must enact and implement a comprehensive national mandatory market-based program to progressively and significantly reduce U.S. greenhouse gas emissions in a manner that contributes to sustained economic growth. Second, . . . the United States must also work with other countries to establish an international framework that engages all the major greenhouse gas–emitting nations in a fair and effective long-term effort to protect our global climate. Third, we must strengthen our efforts to develop and deploy climate-friendly technologies and to diffuse those technologies on a global scale." [7]

In short, both the business community and environmental organizations now fully anticipate carbon emissions will soon have a price and, thereby, gain a set of market signals to drive investment decisions, behavior changes, and technology options. AFL-CIO president John Sweeney urges investors not to wait for governments to act. Sweeney suggests that a 10-year program of investments in solar, wind, geothermal, hydro, nuclear, and carbon capture and sequestration to bring 18,500 megawatts of renewable energy on line annually could generate 2 million full-time equivalent jobs. [9] Businesses that have prepared well to engage in the existing

TABLE 10.3 Climate Action Steps for Business: Preparing for Carbon Markets

What should business be doing?
• Pricing carbon through investment and trade
• Implementing comprehensive, corporate-wide, energy efficiency initiatives
• Asking more of suppliers
• Activating employees
• Disclosing climate performance
• Tying compensation to corporate climate impact

Source: Adapted from [4]

and emerging carbon markets have undertaken a number of specific assessments and behavior changes. These are summarized in Table 10.3.

Of course, many groups are organizing themselves to support what they hope will be a tidal wave of climate protection initiatives under the Obama-Biden administration. These organizing efforts increasingly involve growing alliances between big and small business, nonprofits, and foundations. Some such efforts even have clever names like "BICEP," which stands for Business for Innovative Climate and Environmental Policy. [3] Who knows where all this organizing will lead, but it surely will lead us somewhere different from where the nation has been for the previous decade!

CONNECT THE DOTS

- Banks and insurance companies are beginning to scrutinize the carbon risk inherent in their clients' portfolios. Investors have been engaging with a broader range of companies on issues including disclosure of climate risks and actions to address those risks.

- The United States has several ongoing regional or voluntary greenhouse gas emission markets, such as CCX, RGGI, and WCI.

- A mandatory cap and trade market for managing greenhouse gas emissions is working well in Europe and could be effective in the United States, if part of a clear and long-term federal policy.

III: How We Work Together Now

- Emission reduction scenarios include energy efficiency strategies (that may involve ESCOs) as well as shifts to low- or no-carbon energy technologies.

- Businesses that fail to adapt may be exposed to the risk of asset price declines, capital access difficulties, or insurance shortfalls.

- Climate change presents opportunities for economic growth and job creation.

Online Resources

http://ec.europa.eu/environment/index_en.htm
http://pewclimate.org
www.ceres.org
www.chicagoclimatex.com
www.eia.doe.gov
www.eoearth.org/article/Climate_change_impacts_
  on_non-market_activities
www.eoearth.org/article/Limitations_of_markets
www.eoearth.org/article/Market
www.eoearth.org/article/Market_failure
www.eoearth.org/article/
  Market_impacts_of_climate_change
www.eoearth.org/article/Market-based_instrument
www.globalreporting.org
www.rggi.org
www.us-cap.org
www.westernclimateinitiative.org
See also extra content for Chapter 10 online at http://
  ncseonline.org/climatesolutions

Climate Solution Actions

Item 13: Policy Challenges of GHG Rule Making—
  Where the Rubber Meets the Road
Item 24: Counting Carbon—Tracking and Com-
  municating Emitted and Embodied Greenhouse
  Gases in Products, Services, Corporations, and
  Consumers

Works Cited and Consulted

[1] 1Sky.org (2008) About 1Sky. *1Sky.org* (read October 10, 2008). www.1sky.org

[2] Berst J, Bane P, Burkhalter M, Zheng A (2008) The Electricity Economy: New Opportunities from the Transformation of the Electric Power Sector. www.globalenvironmentfund.com; www.globalsmartenergy.com

[3] Ceres BICEP (2008) Business for Innovative Climate and Environmental Protection. *Ceres.* http://www.ceres.org/

[4] Ceres (2007) Ceres Principles. *Ceres* (read September 24, 2008). www.ceres.org

[5] Chicago Climate Exchange (CCX) (2008) *Chicago Climate Exchange* (read September 24, 2008). http://www.chicagoclimatex.com/

[6] Citi JP Morgan Chase Morgan Stanley (2008) Leading Wall Street Banks Establish the Carbon Principles. *Press Release.* www.jpmorganchase.com/pdfdoc/jpmc/community/CarbonPrinciples-PressRelease_FINAL.pdf

[7] Claussen E (2007) Testimony by Hon. Eileen Claussen, President, Pew Center on Global Climate Change, February 13, 2007, at the US House of Representatives, Committee on Energy and Commerce. *Pew Center on Global Climate Change* (read September 24, 2008). www.pewclimate.org/what_s_being_done/in_the_congress/testimony_feb2807.cfm

[8] Goldman Sachs Global Investment Research (2007) Introducing GS SUSTAIN (read September 19, 2008). www.unglobalcompact.org/NewsAndEvents/news_archives/2007_07_05d.html

[9] INCR (2008) 2008 Investor Summit on Climate Risk. www.incr.com

[10] Inslee J (2007) The New Apollo Act (H.R.2809). *Thomas.* http://thomas.loc.gov/cgi-bin/bdquery/z?d110:h.r.02809

[11] Inslee J, Hendricks B (2007) *Apollo's Fire: Igniting America's Clean Energy Economy.* (Island Press, Washington, DC). http://islandpress.org

[12] Kuznets S (1934) National Income, 1929–1932. *Thomas.* http://thomas.loc.gov/

[13] Litz F (2008) Ten States Formally Begin Carbon Trading. *World Resources Institute.* September 24, 2008 (read October 1, 2008). www.wri.org/stories/2008/09/ten-states-formally-begin-carbon-trading/

[14] Llewellyn J, Chaix C (2007) The Business of Climate Change: Challenges and Opportunities II: Policy Is Accelerating, with Major Implications for Companies and Investors. *Lehman Brothers* (read September 29, 2008). www.lehman.com/who/intellectual_capital/climate_change_ii.htm

[15] NAESCO (2008) What Is an ESCO? *National Association of Energy Service Companies* (read October 28, 2008). www.naesco.org

[16] Oppenheim J, et al. (2007) Shaping the New Rules of Competition: UN Global Compact Participant Mirror (read September 19, 2008). www.unglobalcompact.org/NewsAndEvents/news_archives/2007_07_05d.html

[17] Pew Center (BELC) (2008) Business Environmental Leadership Council. *Pew Center on Global Climate Change* (read November 24, 2008). www.pewclimate.org/companies_leading_the_way_belc

[18] Regional Greenhouse Gas Initiative (RGGI) (2008) Home Page. *RGGI* (read September 23, 2008). http://www.rggi.org/

[19] Remmen A, Jensen AA, Frydendal J (2007) *Life Cycle Management: A Business Guide to Sustainabil-*

ity. (Life Cycle Initiative, United Nations Environment Programme). www.unep.fr/scp/lcinitiative/publications/

[20] Specter M (2008) Big Foot: In Measuring Carbon Emissions, It's Easy to Confuse Morality and Science. *New Yorker*. February 25, 2008. http://www.newyorker.com/reporting/2008/02/25/080225fa_fact_specter

[21] Stern N (2007) The Stern Review Report on the Economics of Climate Change. HM Treasury London. www.hm-treasury.gov.uk/sternreview_index.htm

[22] Swiss Re (2008) 2007 Annual Shareholder Report. *Swiss Re*. www.swissre.com

[23] Talberth J (2008) *A New Bottom Line for Progress*, p 19 in *State of the World 2008*. (Worldwatch Institute, Washington, DC). www.worldwatch.org

[24] Talberth J, Cobb C, Slattery N (2007) *The Genuine Progress Indicator 2006: A Tool for Sustainable Development*. (Redefining Progress, Oakland, CA). www.rprogress.org/publications/

[25] UNCED (1992) United Nations Conference on Environment and Development. www.un.org/geninfo/bp/enviro.html

[26] UNEP FI (1997) UNEP Statement by Financial Institutions on the Environment and Sustainable Development. *United Nations Environment Programme*. www.unepfi.org/signatories/statements/fi/

[27] UNEP FI (1999) Financial Initiative's Climate Change Working Group. *United Nations Environment Programme*. www.unepfi.org/work_streams/climate_change/working_group/index.html

[28] USCAP (2007) US Climate Action Partnership Call to Action. www.us-cap.org

[29] Vine E (2005) *An international survey of the energy service company (ESCO) industry*. Energy Policy 33(5):691–704. http://linkinghub.elsevier.com/retrieve/pii/S0301421503003008

[30] WCED (1987) *Our Common Future (World Commission on Environment and Development)*, p 43 in *World Commission on Environment and Development*. www.unepfi.org/about/background/index.html

[31] Western Climate Initiative (WCI) (2007) Western Climate Initiative (read September 25, 2008). www.westernclimateinitiative.org/

CHAPTER 11

The Climate Message Starts to Stick

We affirm that God-given dominion is a sacred responsibility to steward the earth and not a license to abuse the creation of which we are a part. We are not the owners of creation, but its stewards, summoned by God to "watch over and care for it" (Genesis 2:15). This implies the principle of sustainability: our uses of the Earth must be designed to conserve and renew the Earth rather than to deplete or destroy it. [13]

NATIONAL ASSOCIATION OF EVANGELICALS, 2008

Public opinion on climate change is changing even more rapidly than the climate. In 1997, just 27% of Americans said global warming was important to them personally. A decade later, the number of Americans with this level of concern had nearly doubled to become a small majority of 52%. More than 6 out of 10 Americans now feel that they know a good deal about global warming, again well up from a decade ago. The number of Americans who identify global warming as the single biggest environmental problem is double now—up to 33%—what it was a year ago. [3] In 2008, there were 8 out of 10 Americans who said they believed global warming is happening and will be a serious problem if uncorrected. Seven out of 10 Americans say that they are trying to reduce their use of energy or goods that create greenhouse gas emissions. Almost 3 out of 5 Americans said they were using less gasoline in 2008, a dramatic shift from past years. [2] This shift was probably due to the price of gas, which reached $4 per gallon, more than due to concerns about climate change.

What Do We Think about Climate Change?

Nearly all Americans—94%—say they're willing to make changes in their lives in order to help the environment generally; 80% say so even if it means some personal inconvenience. [3] But the American public is divided along partisan and ideological lines over global warming. For instance, 53% of Democrats call it a very serious problem, compared with 20% of Republicans. Concern also is higher among women, younger adults, and nonwhites and lower among men, whites, and evangelical Protestants. [2] In 2007 more than half of liberals (54%) and Democrats (51%) believed people are the main cause of global warming, while only about half as many conservatives or Republicans (29% and 24%, respectively) agreed. [3]

Yet, by 2008, ABC News and its poll partners found that gap closing when voters were asked about specific potential policy remedies. For example, a majority favored a cap and trade program to limit greenhouse gases. Cap and trade was more popular among Democrats, but "in the most basic measure, 52 percent of Republicans supported cap-and-trade, vs. 66 percent of Democrats and 60 percent of independents." [2]

In addition, age and education seem to make a difference in how people view climate change. Adults younger than 40 are more likely than their elders to think that global warming will be a very serious problem if left unchecked (65% vs. 52%), to think it actually can be addressed (70% vs. 58%), and to say the government should be doing more about it (75% vs. 66%). Younger adults also are more likely to think that most scientists agree that global warming is occurring. [3]

A survey by the Pew Research Center for the People and the Press in 2008 examined the effect—if any—of education level on attitudes about climate change. Pew found a partisan divide similar to that found by ABC: Fewer Republicans think global warming is serious or caused by humans. Within Democrats or independents, Pew found that college graduates have higher levels of agreement that global warming is happening because of human activity than those who are not college graduates. But the opposite is true about Republicans! To quote the Pew survey of May 2008, "Yet for Republicans, unlike Democrats, higher education is associated with greater skepticism that human activity is causing global warming. Only 19% of Republican college graduates say that there is solid evidence that the earth is warming and it is caused by human activity, while 31% of Republicans with less education say the same." [9]

In 2007, 71% of Americans felt the federal government should be doing more to deal with climate change. The number dropped to 60% in 2008. Stanford University researcher Jon Krosnick comments on such a drop as possibly affected by local weather, "People are saying the weather is less variable now than it was a year ago, and this has led some people to become more skeptical about the existence of global warming and humans' role in causing it." However, for 3 straight years from 2006 through 2008, most people thought the federal government should be doing a lot more to address climate change. [2]

A Boxing Match Attracts an Audience

The certainty about climate disruption and the human role in it is much greater among climate scientists than the public realizes. Krosnick reports that the number of Americans who think with certainty that climate scientists agree on climate change has risen to 40%—up from 35% the prior year. So Krosnick wondered whether public perception was being influenced by the media's treatment of the subject. It is common journalistic practice to "balance" news stories by including an "expert" who has a contradictory opinion. The news media often impose such a counterpoint in an attempt to reduce the public perception of bias and, quite frankly, because a boxing match attracts an audience. The public loves controversy. That interest sells newspapers and TV news shows. [10]

But does such "balance" in climate news stories influence people's thinking? With a sample size of 2,617 respondents, Krosnick tested a number of TV news stories about climate change. Each story was shown to viewers either with or without a climate change skeptic's view to balance it. Not surprisingly, adding a skeptic increased the viewers' interest rating for the story significantly without considerably increasing the difficulty of the story. However, adding a skeptic to the story also significantly lowered the percentage of viewers who felt scientists agreed that human-induced climate change is happening (from 58% to 47%). Moreover, Krosnick found, adding a climate skeptic to a news story made the audience feel less certain that climate change is caused primarily by humans, and less likely to perceive climate change as a problem. In other words, the public is less likely to believe climate change causes harm when a story includes "balance." [10]

What message is most likely to convince the public that climate change is happening? Krosnick advises that it is more effective to bypass any debate and to base any communications about climate change on the assumption that it is happening: "Don't argue about whether it's happening. Assume it's happening and argue about its consequences." [10]

Krosnick and his colleagues tested this approach and found evidence to indicate that it works: Articles focusing on consequences of climate change produced a 6% higher rate of conviction that climate change is happening than articles discussing the evidence about climate change. [10]

Sticky Messages

Should government act, either by offering tax breaks to encourage people to address climate change or by requiring businesses to do so? Yes! Huge majorities of Americans favor government action to offer incentives or require mandates for more efficient automobiles, appliances, and buildings and for reducing power plant emissions. Policymakers are discussing many basic options of action on climate change: emission mandates, consumption taxes on emission sources, emission trading schemes, and tax breaks for emission reductions.

In a survey sponsored by *New Scientist*, Resources for the Future, and Stanford University that picked up where the ABC 2007 survey left off, Krosnick and his colleagues tested public opinion toward three different climate action options: standards, incentives, and a cap and trade scheme. Standards would involve a government mandate requiring that changes be made to the way energy is produced. For example, government could specify that more electricity be generated using certain energy sources (such as sunlight or wind) or that gasoline be blended with non–fossil fuels such as

ethanol that might yield lower net carbon emissions. Incentive-based policies would reward companies for reducing carbon emissions or impose costs on high carbon emissions, such as a tax on emission by volume, or both. While standards specify how greenhouse gas reduction would be achieved, incentives allow the industries to decide whether and how to achieve reductions, a flexibility that may cost society less than standards would for equal reductions in greenhouse gas emissions. A cap and trade scheme would require emission cuts but allow companies to trade permits to emit, rewarding those that reduced emission most. [5] Specifically, the *New Scientist* survey asked about 1,500 households whether they would vote for (1) a government-mandated emission limit; (2) a corporate tax on the amount of greenhouse gases companies emit while generating electricity, or that are emitted from the vehicle fuel they sell; or (3) a cap and trade policy in each of two different energy uses—gasoline and electricity. For each policy option and each fuel, the survey offered a low, medium, or high price estimate on how that policy would affect a monthly electricity or per-gallon gasoline cost. Not surprisingly, Krosnick and his colleagues find people are generally less favorable to policies they think will be more expensive. People tend to favor a limitation standard over an increased tax policy, and they favor an increased tax policy over a cap and trade policy. [5]

The *New Scientist* poll results challenge some common preconceptions: "They show clearly that policies to combat global warming can command majority public support in the US, as long as they don't hit people's pockets too hard. Americans turn out to be suspicious of policies that use market forces to help bring down emissions, and are much more likely to support prescriptive regulations that tell companies exactly how they must achieve cuts." [4]

Policymakers will have to wrestle with the survey's finding that the policy options preferred by the public are likely to be more expensive (e.g., mandated limits) than others that

The American public can be divided into two groups: those who already have strong, solid opinions (one way or the other) about whether climate change is happening and is exacerbated by human activities, and those who are uncertain and therefore susceptible to being swayed one way or the other. While the latter group can be converted, these citizens are ironically the least likely to act upon their changed views and most prone to sliding back to earlier views. It is hard to convert them permanently to their new perspective. The trick is to change their views and to make them stick. Jon Krosnick and others who study the psychology of public opinion feel that focusing news stories on climate consequences and solutions, and the relative effectiveness of those solutions, is the best way to make the message stick. [10]

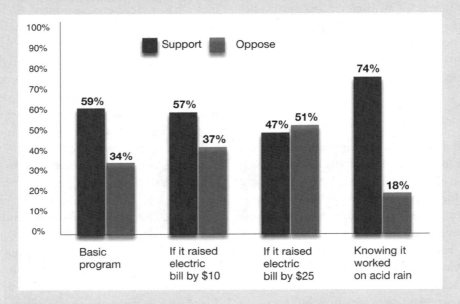

FIGURE 11.1 Public support for greenhouse gas cap and trade policy in 2008

Public support for cap and trade jumps when respondents learn that this same technique worked to dramatically reduce acid rain emissions at a lower cost than would have occurred with mandates. Source: [2]

currently command less support from the public (e.g., a consumption tax or emission trade scheme). When presented with likely costs on gas or electricity for meaningful emission reduction, respondents were more likely to support mandates or incentives that targeted electricity consumption than gasoline consumption. As

Aldhous writes, "Even the least popular electricity policy—cap and trade—won more support at all three prices told to respondents than the most popular vehicle fuel policy." [4]

The most striking finding of the *New Scientist* survey is that respondents would prefer to address the climate change via the electricity sector through the setting of a low carbon emission standard or mandate, and not through an emission tax or trade scheme.

Policymakers expect that a good cap and trade system would be more cost-effective than mandated emission standards. In practice, cap and trade gives companies an incentive to innovate ways to reduce their emissions, rather than imposing a solution that may be more expensive. Upon probing the results further, Krosnick found that the public's main reservation about a cap and trade policy is its effectiveness. In short, the public lacks confidence that it will work. Krosnick concluded that to sell the American public on a cap and trade policy, education about the policy's effectiveness is needed. [5]

> Policies that hit people's wallets hard will be tough to sell to American voters, and those that may prove cheapest seem inherently unpopular. [4]
>
> *Peter Aldhous*

It may behoove us to examine the "issue public"—those who are the most committed to action on a given policy domain. Citizens most likely to vote on climate change and become activists for solutions to climate change are those members of the public most concerned with it. As Jon Krosnick and other researchers use the concept, the issue public is that segment of the population that says an issue is personally important to them. By 1998 during the course of the initial Kyoto Protocol debate, the issue public on climate change grew to about 11% of the total population. [1]

In the past decade the issue public for whom climate change action is personally important has climbed steadily to include almost one in five Americans—now at about 18% and rising. Therefore, within the larger body politic, some 43 million Americans form a nearly unanimous group in believing that global warming is real, that it is caused by humans, and that government should take action immediately. Currently, within the overall citizenry, about 10% of Republican voters and 17% of independent

voters are passionate about climate change, versus 23% of Democrats who are passionate about climate change. As a point of optimism, Krosnick points out that this is a sizable pool to draw upon for the agenda ahead, and it mirrors the public support a generation ago for women's equality and civil rights. [10]

Strange Bedfellows

In early 2007, Bill McKibben, whose longest involvement in the climate change arena has been as a writer, decided to do something more direct. He and a group of Middlebury College students formed Step It Up 2007 (see Stepitup2007.org). Twelve weeks later, on a budget of less than $200,000, demonstrations took place in 1,400 places around the country—a testament to the scalability of social organizing operations and to the passion for this issue among college students, who were the backbone of Step It Up. McKibben argues that climate change is not a second-tier issue for an enormous number of people. They just have not acted, because the issue seems too large. According to McKibben, people need to "screw in the new lightbulb, but then screw in a new senator."

> It is incredibly important for everyone to tell the truth all the time about the state the planet is in as very dire and very dark. We need to tap into that sense of reality. [14]
>
> *Bill McKibben, 2008*

McKibben also stresses that climate activism needs to be nimble and lightweight. Large organizations are not as necessary, especially with the social networking power of the Internet. The politics for climate action have improved strongly in the last few years, even as the scientific findings have gotten worse. According to McKibben, climate change is not a future problem but a deep current contemporary issue: "We're in a situation that requires a kind of honesty, that puts us on 'war footing,' that allows us to make fast changes like we have in the past." McKibben points out that it does not take 51%

of Americans to change the politics around an issue like this.

In an echo of Krosnick's issue-public data, McKibben asserts, "If you can get fifteen or even five percent of Americans deeply engaged, that's more than enough to transform politics. Only politics, in the end, is going to deal with this issue in the time we need it to happen."

McKibben states emphatically, "Look at paleoclimatic data for carbon dioxide! 380 parts per million, where we are now, is too much," he argued. "We need to scramble to a safety zone, somewhere south of where we are now. If we're honest in saying this is an emergency, then we'll have to deal with it. Possibilities for much more dramatic thinking are where we need to be going right now." [14]

In mid-2008, McKibben and his collaborators founded 350.org to "make sure everyone knows the target so that our political leaders feel real pressure to act." The most recent science tells us that unless we reduce the atmospheric carbon dioxide levels to 350 parts per million (ppm), we will cause large and irreversible damage to life on Earth. The UN is in the midst of negotiating a treaty that could put us on the path to reduce carbon levels. But the terms are too weak to get us back to 350 ppm. So who is leading us to take these bold steps? In alliances that would have seemed quite improbable just a few years ago, faith communities, universities, and labor unions are all joining forces with the scientific and environmental communities. [12]

Reverend Richard Cizik of the National Association of Evangelicals feels the need to communicate not just the scope and depth of the crisis we face but also its moral magnitude. The climate crisis is ultimately an issue of meaning. The climate change challenge is one of cosmological nature. "We are all perpetrators and victims," he stated "You can't change the environment in this town unless you change the politics." Cizik believes that evangelicals can change the politics. Indeed, numerous evangelical church leaders have reached out to, not only each other, but like-minded leadership of other

faith communities for a united front on climate protection and stewardship of planet Earth. [13]

Twenty-five percent of the voting public goes to evangelical church on Sunday morning. According to Cizik, "Hope is about having a vision." This faith community has the ability to dream about that which does not exist, but can. And then the community needs a strategy. "A dream without a strategy is a hallucination," notes Reverend Cizik. The scientific community has strategies for climate solutions but needs partners. Evangelicals think global warming is very important, and they shouldn't be cast aside any longer. Cizik told the Eighth National Conference on Science, Policy, and the Environment, "We would benefit from a partnership of these two unlikeliest of bed fellows: the faith and scientific communities."*

But the motivation is not simply stewardship of the planet. At the end of the century, some 1.1 billion people (more than 1 in 6 worldwide) did not have access to safe drinking water, while 842 million (nearly 1 in 7) were classified by the United Nations as "chronically hungry." "The century set records for organized violence, mass poverty, and environmental decline," Worldwatch Institute's Gary Gardner writes. "People of faith need to take seriously the power of their own teachings and acknowledge their value in the realization of a better world. Religious leaders and communities of faith need to bring their social voice to the public square on these issues." [8] Indeed that stirring among the faith communities for more sustainable—and in essence more ethical— ways of living is underway. Other examples include the Interfaith Power and Light projects in the United States to "green" congregations, the efforts of Buddhist monks to protect forests by "ordaining" trees, and the work of the World Council of Churches to help island nations adapt to climate change.

Another unlikely partnership is emerging

---

*For more from the NCSE conference participants, see their complete talks at http://ncseonline.org/climatesolutions.

between labor and the environmental movement. Dan Seligman of the Apollo Alliance, which brings together labor unions, businesses, and environmentalists, stresses the importance of viewing the opportunities in job creation and enhancement of our nation's competitiveness that come with environmental change. "Global warming, despite efforts from Al Gore and many others in the room, is simply not a top tier priority for most in the US,"* he told the National Conference, listing issues such as the economy, national security, Iraq, and health care as current high priorities for the public.

Also according to Seligman, if Congress moves with a robust cap and trade bill, the fossil fuel industries tied to coal and oil will lobby strongly against it. But the Apollo Alliance and its labor allies will be there to argue that environmental protection creates good jobs. Energy production and consumption ties together many top-tier concerns, and the general public is aware of this. Retooling the energy industry is a huge potential source of good, new jobs.

Even as fear presents a challenge around global warming, Seligman thinks hope is in the genetic makeup of the American people. We need to invoke the idea that we have the technological prowess and the can-do spirit to take it on. Seligman argues that we need to link the challenge to job security, because the larger framework helps. For average Americans, global warming isn't being felt as the peril that it is. When people see smart meters, other new clean technology, and the jobs that these innovations bring, they will be seen and felt and anticipated by the middle class, which will create hope.

At Arizona State University, all 9,300 freshmen are required to attend a course module in sustainability. Environmental scientist Michael Crow, the university's president, believes that the basic structure of American universities is a flawed one. As Crow puts it, "Students are usually smarter and more creative than faculty."*

Crow leads a coalition of more than 600 college and university presidents committed to taking campuses to carbon neutrality. Higher-education presidents who sign the American College and University Presidents Climate Commitment pledge to take leadership on eliminating greenhouse gases. Crow states, "American universities are still in the Stone Age, when it comes to environmental and energy-savvy design. We need to re-think the way universities themselves are designed." Crow also stresses the merit in some university endeavors on the local level. Universities can work to solve problems where the effects can be seen locally. For example, Arizona State University has taken on the goal of lowering the nighttime heat index in Phoenix. According to Crow, "a 'Stone Age university' doesn't have such objectives, whereas a 'new university' can take on a broad problem like sustainability. In a Stone Age university, professors teach what they know. Perhaps, though, it's better to go and learn what we don't know—sustainability, urban planning, etc."* This could stimulate more excitement and engagement in the learning process.

Crow states that carbon dioxide in the atmosphere is just one part of the problem. Currently, 80% of scientific funding goes to vaporizing our enemy and extending our individual lives. "Where are the medical school equivalents focused on sustainability, on the environment?" Crow asks. "One stumbling block is: We're overly obsessed with the individual, and insufficiently obsessed with the collective."*

Peter Senge, the founding chair of the Society for Organizational Learning, views the global nature of the challenge with a philosophical bent. One of the subtleties of globalization is multiculturalism. We live with each other and are in each other's backyards, and the "each other" are very different people. Multiculturalism is not a homogenization process. Multiculturalism means learning to live together respectfully despite deep differences. This process of understanding each other across cul-

---

*For more from the NCSE conference participants, see their complete talks at http://ncseonline.org/ climatesolutions.

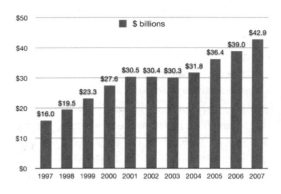

FIGURE 11.2 Foundation giving: 1997–2007

In 2007, total giving by the nation's more than 72,000 grantmaking foundations increased by an estimated 10%, from $39 billion to a record $42.9 billion. Source: [11]

tures, borders, faiths will be critical to addressing the global nature of climate disruption. A second subtlety of globalization is the growing sense we have of contradictions between the natural world and society. Never before have we had to think about how our actions in altering the planet's climate might affect a person on the other side of the planet. "There's only one future for all of us. There's one problem, only one atmosphere. What does it actually mean to all live on Earth together well?" Senge asks. [16]

Philanthropists Filling the Gaps

In the fight against global climate change, philanthropic organizations are uniquely poised to tackle some of the most challenging aspects of the problem. Climate change affects many different sectors of society and therefore encompasses virtually every cause that foundations support. Foundations own at least $670 billion of the global economy. [11] As foundation president Stephen Viederman writes, "To use the endowment to support market-based solutions to climate change does not require a change in program guidelines. It does require a change in thinking about the way that financial assets can and should be used constructively." [17] In addition to the endowment assets to invest in climate protection, foundations also have a critical role to play in creating the solutions.

A study led by the California Environmental Associates (CEA) explores how philanthropic investment can turn the tide against global warming and cites philanthropy's comparative advantages. "Politicians are fixated on the next election; CEOs are focused on next quarter's numbers. Philanthropists, by contrast, have longer time horizons and can tolerate more risk. Besides being more patient investors, philanthropists have a strong tradition of filling gaps, spurring step-changes in technology and pursuing programming that transcends both national boundaries and economic sectors. Such capacities are exactly what are needed to tackle global warming." [6]

> To prevent the planet's mild fever from becoming a life-threatening illness, we must reduce annual greenhouse gas emissions by a staggering 30 gigatons (Gt) by 2030. That's about how much carbon the world emits today, and about half of what's expected by 2030 if development and energy consumption continue apace. [6]
>
> *California Environmental Associates,*
> *Design to Win, 2007*

Yet climate change solutions are a relative newcomer to the world of philanthropic investments, and foundations are considering their approaches carefully, examining the extent to which initiatives fit under current giving schemes and developing innovative and forward-thinking new initiatives to affect change where it is needed most. Large US foundations invested $436 million to address climate change in 2007, a doubling since 2004. Of the $436 million, 41% was spent internationally, mostly in

FIGURE 11.3 Policy spurs carbon markets

Strong financial signals are necessary to spark real collective action. Either through an emissions cap or through other means, we must put a price on carbon to force businesses, consumers, and governments to pay for their pollution. In turn, investment will shift to cleaner options. Source: [6]

developing countries. Foundation giving for climate change has increased nearly five-fold domestically since 1997 and nearly eight-fold internationally. By comparison, US philanthropy devoted $3.2 billion to health, $3.1 billion to education, and $1.5 billion to the arts in 2004, according to the Giving USA Foundation. Only 100 of the funders who underwrite global warming work have assets over $100,000. [6] (See Figure 11.2.)

Thus, major philanthropic funders of climate change solutions—such as the Surdna Foundation, the Doris Duke Charitable Foundation, the Pew Charitable Trusts, the John D. and Catherine T. MacArthur Foundation, and the Energy Foundation—are unique in their ability to advance solutions and are working to carefully carve out the niche in which their efforts might be the most beneficial. These foundations are consulting with scientific experts, policymakers, academics, and activists, not to mention their fellow foundations, to develop their focus and programs.

The Doris Duke Charitable Foundation (DDCF) took this approach in 2005 when they identified climate change as the most pressing issue facing society, outside of the areas they were then funding. Quickly, their consultations led to the consensus to support "technology policy"—the policy framework needed to move new low-carbon technologies quickly from expensive

new innovations to affordable norms—and thus help to reduce the energy we use and improve efficiency overall.

In 2007, DDCF created a new $100 million 5-year Climate Change Initiative focused on two strategies: carbon pricing policies and technology policies to help less-polluting technologies become more widely and quickly adopted across the economy. The first strategy supports the development of optimal pricing policies for carbon dioxide and other greenhouse gases. These policies are surprisingly complicated. Design work is needed to work out all of the details so that the policies are effective at helping society efficiently achieve emission reductions in an amount and on a schedule that science indicates is necessary. The policies must also be designed with an eye toward what is politically possible. One of the DDCF pricing policy grants supports the Harvard Project on International Climate Agreements, which strives to attain all of these goals in the design of an international climate accord that will replace the Kyoto Protocol once it expires in 2012. The second DDCF strategy supports the development of policies that will bring already available clean energy technologies, as well as new technologies, to market more quickly. This speed to market is critical for technologies related to energy efficiency, renewable energy, and low-emission uses of coal, such as carbon capture and storage technologies. More than $30 million in grants were approved by the end of 2007. [15]

One of the nation's largest scientific and environmental advocacy organizations, the Pew Environment Group, has worked for nearly two decades on climate change solutions, including advancing national and global policy regarding energy efficiency and clean energy. Part of the powerful Pew Charitable Trusts, the Pew Environment Group comprises more than 80 staff—with a presence throughout the United States as well as in Canada, Europe, Australia, New Zealand, the western Pacific, and the Indian Ocean.

The Pew Environment program includes climate change, protecting wilderness and public lands, and protecting ocean life. Pew uses two approaches to address climate change: science and policy analysis and advocacy campaigns. Moving forward on global warming is now primarily a political problem, says Kathleen Welch, deputy director of the Pew Environment Group. Substantial, permanent, and mandated reductions in US emissions are essential to a global strategy for climate change, Pew believes. The Pew Campaign on Global Warming is aimed at adoption of a national policy to reduce emissions throughout the economy. This work at the national level, and more recently the state level, revolves around building the political will for a robust national agenda for climate change that would involve the establishment of mandatory emission limits, coupled with a market-based system that would allow reductions to be achieved as cost-effectively as possible. Complementary energy policies are also needed, Pew says, including more-stringent fuel efficiency standards for vehicles, a national renewable energy standard, energy efficiency measures, and other short- and long-term strategies to speed the transition to low- and zero-emission technologies.

The Pew Campaign for Fuel Efficiency addresses an important portion of these complementary policies as it seeks more-stringent fuel efficiency standards for the nation's cars and trucks. A massive public education campaign aims to build public support and move elected officials toward stronger energy efficiency standards, which can reduce our dependence on oil, enhance security, save consumers money, and stimulate investments in vehicle technologies that help solve global warming. The campaign is seeking support in Congress for stronger US standards, conducting public education efforts in 15 to 20 key states, and coordinating a coalition of environmental groups at the national level. In addition, the campaign is conducting media outreach and public opinion polling and nonpartisan research and analysis. Pew also supports the independent Pew Center on Global Climate Change, a major research and policy center that brings better science into the legislative and regulatory arenas. Launched in 1998, the center advances debate through analysis, public education, and a cooperative approach with business.

The Energy Foundation also focuses on advancing policy on climate change, but it takes a broad, sector-by-sector approach as well as an international approach. The Energy Foundation is a partnership of major donors including, but by no means limited to, DDCF, the MacArthur Foundation, and the Pew Charitable Trusts. The Energy Foundation acts as a grantmaker on behalf of these foundations to advance energy efficiency and renewable energy. The foundation supports policy advocacy in the largest and fastest-growing energy markets in the world, the US and China, that will help to grow their economies while dramatically reducing carbon emissions and air pollution. The foundation awards for policy solutions in four program areas in the United States: power, buildings, transportation, and climate. The foundation also funds the China Sustainable Energy Program, with programs in low-carbon development paths, transportation, renewable energy, electric utilities, buildings, and industry.

With policy advocacy as their focus in each of these sectors, the Energy Foundation is closely following and helping to shape the development of cap and trade systems for carbon management. But opportunities are not limited to cap and trade only, and the foundation is working to develop an additional suite of policies with cap and trade that would enhance the benefits, both in efficiency and economics, of this system. Eric Heitz, president of the Energy Foundation, says there are other mechanisms through which low-cost carbon can be pulled from the economy at a profit. Strategies such as appliance standards, which are federally regulated efficiency standards on household appliances such as refrigerators and microwaves, are not going to be covered in a cap and trade scheme. Also, the Energy Foundation is working to make investing in energy efficiency profitable for utilities, by enabling them to get a return on investment in efficiency improvements.[†] This reverses a perverse incentive in the energy sector, where profits are usually tied directly to energy production. In the traditional market, the more energy utilities sell, the more money they make. In a market with efficiency incentives, the more efficiency utilities achieve, the more money they make.

One of America's oldest family foundations, the Surdna Foundation, has approximately $1 billion in assets and gives annually more than $40 million in grants in program areas including environment, community revitalization, effective citizenry, arts, and the nonprofit sector. The Surdna Foundation's Environment Program provides close to $9 million in grants each year to US nonprofit groups that advance solutions to climate change, improve transportation systems and patterns of land use, and safeguard oceans.[‡]

The Surdna Foundation's Environment Program is its largest program area and is national in scope. Mitigating the interrelated threats of global climate change, biodiversity loss, and unsustainable levels of resource consumption are at the core of this program, and the central goals of the program are the following:

- Build support for programs to stabilize climate change at the local, state, and national level.
- Improve transportation systems and patterns of land use across metropolitan areas, working landscapes, and intact ecosystems.
- Safeguard the biological diversity and productivity of US domestic oceans.

Like many of the major players in climate change philanthrophy, Surdna's grantmaking interests are more policy related than science related, but Surdna is unique in that their work revolves primarily around public education and awareness campaigns. This includes building the understanding of science and impacts of global climate change and generating support for programs to stabilize climate change by

- mobilizing new constituencies to make the case that climate change is more than an environmental issue and promote action to address it
- advancing state, regional, and city policy and leadership to create and implement plans that address climate change
- accelerating energy-efficient solutions to conserve energy, reduce emissions, and promote a green economy

Specific policies and activities Surdna is working to promote include carbon pricing policies and markets, the deployment of clean energy and energy efficiency technologies (in conjunction with the Energy Foundation), and research and development policies for understanding and expediting new technologies. Additionally, Surdna is beginning to explore its role in climate change adaptation strategies, which is emerging as a major consideration in their work in wildlife and biodiversity conservation as well.[‡]

Funding Climate Adaptation Work

Other environmentally oriented philanthropies are also realizing that climate change impacts are beginning to blur the lines be-

---

[†]For more, see Energy Foundation, www.ef.org.

[‡]For more, see Surdna Foundation, www.surdna.org.

tween program areas, and mitigation policy solutions may not be enough. Adaptation—the process of adapting our society to the vicissitudes of climate disruptions—is becoming a central concern for foundations, as scientific models consistently predict the worst impacts of climate change occurring in areas contributing in disproportionately small amounts to the problem. These places lack the resources to deal with the climatic changes that have already begun to affect them, further exacerbating problems of poverty, public health, and a lack of development. Foundations such as MacArthur and the Pew Charitable Trusts are investing seriously into these much-needed adaptation solutions.

Elizabeth Chadri, program officer for conservation and sustainable development for the Program on Global Security and Sustainability of the MacArthur Foundation, says climate change adaptation solutions are necessarily much more international in scale than mitigation solutions tend to be.§ Working internationally on coral reefs in 2004, the MacArthur Foundation staff began to notice the impacts climate change was already having on these environments, and they began to raise the issue no one wanted to discuss: If some climate change in inevitable, what can be done to minimize its impacts?

This discussion may have been taboo just a few years ago, but consultations with scientists working on mitigation confirmed their suspicions: There is a need to understand the damage that has already been done and what the implications are. Adaptation strategies and policies have a critical role to play in minimizing the worst effects of unavoidable climate change, and the MacArthur Foundation and others are breaking new ground in philanthropy to make sure this role is fulfilled in time.§

Many developing countries are seeing greater impacts from climate change, despite low contributions to carbon emissions. There

are indications that the already harsh climate of sub-Saharan Africa will be particularly vulnerable to climate change. Yet adaptation initiatives often stall because of hesitations and international tensions regarding who should be held responsible for the impacts of the Western world's carbon on this particularly sensitive region. The impacts are already being noticed throughout communities, though the exact cause and concept of climate change may not be fully grasped. Communities sense this change is different from droughts they have previously experienced, and they are anxious to start talking about it.

With the help of prominent advisors such as former president of Stanford Donald Kennedy, then editor in chief of the journal *Science*, and Peter Hayes, executive director of the Nautilus Institute for Security and Sustainable Development, the MacArthur Foundation began to explore the challenge of how to facilitate climate change adaptation.

Adaptation in developing nations is, after all, as much a development issue as it is an environmental issue as it is a public health issue. The multidimensional, multidisciplinary nature of this challenge cannot be overstated. And as critical as modeling climate change impacts and predicting their implications is, there is the very pressing issue of finding solutions quickly, both in time for them to be effective as forecasts become reality, and in time for communities and governments to wrap their collective heads around this staggering issue before a sense of desperation sets in. Akin to preparing for the onslaught of a hurricane, the more safeguards put in place before the storm is bearing down, the less a sense of desperate urgency will guide the process. Therefore, MacArthur has brought together a broad stakeholder group, including climate specialists, biologists, agriculture experts, and even other foundations, to influence how they will look at climate impacts, implications on natural and human systems, and adaptive solutions.

In fact, interdisciplinary approaches, cross-

---

§For more, see MacArthur Foundation, www.mac found.org.

sectoral solutions, and broad stakeholder coalitions are characteristic of the activities of all of the major philanthropic players in climate change solutions. The need to collaborate and think beyond narrow program areas in order to tackle the far-reaching problem of climate change is changing how philanthropies do business. As the MacArthur Foundation considers the role of biofuels in the spectrum of renewable energies, they are thinking a lot about the potential unintended consequences on ecosystems. Surdna keeps a broad focus by connecting climate change and land use, working to develop smart growth polices that view open spaces through a carbon-oriented focus.

Looking forward, while most climate change philanthropy today tends to focus on mitigation, especially carbon pricing and energy policies, philanthropies such as DDCF and the Energy Foundation cite adaptation policies and city- and state-level policy advocacy as major future priorities. But these organizations acknowledge that gaps still exist and acknowledge a sense of urgency to move the political ball forward, knowing that these solutions will not mature in a single step forward.

The science says we have to go much further than anything currently on the political table, warns Eric Heitz, president of the Energy Foundation. Implementing the best strategies for removing carbon from the economy now is essential to beginning work on the next phase of solutions. Setting an economy-wide cap on carbon opens the door to international negotiations, which are essential for bringing in the major emitters, China and India. All of this will require a tremendous amount of work—by science and to determine which policies will facilitate the right kind of implementation, notes Heitz.*

For philanthropies working in developing nations, climate change adaptation strategies will be essential to development strategies, says Elizabeth Chadri. Opportunities are emerging in developing new technologies, energy efficiency strategies, and renewable energy sources that create more opportunities for development and more incentives for mitigation in climate-sensitive areas such as sub-Saharan Africa.*

Clearly climate leadership and funding must come from beyond philanthropy's current environmental portfolio. Instead of replacing other funding areas with climate funding we encourage funders to integrate a climate lens (a perspective that considers climate change in all strategic decisions) into their grantmaking process. [7]
Environmental Grantmakers Association

All things considered, the future of climate change philanthropy is promising, with the number of organizations funding climate change solutions and the funding available to grantees increasing. Climate change has emerged as a major priority for philanthropic foundations, and as the interdisciplinary, collaborative approaches being taken by organizations suggest, one that has the potential to redefine how charitable giving is done in the 21st century. With a strong tradition of "filling gaps, spurring step-changes in technology and pursuing programming that transcends both national boundaries and economic sectors," philanthropic organizations are uniquely poised to advance climate change solutions in a way no other sector may have the capacity to. [6]

CONNECT THE DOTS

- Many Americans oppose taxes on use but support incentives and requirements for citizens and businesses to take action to reduce emissions.

- A growing number of religious faith communities are focusing on the moral obligation to protect the Earth's climate from human-induced disruption.

---

*For more from the NCSE conference participants, see their complete talks at http://ncseonline.org/climatesolutions.

- A partnership is emerging between labor and the environmental movement around retooling the economy for green jobs.

- Climate change has emerged as a major priority for philanthropic foundations.

Online Resources

1 Sky, www.1sky.org

350.org, http://350.org

Apollo Alliance, http://apolloalliance.org

American College and University Presidents Climate Commitment, www.presidentsclimatecommitment.org

Climate Center, www.climatecenter.org

Doris Duke Charitable Foundation, www.ddcf.org

Energy Foundation, www.ef.org

Environmental Grantmakers Association, www.ega.org

Foundation Center, http://foundationcenter.org/focus/gpf/climatechange

MacArthur Foundation, www.macfound.org

National Association of Evangelicals, www.nae.net

Pew Charitable Trusts, www.pewtrusts.org

Chronicle of Philanthropy, http://philanthropy.com

Surdna Foundation, www.surdna.org

The We Can Solve It Campaign, www.wecansolveit.org

www.eoearth.org/article/Post-Normal_Science

See extra content for Chapter 11 online at http://ncseonline.org/climatesolutions

Climate Solution Actions

Action 29: Mass Action—How Scientists Can Engage the Public in Global Dialogue Toward Shared Policy and Behavior Change Solutions for Global Climate Change

Works Cited and Consulted

[1] Abassi D (2006) *Americans and Climate Change: Closing the Gap between Science and Action.* (Forestry and Environmental Studies, Yale University, New Haven, CT). www.yale.edu/environment/publications

[2] ABC News/Planet Green/Stanford Poll (2008) Fuel Costs Boost Conservation Efforts: 7 in 10 Reducing "Carbon Footprint." August 9, 2008 (read September 29, 2008). http://woods.stanford.edu/research/surveys.html

[3] ABC News/Washington Post/Stanford Poll (2007) The Environment: Concern Soars about Global Warming as World's Top Environmental Threat. April 20, 2007 (read September 24, 2008). http://abcnews.go.com/images/US/1035a1Environment.pdf

[4] Aldhous P (2007) *Global warming: the buck stops here.* The New Scientist 194(2609):16–19. www.newscientist.com

[5] Bannon B, DeBell M, Krosnick JA, Kopp R, Aldhous P (2007) Americans' Evaluations of Policies to Reduce Greenhouse Gas Emissions. Resources for the Future, New Scientist Magazine Technical Report (read October 10, 2008). http://woods.stanford.edu/research/surveys.html

[6] California Environmental Associates (2007) Design to Win: Philanthropy's Role in the Fight Against Global Warming. www.ef.org/documents/Design_to_Win_Final_Report_8_31_07.pdf

[7] Environmental Grantmakers Association (2008) EGA Intersections: Confronting the Climate Challenge. *Environmental Grantmakers Association* (read September 10, 2008). www.ega.org/

[8] Gardner G (2006) *Inspiring Progress: Religions' Contributions to Sustainable Development.* (Worldwatch Institute, Washington, DC). www.worldwatch.org

[9] Kohut A (2008) A Deeper Partisan Divide over Global Warming. May 8, 2008. *The Pew Research Center for the People and the Press.* http://people-press.org/report/417/a-deeper-partisan-divide-over-global-warming

[10] Krosnick J (2008) American Perspective on Climate Change. *National Conference on Science, Policy, and the Environment: Climate Science and Solutions.* http://ncseonline.org/2008conference

[11] Lawrence S (2008) *Foundation Growth and Giving Estimates 2008.* Foundations Today Series. (Foundation Center, New York). http://foundationcenter.org

[12] McKibben B (2008) 350.org (read August 15, 2008). 350.org

[13] NAE (2008) For the Health of the Nation: An Evangelical Call to Civic Responsibility. *National Association of Evangelicals* (read September 30, 2008). www.nae.net

[14] NCSE (2008) Climate Science and Solutions. *National Conference on Science, Policy, and the Environment.* www.ncseonline.org/2008conference/

[15] New York Regional Association of Grantmakers (2008) Doris Duke Going Green, p 9 in NYRAG Memo: Environmental Grantmaking. www.nyrag.org

[16] Senge P (2008) Peter Senge "Impact of Globalization" QuickTalk. www.solonline.org

[17] Viederman S (2008) "How Grant Makers Can Curb Global Warming." *The Chronicle of Philanthropy,* February 13, 2008. http://www.philanthropy.com

CHAPTER 12

Think Globally, Incubate Locally

All our energy problems—price instability, energy security, and climate change have the same solutions—conservation, efficiency, and clean, renewable energy.

ROSS C. "ROCKY" ANDERSON,
Mayor of Salt Lake City, 2000–2008

Registered Republicans in Utah outnumber registered Democrats almost 5 to 1—639,161 to 136,891 according to state figures for 2008. In 2004, Utah gave Republican presidential candidate George W. Bush his widest state margin of victory over his Democratic challenger: Utah voters gave Bush 71.5% of the vote for a second term, up from 66.8% four years earlier. In 2008, Republican presidential candidate John McCain won the state convincingly, 63% to 34% over the eventual national victor, Barack Obama. Only one of the five members of Utah's congressional delegation is a Democrat. So why would the largest city in this state be such a leader in climate policy?

Salt of the Earth: Climate Leadership from Unlikely Utah

While Utah may be solidly conservative and politically Republican in federal elections, Salt Lake City, its largest city and county, with over one-third of the state's population, has been tipping toward green under the impetus of an unusual leader and coalition of civic organizations. We recall from the public opinion research discussed in Chapter 11 that the "issue public" for climate change skews toward Democrats, younger voters, and women. So why has Salt Lake City, located by many measures in the nation's most conservative state, become a climate action leader and environmental vanguard? Maybe it has something to do with seeing the snow season shorten on the peaks just miles from downtown or watching the Great Salt Lake shrink to become a lesser salt lake. The success in Salt Lake City appears to stem from binding the economic reasons to be environmentally more sustainable with the climate actions available to citizens and businesses. It also is the result of local government leadership.

While mayor of Salt Lake City from 2000 to 2008, Rocky Anderson instituted a suite of policies to implement these solutions. "Leadership is the need to choose wisely in the face of sometimes very strong opposition," he stated. Issues that proved to be key for Salt Lake City included addressing sprawl and stopping reliance on coal power. Mayor Anderson called for a moratorium on all new coal-fired power plants in the region, in direct confrontation with a key state industry. The coal phase-out movement has contined to gather steam with a large demonstration taking place in Washington, DC, in early 2009. The decommissioning of operating coal-fired power plants remains a rallying point for both those concerned about climate disruption and those concerned about supplying adequate electricity generating capacity.*

The first step for any local government seeking to put itself on a low-carbon diet is to do a sustainability inventory, including water, energy use, and air quality. Low-hanging fruit can be identified, such as replacing lightbulbs in City Hall (saving $33,000 a year) and changing traffic lights to use light-emitting diodes (LEDs). City workers were told to turn off lights and computers when not in use. The city offset its air travel on official business by paying to sequester carbon in Costa Rican rain forests. It transitioned to a fleet of city vehicles using compressed natural gas, and eliminated all SUVs. One of the most popular plans was to allow drivers of certified low-polluting vehicles to park for free at city meters. Bike racks and bike paths were extended; walking was made safer by installation of countdown signals for pedestrians and mid-block crosswalks that light up when people are using them. Leaders who push new plans may come in for derision, such as when the city put orange flags on the streets for crossing

*The array of Web sites devoted to the topic of burning coal for electrictiy is a good example of the diversity of opinion that arises on climate-related topics: Pro-moratorium: http://cmnow.org, http://coalswarm.typepad.com, www.coal-is-dirty.com; Anti-moratorium: www.coalisclean.com, www.cleancoalusa.org, www.iea-coal.org/site/ieacoal/home.

pedestrians to carry. However, Anderson said that now people use the flags eagerly.

Some projects were larger and more difficult but yielded significant results. An executive order required all new buildings to meet green standards. The green buildings have higher occupancy and command higher rents. Methane recovery and cogeneration have been successful at Salt Lake City public utilities and the town landfill. Smart growth regulations that plan for greater density in some areas, yet provide high livability with recreational green space and walkways, have been adopted in many cities.

Salt Lake City instituted programs that provide energy audits and green certifications for businesses and private homes. Peer pressure as well as economic advantages for businesses can energize these programs. Sometimes, political leadership is required. Programs that initially met resistance, such as implementation of light-rail lines, are now very popular. In fact, based upon Salt Lake City's success, four other towns in Utah have voted for sales tax increases to build light-rail lines.

A key feature of Salt Lake City's success has been engaging businesses, citizens, and students directly. The Salt Lake City e2 Citizen program ("e2" stands for environmentally and economically sustainable) educates and supports city residents who take steps in their own lives to address climate change and the further degradation of our planet. Similar programs exist for businesses and for students—e2 Business and e2 Student, respectively. In each case, the participant registers with Salt Lake City Green and commits to at least three new goals from the following behaviors that align with economic sectors: transportation; energy conservation; reduce, reuse, recycle waste management; water conservation; food; health; and community education. Salt Lake City alone has reduced greenhouse gas emissions by 31% and has saved money in the process. These successes on the local level inspire people and let them know not only that there is a problem but also that there are solutions, Anderson stressed. Seeing these

TABLE 12.1 Salt Lake City Green Accomplishments
Highlights of Climate Change Initiatives: 2000–2008

City fleet

- Since 2003 SLC has decreased its light-vehicle fleet by 143 vehicles.
- Biodiesel is now being used at all fueling locations throughout the city.
- All diesel fuel now being received from the refineries has gone from 550 ppm to 15 ppm ultra-low sulfur content, which complies with recent EPA mandates.
- 62% of all city-owned off-road vehicles are being operated on alternative fuels (biodiesel, CNG, hybrid, etc.).

Bike and alternative transit

- SLC provides free bus passes for city employees.
- Employees who carpool to work receive permits for reserved parking spaces at the city-county building.
- There is free parking for low-emission and alternative-fuel vehicles at all city parking meters.
- SLC has added 15.3 miles of bike lanes on city roadways.
- Over 1,000 pedestrian countdown timers have been installed at intersections to boost safety for walkers.

Energy and buildings

- Salt Lake City Corporation has reduced CO_2 emissions from energy use at its municipal operations by 31% since 2001, surpassing the city's goal to meet the Kyoto Protocol standard by 148% and doing it 7 years early.
- City-owned building projects and renovations will be certified energy and environmentally efficient pursuant to the US Green Building Council's LEED silver standard.
- Salt Lake City purchases 12,960 kWh of wind power per year, reducing CO_2 emissions by 796 tons per year.
- 100% of the electricity used at the city-county building is renewable energy.
- LEDs (light-emitting diodes), which use 90% less energy and require less maintenance than standard lightbulbs, have been installed on all city-owned traffic signals for an estimated savings per year of $55,000.
- Landfill methane produced and captured at SLC's municipal waste site provides enough renewable electricity to power over 2,500 homes.
- Digester gas from the wastewater reclamation plant is captured to produce electricity, saving the city $160,000 per year.

Other successes

- All future carbon emissions from city-related air travel will be offset through Pax Natura.
- SLC residents recycle 11,000 tons of materials each year, an 85% increase since 2000.
- No city-sponsored meetings, interoffice functions, or events will provide bottled water. Instead, pitchers and refillable containers will be made available.
- Water use by city residents has dropped by 20% since 2000.
- A walkable-communities ordinance has been passed to promote transit-oriented development.
- SLC is now using 60% less chemical intervention to abate insect problems in the urban forest.

Source: Adapted from [24]

achievements will empower individuals to make the changes that are necessary. Anderson said that this individual action needs to be converted into political action and that citizens need to insist that elected officials rise to this challenge. The solutions we develop and implement to address climate change, he stressed, will have additional benefits for enhancing national security, increasing energy independence, and improving local air quality.

Mayor Anderson notes that the story of climate action is not all gloom and doom but that "we need to show people that we have the means to make the changes necessary. We can develop the technologies and export them to places like India and China, creating markets for these technologies especially in the area of electricity generation."[†] In other words, strong incentives exist for economic development in the area of energy efficiency technologies.

Cities Take Measure

It is hard to go on a diet without stepping on a scale. In the same fashion, communities serious about reducing greenhouse gas emissions need to inventory all emission-producing activities before figuring out where to shed some gas. Thankfully, excellent tools exist to help municipalities do so.

So what can cities do? We opened this discussion by examining Salt Lake City. But Salt Lake is not alone. The United States Conference of Mayors launched a well-received Climate Protection Center in early 2007. In adopting the center's Climate Protection Agreement, mayors commit their cities to reduce emissions by 2012 to 7% below 1990 levels. In the first year and a half, well over 884 mayors committed to this goal, representing a total population of over 80,950,895 citizens from all 50 states and Puerto Rico. This rapid mobilization of urban centers large and small is a sign of the broad sense of urgency around climate protection. From Albuquerque to Waukesha, cities are enacting local and specific climate protection measures across all sectors of urban life, from transportation and waste management to purchasing policies and water use, as we can see from the sample of accomplishments for Salt Lake in Table 12.1. [27]

More than 800 local governments around the globe have adopted tools and policies to reduce emissions by joining ICLEI—Local Governments for Sustainability—in order to participate in the Cities for Climate Protection (CCP) Campaign. Founded in 1990 as the International Council for Local Environmental Initiatives, ICLEI focuses on improving local

sustainability. The CCP Campaign helps cities adopt policies and implement quantifiable measures to reduce local greenhouse gas emissions, improve air quality, and enhance urban livability and sustainability. The goal is for local governments around the globe to integrate climate change mitigation into their decision-making processes. This highly successful and widely recognized campaign is based on an innovative performance framework structured around five milestones that local governments commit to undertake. The milestones allow local governments to understand how municipal decisions affect energy use and how these decisions can be used to mitigate global climate change while improving community quality of life. The CCP methodology provides a simple, standardized way of acting to reduce greenhouse gas emissions and of monitoring, measuring, and reporting performance. As ICLEI explains, "Communities benefit from the actions that they take to reduce greenhouse gas emissions through: (1) Financial savings in reduced utility and fuel costs to the local government, households, and businesses. (2) Improved local air quality, contributing to the general health and well being of the community. (3) Economic development and new local jobs as investments in locally produced energy products and services keep money circulating in the local economy." [10]

In addition to the local action plans mentioned above, communities can use the resources of organizations such as Climate Action Network (CAN), a worldwide network of over 365 nongovernmental organizations from 85 countries working to promote government, private-sector, and individual action to limit human-induced climate change to ecologically sustainable levels. The US affiliate was established in 1989 "to provide groups working on global warming with a forum for joint strategy development and advocacy to affect change in a coordinated way at the United Nations and in Washington, DC." [26]

Creating a shared vision that reflects the entire community's desired future that excites and engages citizens may be easier at the local

---

[†]For more from the NCSE conference participants, see their complete talks at http://ncseonline.org/climatesolutions.

scale of a city or neighborhood than at the national scale. Yet, establishing this common ground is infectious and can bring out larger-scale awareness and ultimately improved decision making. Many communities have successfully used Natural Step workshops to bring a community strategically closer to environmental and social sustainability in an economically sound way. For example, Whistler, a mountain resort community in British Columbia, was the first community in North America to adopt the Natural Step Framework. This framework is a series of four steps that a community or business can use to identify common goals and a process for working toward them. With its stunning natural beauty, tremendous growth challenges, and upcoming 2010 Olympic Games and Paralympic Games, Whistler has spent the last decade understanding what sustainability means in its specific context. [12]

Who Took Our State's Snow?

Utah is not the only state concerned about the impact of less snow bringing fewer ski dollars to its economy. Increasingly, cities, states, and provinces are hiring sustainability advisers or conducting research on sustainability measures by bringing on consultants with the appropriate expertise from the nonprofit world. Utah's population jumped 40% between 1990 and 2003. When Utah wanted to examine the ecological footprint that its burgeoning population placed on the region's natural resources, the Utah Population and Environment Coalition collaborated with Global Footprint Network to prepare the Utah Vital Signs Project 2007. The goal of the report was quite simple, "an informed basis for reflection on what we Utahns require of the earth." This study was the first state-level ecological footprint study in the United States, and it compared Utah's ecological footprint in 1990 and 2003. The data in the Utah Vital Signs report will help Utah residents and policymakers meet the goals in implementing ICLEI's Local Governments for Sustainability initiative. [28, 29]

In the United States, cities, counties, and states have provided leadership to combat climate change. The National Council for Science and the Environment conference in January 2008 featured several elected officials commenting on the political landscape for climate action, including Jay Inslee, a member of Congress from Washington State since 1999, and former Mayor Anderson of Salt Lake City. Representative Inslee has long been a proponent of a comprehensive national energy policy in which renewables play a key role. And Anderson, as described above, led Salt Lake City to an number of groundbreaking energy efficiency measures, long before it was in vogue. What did they have to say?

Calling on political leaders to create a "vision of opportunity and optimism for Americans," Representative Inslee pointed out that the tools needed to combat climate change already exist. People are coming up with solutions in labs and small businesses all over the country. Inslee noted, "Global warming is as much an economic opportunity as an environmental challenge." He is pushing for government support for clean energy technologies. He is the lead sponsor of the New Apollo Energy Act, aggressive clean energy legislation in the US House of Representatives. [11]

Representative Inslee and Mayor Anderson agree that the federal government has failed to provide adequate leadership on this issue. While individual agencies have worked hard on issues such as energy efficiency and curbing greenhouse gases, the marching orders from Washington on addressing climate disruption thoroughly and effectively were largely absent during the Bush administration. Fortunately, action at the state and municipal levels has helped to fill the void at the federal level. Municipalities and states are great laboratories for innovation.

Many groups have sprung up to help communities develop meaningful and effective climate change strategies. For example, the Center for Climate Strategies (CCS) brings together experts

TABLE 12.2 American States as Environmental Policy Pioneers

State action	When	Corresponding federal action	When
State acid rain laws	1985	Federal acid rain program*	1990
State air toxics laws	1987	Federal air toxics program[†]	1990
State NO_x trading (OTC)	1995	Federal NO_x > SIP call[‡]	2004
State mercury laws	1998–2002	Federal clean air mercury rule[§]	2005
State renewable portfolio standards (RPS) laws	1997–2007	Federal RPS law	Introduced
State "4-P" laws for power plants	1997–2002	Federal "4-P" law	Introduced
Statewide GHG reduction laws	2003–2006	Federal GHG reduction law	Introduced
State GHG reductions from vehicles	2002	Federal vehicle GHG standards	?

*www.epa.gov/acidrain/

[†]www.epa.gov/ttn/atw/

[‡]www.epa.gov/airmarkt/progsregs/nox/index.html

[§]www.epa.gov/camr/

Source: Adapted from [7]

Note: "4-P" stands for four pollutants ideally integrated in a regulatory package: sulfur dioxide, nitrogen oxides, mercury, and carbon dioxide.

in public policy, business, economics, finance, management, law, science, engineering, and communications to help states, regions, and national governments tackle climate change. Kenneth Colburn, codirector at CCS, notes that states are often originators of innovative programs to address the most important issues. State governments use a consensus-building approach to resolve conflicts. States are proactive on climate change because they see both economic opportunities and political opportunities to address the issues. They also see firsthand the impacts of climate change such as a shrinking length of the economically critical ski season in Utah—"Who took our snow?!" [5]

States consistently shape federal policy, as we can see in Table 12.2. Colburn cites the work of two states in developing climate change policies. In Arizona, Governor Janet Napolitano created a multi-stakeholder Arizona Climate Change Advisory Group. The group expected that 285,000 jobs could be created as the state addresses climate change. CCS has calculated the net economic cost savings of the Arizona proposals to be $5.5 billion for the period between 2007 and 2020. That much money represents a powerful financial incentive for enacting climate protection without delay. Of the

group's 49 recommendations, 45 were adopted unanimously. In Arizona, the following measures saved more money than they cost: clean cars, appliance efficiency standards, electricity pricing, distributed generation and combined heat and power, demand-side management, and building codes. The following is a sample of the abatement strategies that required additional investment above the immediate savings through 2020: renewable portfolio standards, truck speed limits, increased reforestation, decreased carbon intensity, and reduced land conversion.

In neighboring New Mexico at the same time, Governor Bill Richardson asked a multi-stakeholder advisory group to come up with ways to meet an ultimate goal of reducing greenhouse gas emissions to 75% of 2000 levels by 2050. The resulting New Mexico Climate Change Advisory Group determined that New Mexico could not only meet but exceed the governor's goal, and realize a savings of $2.1 billion.[‡]

As Michael Northrop of the Rockefeller Brothers Fund and his colleagues write, "States

[‡]For more details of the policy recommendations, see www.azclimatechange.us and www.nmclimatechange .us.

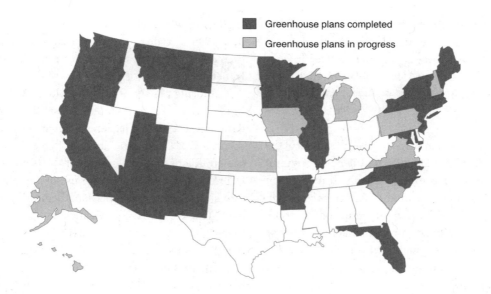

Greenhouse plans completed

Greenhouse plans in progress

FIGURE 12.1 State-level climate action plans underway or in development in 2008

Since 2000, some 30 states have developed or are developing climate mitigation action plans through open, democratic, and bipartisan consensus-building processes. These statewide climate action plans include goals of increasing energy efficiency and reducing greenhouse gas emissions. Source: [20]

have considered such policies on the basis of their ability to reduce greenhouse gases, their cost effectiveness, their feasibility and their potential co-benefits." [17] By early 2008, 13 states had taken leadership roles with significant statewide initiatives on climate protection policy: Arizona, California, Colorado, Connecticut, Maine, Minnesota, Montana, New Mexico, New York, North Carolina, Rhode Island, Vermont, and Washington. At least 18 more states are in the process of developing comprehensive climate action plans. The map in Figure 12.1 illustrates this national activity. (Compare this map to the one in Figure 10.3 showing the states with carbon markets.)

Scaling Up

In 2007 McKinsey, the worldwide consulting firm, asked what the costs might be for greenhouse gas reduction options. McKinsey came

to very similar conclusions as the CCS staff: Marginal costs are very favorable for immediate climate action. In plain English, the first 15% of the greenhouse gas reductions through such aggressive policies as already in motion in these 13 states actually cost no new money, and they reduce costs per ton of gas avoided. [14] Figure 12.2 shows this cost curve for 2020. Each step from left to right in the curve represents the impact of one greenhouse gas mitigation policy or action. Properly implemented, sector-based climate change mitigation policies can reduce pollution, lower emissions, save money, and create jobs. In short, state opportunities can be scaled to the national level.

FIGURE 12.2 online at ncse.org/climate solutions

Economy-wide marginal cost/savings curves for greehouse gas removal

> Either we can be a part of the solution and get ahead of the curve, or we can get run over by whatever is proposed in Washington.
>
> Governor Mark Sanford, April 27, 2007,
> South Carolina Climate, Energy and Commerce
> Advisory Committee meeting

Climate action plans involve all sectors of the economy and many different behaviors within each sector. For example, state climate action plans are always a product of intensive stakeholder and technical work group collaboration. They are designed to reduce state greenhouse gas emissions through a wide range of specific policies and programs in the following sectors: energy efficiency and conservation; clean, advanced, and renewable energy supply; transportation and land use improvements; forest and farm conservation; waste management; and industrial processes improvements. [4]

In each of the above sectors, greenhouse gas abatement could come from any number of process or behavior changes, such as codes and standards; market-based systems (e.g., cap and trade); funding and technical assistance (e.g., state grants); reporting and disclosure (e.g., toxic waste inventories); voluntary agreements; information and education programs (e.g., Energy Star outreach); and other implementation methods. A typical climate plan has 40 to 70 actions that are combined into a balanced and comprehensive "portfolio" that covers all sectors and uses a combination of implementation methods, with elements of both traditional and innovative policy mechanisms. [4]

California and Beyond

Dan Kammen, from the Renewable and Appropriate Energy Laboratory at the University of California, Berkeley, offers an analysis of California's Assembly Bill 32, which requires the greening of energy sold to California and a major reduction of carbon emissions by 2020. Professor Kammen points out that California's laws affect other states, such that there is now

a growing group of western states with commitments to reducing carbon emissions over their whole economies (the Western Climate Initiative).

California's enactment of the Global Warming Solutions Act of 2006 (AB 32) set the standard as the US state leader in climate change action. AB 32 calls for reductions of greenhouse gas emissions from all sectors of the economy to 1990 levels by 2020, with the use of a mandatory statewide cap on emissions beginning in 2012. The greenhouse gases include all of the following gases: carbon dioxide (CO_2), methane (CH_4), nitrous oxide (N_2O), hydrofluorocarbons (HFCs), perfluorocarbons (PFCs), and sulfur hexafluoride (SF_6). These are the same six gases listed as greenhouse gases (GHGs) in the Kyoto Protocol. In addition, an executive order by Governor Schwarzenegger sets state targets seeking to reduce greenhouse gas emissions to 80% below 1990 levels by 2050.

Standards for building construction will play an important part in the control of carbon emissions. California's goal is for all residential construction to be zero net energy by 2020, and all commercial new construction is to reach this same goal by 2030. One action taken by the city of Berkeley has proven so popular that the state government of California may facilitate its adoption by other cities. Residents who are willing to undergo a home energy audit and who make necessary upgrades are eligible for loans from the city to purchase solar energy systems. Payment is spread over 20 years and paid as part of the city property tax bill. The city floats bonds to pay for the program, gaining a better rate than an individual citizen could get and hence making solar energy competitive with the commercial power grid.

Further, building codes will need to be adjusted to remove unnecessary barriers for the low-energy and low-emission building projects. We need to rethink state building codes and local zoning rules to avoid penalizing safe, high-efficiency design or, even better, to provide incentives for high-efficiency design. Right now, for example, new, high-efficiency water treatment systems that recycle wastewater from dishwashers for use in flushing toilets may not be allowed in most states, because of arcane building codes. The Passive House Institute in Germany pioneered homes with passive heat exchangers that consume only 5% to 15% of the energy that a typical home uses. [8, 23] The National Renewable Energy Laboratory is collaborating with a number of state-level organizations of zero-emission homes to collect detailed performance data that will allow refinement of the approach for integrating energy efficiency measures with photovoltaic and solar thermal systems. [18]

Nationwide, 29 states and the District of Columbia have renewable portfolio standards (RPSs) that require a designated percentage of the energy mix to come from renewables by a certain year. The consequence is a reduction in carbon dioxide emissions. Texas was the first to initiate such a requirement and has already exceeded its targets by using wind generation on its farmlands. Kammen pointed out that other states, such as North Dakota, have wind resources that exceed those in Germany, where wind power is a major industry. Other ideas from nations abroad are relevant to state governments, Kammen reminds us: "In Denmark, waste heat from power plants is used for home heating." [22]

The midwestern states are also rising to the energy challenge. Gary Radloff, director of policy and strategic communication for Wisconsin's Department of Agriculture, Trade, and Consumer Protection, talked about the linkage between energy security and climate stewardship from the perspective of the north central states. "State and local governments are incubators for innovation," he declared. Models include regional collaborations for biofuels, bioenergy, and bioproducts. Corn-based ethanol, which has received so much attention of late, is only the first step, opening the door to go further with integrated biorefinery, such as forest biomass conversion. "If you Google 'venture capitalists' and 'cellulosic,' you will see the opportunities!" he enthused. [21]

"For the Midwest, biofuels are the key, the venture capitalists are our friends, and we're just not giving up," Radloff declared. The Midwestern Governors Association held a summit in 2007 that led to an accord for comprehensive energy policy planning to reduce the carbon footprint. Participation by Manitoba, Canada, makes this an international-regional effort. Challenges at the regional level include a transmission adequacy initiative, that is, an initiative to link wind power to an electric grid and provide hydropower as backup. State climate change committees will be an important part of implementing the regional vision. Partners include land grant entities, state agricultural departments, entrepreneurs, nongovernmental organizations, and private foundations. This work continued into 2008 with the midwestern states adopting guidelines for implementing a regional bioproduct procurement system with common definitions and standards for bio-based products and setting out methods for midwestern states interested in procuring those products. [15]

Radloff expresses some concern that, all too often, state legislatures are not moving fast enough to implement state and regional plans. Items agreed to by the governors' summit, such as the Greenhouse Gas Accord, require state-level legislation that has not yet been forthcoming. Nongovernmental organizations such as the Midwest Ag Energy Network and consortia such as the North Central Bioeconomy Consortium have aligned with state and local administrations to move ahead despite these regional roadblocks.§

Consortia can leverage regional research capabilities. They can also implement and link regional cap and trade programs. "If we don't have cap and trade nationally in the next two years, I'll be surprised, but if not, the states may just surround DC and take over," Radloff declares. [21]

---

§For more on these regional cooperatives, see www.ncbioconsortium.org and www.midwestagenergy.net.

National roadblocks are another problem. Radloff cites the US Environmental Protection Agency's failure under the Bush administration to allow California and other states to implement their own tougher low-carbon fuel standards. In addition, some state and national regulations may need to be softened temporarily, to allow for biofuels development. Another hindrance to progress in Wisconsin has been the structure of the Governor's Task Force on Global Warming. Its major work was assigned to various work groups, which became bogged down as "industry groups punted." Radloff suggests that such task forces need to "have a core that can hold together" to discuss the tough issues.

Do the Feds Get It?

Most observers of climate legislation believe that—one way or another—a cap on carbon emissions is just around the corner. In the fall of 2007, Senators Joseph Lieberman and John Warner introduced a historic "cap and trade" bill that would require the country to reduce its carbon dioxide emissions by 70% before 2050. A number of federal bills have been deliberated by the US Congress, the most well known of which—the Lieberman-Warner Act—failed to be adopted in 2008. The dozen states with existing greenhouse gas reduction plans project greater greenhouse gas emission reductions than even the most aggressive congressional reduction plan under consideration in 2008 (the Sanders-Waxman-Boxer plan). For example, reducing greenhouse emissions to 1990 levels by 2020—as most of the 13 early states have targeted—would result in a reduction of greenhouse gas emissions by 58% and a net economic savings of over $25 billion in that time frame. In 2009 Nancy Pelosi has made legislation on climate change a priority, and there may be legislation to control carbon emissions signed by President Obama by the time you are reading these words.

In one sector alone—residential, commercial, and industrial buildings—the net eco-

TABLE 12.3 New Potential for US Net Economic Cost (Savings) by Sector by 2020

Sector implementing greenhouse gas reductions	$ billion
Residential, commercial, and industrial	−$43.4
Transportation and land use	−$2.3
Subtotal	−$45.70
Energy supply	$16.2
Agriculture, forestry, and waste	$3.8
Subtotal	$20
Net total economic savings through 2020	−$25.6

Source: [7: p. 19]

nomic benefit of aggressive greenhouse gas abatement would be a savings to the country of over $43 billion by 2020. In Table 12.3, we see that efficiency in the building sector more than pays for abatement in other sectors. [7]

Science tells us that emissions must be cut at least 80% by 2050. Lexi Shulz, deputy director of the climate program at the Union of Concerned Scientists, asserts the best proposals put forth by the 110th US Congress (2006–2008) would only hold future atmospheric concentrations to about 450 parts per million (ppm), well above the 387 ppm of today and too high to prevent severe climate disruption. "To meet the 80% goal, voluntary reductions won't be enough," Shulz stressed. "If the targets are not set at the right level, then there may be a 'crash finish' to meet the goal more quickly at the very end." [25]

James Bradbury from Representative Jay Inslee's office notes, "Even if we put a price on carbon, this would be unlikely to adequately address transportation issues such as the failure of oil prices to offset consumption costs." [16] Petroleum-based fuels represent 95% of all transportation fuels. However, a low-carbon fuel standard would regulate all transportation fuels. Inslee introduced HR 2215, which is consistent with California law. Putting a price on carbon does not necessarily address issues of growth and does not include incentives for state and local governments to implement "smart growth" programs that would maximize conservation and reduction in fuel consumption. Putting a price on carbon does not address the existing complex regulatory structure, the incentives that utilities have to promote consumption, or the low percentage of renewable energy in the power generation sector.

With the changes in the composition of both the US House and Senate, and the congruent changes at the other end of Pennsylvania Avenue in the White House of Barack Obama, the legislative initiatives for climate protection are sure to be in the spotlight. But what about all the younger Americans, whose entire adult lives will play out in the next decades under whatever climate disruption we fail to avoid? Let's take a brief look at climate activism among younger Americans. This younger generation is the one that will have to both implement the policies and live longer with the consequences—good or bad—than the politicians currently ruling the roost in either Washington or the 50 state capitols.

Next Gen Leadership

The leaders of tomorrow have a message for those working on climate change today: "Thank you!" Young movers and shakers want their elders to know that the next generation is poised and ready to accept this important torch. The innovations in climate science, technology, and policy that are the fruits of dedicated careers in climate change have set the stage for taking solutions to the next level. As a self-proclaimed "youth leadership talent scout," Douglas Cohen asserts, "Intergenerational partnerships are exactly what is needed to bring about a sustainable future." Cohen is founder of the Leadership Institute and cochair of the US Partnership for Education for Sustainable Development's national youth initiatives. [6]

Scientific research and an understanding of social change landscapes can lead to systemic change if it is supported by leadership that inspires and awakens, creating a sweeping scope of collective action toward sustainability. This movement is already underway. Climate change awareness campaigns like Focus the Nation are engaging both young and old in demonstration and dialogue on climate change solutions. Youth-led initiatives such as the Energy Action Coalition, the Envirolution, and DoRight show the progress that the next generation of leaders is already making in creating a paradigm shift toward sustainability.

By the time these young leaders reach his age, says economist Eban Goodstein, founder of Focus the Nation 2008, they will have put an end to the fossil fuel era. They will have rewired technologies, redesigned every city on Earth, reimagined the global food system, reinvented transportation, created tens of millions of new jobs, lifted billions out of poverty, and thus will have become the new "Greatest Generation." But for this to happen, the generation currently in power has a responsibility to take actions now. We are standing at a critical moment in history, says Goodstein, with huge implications on the scale and magnitude of the climate challenge that the next generation of leaders will face. Our decisions over the next year or two on whether and how we reduce carbon emissions, invest in green technologies, and support needed research will set the trajectory for a much longer time scale. Those in leadership roles must act now, and this will require a push from stakeholders of all generations. [9] Through awareness-building, collaborative efforts like Focus the Nation 2008 and Power Shift, the next generation of leaders can begin earning the title of Greatest Generation now, by pushing hard to start making the changes we need, from local to global scales. Focus the Nation 2008 was a national educational initiative on global warming solutions, committed to empowering a generation to accelerate the transition to a just and prosperous clean energy future.*

Seizing the momentum of this burgeoning awareness, three young Yale alums, Alex

---

*Focus the Nation is now known as the National Teach-In on global warming (www.nationalteachin .org).

INSIGHT 12: DOING RIGHT BY THE KIDS

An education reformer, Scott Beall, founder of DoRight, has a revolutionary education concept for America. DoRight Enterprises is a sustainability consulting firm, offering pro bono advice to local businesses and organizations to help them raise profits and reduce environmental footprints. What is unique about DoRight's business model is that it is entirely run by middle school students.

DoRight Enterprises was the outcome of Beall's desires to fix the educational shortcomings of American schools, coupled with his own insights from the education field on the lack of buy-in and awareness of environmental issues by Middle America. Without their wholesale support, he says, it will never be possible to achieve the kind of cuts in carbon emissions vital to stabilizing the climate system. Inspired by organizations such as the Natural Step, which provides environmental, social, and economic sustainability advice to major corporations, Beall developed DoRight Enterprises into a model for middle school students, who, he says, are "dying to speak truth to power." DoRight Enterprises was founded in 2005 with the goals of mobilizing youth action to educate and change the behaviors and attitudes of an important and large demographic with respect to environmental and sustainability issues and to reform educational systems with a "textbook example of every best practice pedagogy that schools are clamoring for." [3]

Rigorous training in systems thinking and integrating math, social studies, science, and English prepares students to choose one of three roles within the firm. Advanced applied and holistic thinking is developed through "end of oil" calculations and comparisons between industrial production models and ecosystem models. Once trained, students choose between business consulting, political action, and public relations tracks. They then perform functions within the firm ranging from doing multiple-point sustainability audits on local businesses to running letter-writing campaigns to producing films and publicity materials.

(continued)

Gamboa, Timothy Polmateer, and Antuan Cannon, have created a forum to bring all youth to the climate change solutions table, using Web-based tools of the 21st century activism trade such as social networking and open source platforms, as well as community-based activism in the form of local chapters, events, and leadership training to encourage young people to get involved in the green movement in the way that speaks to them. Aptly christened the Envirolution, this truly collaborative project highlights an important aspect of the next generation of leadership in the green movement: the concept that green is universal and that the trilogy of environmental, social, and economic sustainability can be not only an important part of everything we do but

a common ground for bringing people together. With this in mind, the Envirolution aims to prepare the next generation of leaders to take the torch from those who have laid the groundwork.

Capitalizing on what they view as a commonly overlooked demographic, the Envirolution seeks to harness the enthusiasm, ambition, and drive of high school and university students, as well as "the youth of all ages" (in keeping with their inclusive approach), to empower them to become leaders in the sustainability movement. The three core missions of the Envirolution, to "educate, unite, and take action," are the guiding principles of the chapters, beginning with students' commitment to educate first themselves and then their peers and local communities, and

As Beall says, seventh-grade students are underestimated 90% of the time, regarding the level of complexity that they are capable of understanding. Madeline Skaller is chairwoman of DoRight Enterprises' extracurricular counterpart, the DoRight Leadership Corps, and a freshman at Brewster High School in Brewster, New York. She says her work as the sustainability auditor of the Putnam Hospital Center gave her confidence and empowered her to discover her own voice in society. With her detailed recommendations adopted immediately, she has smartly leveraged this success, talking on radio shows and presenting at conferences, discussing the role of youth in climate change solutions. "Young people can see things in a way that adults cannot," Skaller asserts, referring to the objectivity and fresh perspective these younger minds can offer to old challenges. "We need to change mindsets and mental patterns to enact change in the world." [3]

James Smith, eighth-grade student and DoRight Enterprises consultant, offers an analogy to illustrate this fresh perspective regarding generational responsibility and the current state of the environment: "It's like when your mom is going to drive you and your little brother to a party. You are all set to go, when you go in the kitchen and find that your older brother has made a huge mess and he's leaving. At this point, you've got two options: You could sneak out, go to the party and have a good time, leaving your little brother at home to take all the heat for it. Or, you can stay, clean up, and even if you can't go to the party, you know your little brother can. The moral of the story is: if you can't fix it, stop breaking it. Many people in my generation can't decide what choice to make, but if the older brother stays, maybe we will have an easier time choosing." [3]

There are nearly 30 million people in the United States between the ages of 12 and 19, Beall asserts—a massive untapped resource. The long-term, four-year mission of DoRight Enterprises is to populate all the cities in this country with DoRight consultants. People will listen to this age group, Beall says; it breaks the psychology of denial in adults to hear them speak this truth. As Suzuki Roshi said, "In the mind of the beginner there are many possibilities, in the mind of the expert there are few."

even their younger counterparts via after-school education for elementary and middle school students, on the concepts and applications of sustainability. Students learn to build coalitions and engage stakeholders by actively pursuing alliances with other clubs, local businesses, and nonprofit organizations. Finally, students take action within their communities and through the Envirolution's Web-based community, using their individual skills and talents to create and implement projects that advance sustainability.

The Envirolution pilot chapters and members represent a range of socioeconomic backgrounds, races, and cultures, but all have responded with the same eagerness to get involved. Given the chance, these young people are becoming drivers of change in their communities, and their impact will only continue to grow. The three young leaders of this inspiring organization reach out to the generation in power now. Says Gamboa, "Reach out to one youth. Mentor them. Let them in on what you are doing. Once you incite this passion in them and offer your courage and support, it is amazing to see what they can accomplish."

Jessy Tolkan, the 26-year-old executive director of the 46-member-organization-strong Energy Action Coalition (EAC), speaks from experience when she asserts that this generation is doing something about climate change right now. Tolkan and a handful of other young leaders were the forces behind November 2007's Power Shift conference, which brought together over 6,000 students and youth from across the country and around the world to discuss solutions to what this generation views as the most critical issue of our time. The event culminated in a record-breaking descent on Capitol Hill, with 2,000 young people participating in the single largest lobby day on global warming, including testimony for the Select Committee on Energy Independence and Global Warming. By the end of the day, the students had collectively visited over 300 representatives and nearly all 100 senators. Power Shift 2009 was even larger.

Tolkan says that the youth involved in Power Shift, the Campus Climate Challenge (the EAC's main campaign, which empowers students to demand clean energy and carbon neutrality on 700 campuses across North America), and similar movements are demonstrating that a large, loud, and active movement to demand bold solutions to climate change is well underway and making real strides. Built upon the concept that campuses can be models for a clean energy future, the Campus Climate Challenge has helped presidents from over 600 college campuses in the United States make a signed commitment to carbon neutrality, and hundreds more claim climate victories, including increased wind and solar power, entire transportation systems switched to run on biodiesel, and investment priorities shifted to companies committed to sustainability. The scale of these promises is impressive: The entire University of California system has committed to climate neutrality, with the governor of California further committing four institutions to be run on 100% green energy in the next 5 years. In addition, 12 University of Tennessee campuses are going carbon neutral using a student-innovated system of distributing the cost among all students via "green fees," which has, in turn, inspired the Tennessee state legislature to pursue a similar measure.

Along with the Campus Climate Challenge, the EAC and its member organizations are working with young people to guarantee climate change solutions such as the creation of 5 million new green-collar jobs—technical jobs in emerging markets such as solar panel installation and green construction that will support our economy while also creating pathways out of poverty.

The EAC calls for a moratorium on the construction of all new coal-fired power plants as well as dramatic increases in the use of wind and solar energy. The EAC is leading the call among America's young people for bold, decisive climate action. "We don't just want an 80% reduction in carbon emissions by 2050," says Tolkan. "We want a 30% reduction by 2020—we need to get this show on the road!"

Young people are uniting around the issue of climate change and making their voices heard. In the 2 years since the start of the Campus Climate Challenge, EAC as reached out to 3 million young people, and that number is growing constantly. Young voters are coming to the polls in record numbers, and they are unafraid to demand the actions actually necessary to mitigate climate change. Perhaps most importantly, the young people speaking up on climate change represent a broad array of backgrounds, cultures, and races. The EAC seeks to engage all people, young and old, rich and poor, students, veterans, parents, and so on, believing that diversity is a core virtue of a successful movement. Thus far, it seems they are spot on.

Coalitions, like the one Ms. Tolkan has built, seem to be central to the environmental and sustainability movements of the future that these forward-thinking presenters have offered a glimmer of. If their intentions hold true, we can expect from the next wave of leadership a divergence from the 20th century ideals of competition and specialization and a shift toward collaboration and holistic thinking. By changing the rules of the game, forward-thinking leaders of all generations are designing new ways that we all can win.

The EAC is a founder and supporter of 1Sky, an organization with one goal: bold federal action by 2010 that can reverse global warming.

We will close this chapter with the coda from 1Sky:

> The 1Sky Solutions are grounded in scientific necessity—they are the bottom line of what's needed to dramatically reduce carbon emissions while maximizing energy efficiency, renewable energy and breakthrough technologies. They also represent significant economic promise. By pivoting to a clean energy economy, we can relieve our dependence on foreign oil, unlock the potential of sustainable industry and usher in a new era of prosperity and green jobs. . . . American citizens are building support for the 1Sky Solutions in key Congressional districts on a nonpartisan basis, using cutting-edge communications, Internet and old-fashioned neighbor-to-neighbor outreach. [1]

In sum, most observers expect a vigorous debate in the Congress and the nation as a whole over putting a price on carbon, such as by establishing a federal cap and trade emission policy. [2] By combining the mounting climate action at the state, local, college, and youth levels with nationwide acceptance of the need for climate protection, it may only be a matter of *when*—and not *if*—the United States rejoins the rest of the world in active work on climate protection policies.

CONNECT THE DOTS

- Cities have been in the forefront of enacting climate protection policy in the United States and elsewhere.

- States and regional collaborations among states working on greenhouse gas abatement now span the continent.

- Federal proposals to date have lagged behind those of the subnational units of American government.

- The younger generation of Americans is powerfully energized and organized to push for immediate, aggressive climate protection policy.

Online Resources

www.eoearth.org/article/Climate_leadership_in_northeast_North_America
www.eoearth.org/article/Climate_politics_in_Mexico_in_a_North_American_perspective
www.eoearth.org/article/Communicating_climate_change_motivating_citizen_action
www.eoearth.org/article/Second_generation_climate_policies_in_the_United_States
www.eoearth.org/by/topic/environmental%20policy
1Sky, www.1sky.org
Arizona Climate Change Action Group, www.azclimatechange.us
Center for Climate Strategies, www.climatestrategies.us
Consumer Energy Center Glossary, www.consumerenergycenter.org/glossary
DoRight Enterprises, www.scottbeall.com/doright-summary.htm
Energy Action Coalition, energyactioncoalition.org
Envirolution, www.envirolution.org
Focus the Nation 2008, www.focusthenation.org
Footprint Network, www.footprintnetwork.org
ICLEI—Local Governments for Sustainability, www.iclei.org
Midwest Ag Energy Network, www.midwestagenergy.net
National Teach-In, www.nationalteachin.org
Natural Step, www.naturalstep.org
New Economics Foundation, www.neweconomics.org
New Mexico Climate Change Advisory Group, www.nmclimatechange.us
North Central Bio-Economy Consortium, www.ncbioconsortium.org
Renewable and Appropriate Energy Laboratory, rael.berkeley.edu
Salt Lake City Green, www.slcgreen.com
US Climate Network, usclimatenetwork.org
US EPA Clean Energy Programs, www.epa.gov/cleanenergy/energy-programs/
US Mayors Conference Climate Protection, www.usmayors.org/climateprotection
US Partnership for Education for Sustainable Development, www.uspartnership.org
See also extra content for Chapter 12 online at http://ncseonline.org/climatesolutions

Climate Solution Actions

Action 1: Green Buildings and Building Design
Action 16: Urban Responses to Climate Change in Coastal Cities
Action 31: Communicating Information for Decision Makers—Climate Change at the Regional Scale

Works Cited and Consulted

[1] 1Sky.org (2008) About 1Sky. *1Sky.org* (read October 10, 2008). www.1sky.org

[2] Batten K, Goldstein B, Hendricks B (2008) Investing in a Green Economy. *Center for American Progress* (read October 10, 2008). www.american progress.org

[3] Beall S (2008) DoRight Enterprises: A Youth-Run Consulting Firm. Beacon, NY. www.doright enterprises.org

[4] CCS (2008) Brief Description of State Climate Actions: Residential, Commercial and Industrial Sectors. *The Center for Climate Strategies* (read November 7, 2008). www.climatestrategies.us

[5] CCS (2008) Center for Climate Strategies Publications. *The Center for Climate Strategies* (read September 5, 2008). www.climatestrategies.us

[6] Cohen D (2008) US Partnership for Education for Sustainable Development (read October 10, 2008). www.uspartnership.org

[7] Colburn K (2008) Legislative Agenda for Addressing the Carbon Problem. *National Conference on Science, Policy, and the Environment: Climate Science and Solutions.* http://ncseonline. org/2008conference

[8] Energy Design Update (2004) *An Illinois "Passivhaus."* Energy Design Update 24(5):1–5. www .aspenpublishers.com

[9] Goodstein E (2007) *Fighting for Love in the Century of Extinction: How Passion and Politics Can Stop Global Warming.* (University of Vermont/University Press of New England, Lebanon, NH). www .upne.com

[10] ICLEI (2008) ICLEI: Local Governments for Sustainability (read October 10, 2008). www.iclei .org

[11] Inslee J (2007) The New Apollo Act (H.R.2809). *Thomas.* http://thomas.loc.gov/cgi-bin/bdquery/ z?d110:h.r.02809:

[12] James S, Lahti T (2004) *The Natural Step for Communities: How Cities and Towns Can Change to Sustainable Practices.* (New Society Publishers, Gabriola Island, BC, Canada). www.naturalstep .org

[13] Lash J (2008) Mayors "Get It" on Climate Change. *WRI* (read October 1, 2008). www.wri.org/ stories/2008/09/mayors-get-it-climate-change

[14] McKinsey (2007) Reducing US Greenhouse Gas Emissions: How Much at What Cost? www.mc kinsey.com/clientservice/ccsi/pdf/Greenhouse_ Gas_Emissions_Executive_Summary.pdf

[15] Midwest Governors Association (2008) Status Report on Midwestern Energy and Climate Accords (read November 7, 2008). www.mid westerngovernors.org/EnergyInitiatives.htm

[16] NCSE (2008) Climate Science and Solutions. *Eighth National Conference on Science, Policy, and the Environment.* www.http://ncseonline.org/ climatesolutions

[17] Northrop M, Sassoon D, Colburn K (2008) *US policy: governors on the march.* Environmental Finance, June 2008. www.environmental-finance .com

[18] NREL (2008) Zero Energy Buildings. *National Renewable Energy Laboratory.* www.nrel.gov/ buildings/zero_energy.html

[19] Passey B (2008) Conservative Bent of Utah Favors GOP. *Gannett News Service.* September 21, 2008. www.gannettnewsservice.com/?p=2619

[20] Peterson T, McKinstry R (2008) Integrating State and Federal Action in National Climate Policy: A Case for Partnership. *The Center for Climate Strategies (CCS).* www.climatestrategies.us/ Publications.cfm

[21] Radloff D (2008) Engaging State and Local Government: Developing and Implementing Climate Action Plans. *National Conference on Science, Policy, and the Environment: Climate Science and Solutions.* http://ncseonline.org/2008conference

[22] Renewable and Appropriate Energy Laboratory (2008) Dan Kammen, Congressional Testimony: "Ten Year Outlook for Energy," Feb. 28, 2007, and "Opportunities for Greenhouse Gas Emissions Reductions," Nov. 8, 2007. (read October 1, 2008). http://rael.berkeley.edu

[23] Rosenthal E (2008) "Houses with No Furnace but Plenty of Heat." *New York Times*, December 27, 2008. http://www.nytimes.com/2008/12/27/world/ europe/27house.html?emc=eta1

[24] Salt Lake City (2008) Salt Lake City Green (read November 6, 2008). www.slcgreen.com

[25] Shulz L (2008) Legislative Agenda for Addressing the Carbon Problem: How to Avoid Dangerous Climate Change. *National Conference on Science, Policy, and the Environment: Climate Science and Solutions.* http://ncseonline.org/2008conference

[26] US Climate Action Network (2008) About USCAN. *US Climate Action Network* (read October 15, 2008). http://usclimatenetwork.org

[27] USCOM (2007) Climate Protection Strategies and Best Practices Guide: 2007 Mayors Climate Protection Summit Edition. *US Conference of Mayors* (read October 1, 2008). www.usmayors.org/ climateprotection

[28] Utah Vital Signs (2007) The Ecological Footprint of Utah. www.utahpop.org/vitalsigns

[29] Wackernagel M (2008) Ecological Footprint. *Global Footprint Network* (read August 20, 2008). www .footprintstandards.org

CHAPTER 13

Where Science, Policy, and Public Meet

If there's no action before 2012, that's too late. What we do in the next two to three years will determine our future. This is the defining moment. [18]

DR. RAJENDRA PACHAURI,
Intergovernmental Panel on Climate Change, 2007

In order to address the problem of climate change, three things have to happen. First, scientists must conduct research to understand how the change is occurring and how it can be halted or slowed. Second, the general public must believe that climate disruption is real and that collectively urgent and long-term action is necessary. And third, government officials and others who reflect a desire to mitigate and adapt to climate change must enact sensible policies.

In 2006, Manfred Milinski and colleagues at the Max Planck Institute of Limnology in Plön, Germany, conducted an experiment to test whether people would act to lessen the impact of climate change if their actions benefited someone else. In public goods experiments, participants assume roles in which they benefit from a public good, for example, abundant ocean fisheries, at some relatively low cost to themselves. Most such experiments find that cooperation to avoid depleting the public good erodes as individuals realize that they can

extract more benefit for themselves at very little extra cost, such as spending a few more days fishing, even when the net effect of all players' doing so is that the resource collapses. Milinski and his team tested whether "reliable information on prospects of the global climate can be an incentive for humans to invest private money in sustaining the climate." Milinski and his team wondered whether people would act altruistically, if the direct benefits of their actions did not flow back to them personally but rather to society as a whole. In the game they conducted with undergraduate students, the contents of the public pool were not redistributed among the participating players. Instead, any gains were transferred to a "climate account" after the players had made their contributions and the total amount had been doubled. This condition increased the public-goods-group size to all humans that profit from a potential improvement of the climate. The players were told the "climate account" would be used to fund newspaper advertisements about climate action. The

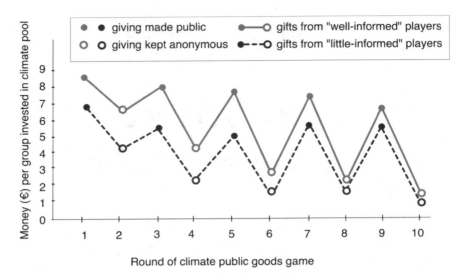

FIGURE 13.1 Climate public goods game

The Max Planck Institute's research suggests strongly that contributions toward climate awareness campaigns that are made publicly are higher than those made anonymously and that the givers who are better informed consistently make higher contributions than the less-informed givers. In this figure, the amount of money gifted by well-informed participants is shown as the solid line. The amount of money gifted by less-well-informed participants is shown as a dotted line. In the experiment's odd-numbered rounds of giving, the giving of each participant, whether well or little informed, was made public. In the even-numbered rounds, all participants' gifts were kept anonymous. Source: [8]

more money in the account, the more advertisements would be published using the scientific information from the Max Planck Institute. The results of this research indicated that players can behave altruistically to maintain the Earth's climate, given the right set of circumstances. Personal investments in climate protection increase substantially when the players can invest publicly, thereby gaining social recognition, thus reinforcing their altruism, as we can see in Figure 13.1. In short, Milinski writes, "Our finding that people reward contributions to sustaining the climate of others is a surprising result. There are obvious ways these unexpected findings can be applied on a large scale." [8]

In a word, under the right approach, the public will support climate protection if they feel the information is reliable and if they feel socially rewarded for acting. In practice, we must under-

stand and improve the use of scientific information in decision making. Decision makers are increasingly aware of scientific conclusions regarding climate change. However, persistent gaps remain between basic awareness and deliberative consideration of climate change in their decisions and policymaking, from the local up to the national and international scales. This chapter will examine the interface between those conducting climate science and those charged with making decisions informed by science. We will also explore how media professionals are communicating climate change to the public.

Do Policymakers Think about Climate Change?

Yes, sometimes policymakers think about climate. In the prior chapter we saw how states and cities have been the policy leaders on climate protection in the United States. In 2007, University of North Dakota researcher Rebecca Romsdahl asked decision makers in that state whether they were considering climate change in making official decisions regarding natural resource and public health management. Romsdahl found there is very little planning taking place in North Dakota to address potential impacts from climate change at the regional and local government levels across the state; 97% of respondents indicated that no plans were being made or they did not know if planning was taking place. Half of respondents said that they see the following as significant barriers to taking action: a lack of public awareness or interest in the issue, monetary constraints, and insufficient staff resources to analyze and assess relevant information. Nearly 30% of respondents also indicated they believe the science is too uncertain, they feel there are not clear decision options, and they lack a legal mandate to address climate change. [14]

In her survey Romsdahl also asked decision makers what would help them take climate change into consideration. Roughly 70% responded that hands-on training and regular conferences would be beneficial, while roughly 60% desired easy-to-understand Internet-based resources. An initial conclusion, according to Romsdahl, is that collaboration between scientists and decision makers can be improved. Just as distributed agricultural extension offices helped improve farming practices in the past century, the placement of extension-like agents for climate mitigation and adaptation in local institutions might raise awareness among local decision makers and assist with their decision support needs. [14]

Researcher Stacy Rosenberg and her collaborators at the Institute for Science, Technology, and Public Policy at Texas A&M University found significant differences between how climate scientists and local and regional decision makers view and use climate science. [17] Rosenberg's team conducted two surveys that each included hundreds of respondents from

across the United States. One survey examined policymakers. A separate survey examined attitudes among climate scientists.

At least four important issues affect whether policymakers incorporate climate science into decisions. First, is the timescale of the science relevant to the policy being decided? Second, is the information on a topic relevant to the policymaker? Third, is the information presented in an understandable and concise manner? Fourth, are policymakers receptive to the science? [6]

Rosenberg's first survey of local and regional decision makers included government officials and interest group representatives in targeted subgroups (e.g., public health, economic development, planning). On initial questions, the majority of government officials and interest groups agreed that climate change was occurring, that climate change was accelerated by human activities, and that the respondents felt they had a good grasp on the science of climate change. Fewer respondents claimed that their organizations considered mitigation and adaptation frequently, though most did to some extent. Climate scientists believed that their work was most relevant to policymakers in land use and agriculture. [17]

Rosenberg and her colleagues also asked the policymaking respondents about how they obtained information on climate science and what further information they desired. Most decision makers trusted journal articles and university scientists to give them accurate information, although about 40% believed the science could be biased. Decision makers reported that they desired strategies to lessen the impact of climate change on humans specifically at a local level. They also wanted reliable predictions, historic data, details about current impacts on a local and regional scale; and information that was concise and understandable. When decision makers were asked about their use of climate science to evaluate policy, 35% said they made modest to frequent use of climate science findings, 35% made limited use, and 30% reported they never made use of science. [17]

But what about attitudes among the scientists? Rosenberg and her colleagues also surveyed US climate scientists who had been authors of certain articles in the peer-reviewed literature between 1995 and 2004. Scientists responded with a stronger degree of agreement on most survey questions than the policymakers or interest group members. The research found that when climate scientists were asked what the media and decision makers should know, however, scientists reported a slightly different set of priorities. Scientists felt policymakers needed an increased understanding of uncertainties, knowledge of future impacts and consequences on local and regional levels, knowledge about paleoclimatic and historic data, a basic understanding of scientific principles, and a grasp of mitigation measures. The similarity in the communication priorities between policymakers and climate scientists bodes well for strengthening the relationship between the two groups. And, importantly, each group felt that its own work was relevant to the other. [16]

Survey results found climate scientists believe that climate change will occur with a combination of gradual changes in all areas of the world and the possibility of abrupt changes in some areas of the globe. All of this underscores the need to include science in the policies that address climate. The vast majority of scientists agree with the findings in the most recent Intergovernmental Panel on Climate Change (IPCC) report that climate change is occurring and that humans contribute to it, and they agree that the public does not understand these findings very well. Climate scientists strongly support a diversity of policy initiatives to reduce greenhouse gases, such as (1) the use of market incentives to reduce emissions, (2) taxing industry and individuals to discourage emissions, (3) promoting public education about climate change, (4) setting higher prices on energy and consumer goods that are not environmentally friendly, and (5) ratification of the Kyoto Protocol, to name a few. [16] With this kind of consensus, the scientific community can play a strong and important role in advising policymakers on ways to implement adaptation and mitigation strategies for climate change.

Know What Your "Ask" Is

How can science help inform policy? As a former science advisor to Congress, Kit Batten notes, "Sometimes the role of science can be negative because it can be misinterpreted or ignored by policymakers." Batten worked with Senator Joseph Lieberman to investigate allegations of climate science censorship at several government agencies, including the National Aeronautics and Space Administration, National Oceanic and Atmospheric Administration (NOAA), Environmental Protection Agency, and US Forest Service.

To keep science from being misinterpreted or ignored, Batten encouraged scientists to reach out to policymakers and to prepare carefully for meetings with them. University government relations offices can be convenient places for university scientists to start to become involved in policy. These offices can help scientists get in touch with local or higher-level politicians and can also provide advice and media training for scientists not familiar with these types of communications. The American Association for the Advancement of Science (AAAS) also offers good basic outreach tools, including short video presentations suitable for showing the public or policymakers.*

Currently, as director for environmental policy at the Center for American Progress, Batten has encouraged scientists "to know what your 'ask' is." In other words, have a clear, definable goal for the meeting. "It might be to present them with a result that you'd like to see a specific policy recommendation made from," she said. In addition, it is important to pay attention to bills or executive orders under consideration in government, to see if your issue could be tied into a

---

*See the AAAS website: http://communicatingscience .aaas.org

TABLE 13.1 California's Desired Regulatory Process under AB 32

Level of government	Existing process	Desired new process
State	AB 32	AB 32 Scoping plan
Regional	(None)	Regional goals Local and regional action plans
Local	New construction projects	New construction projects Existing building projects

Source: [15]

relevant piece of policy. Also pay attention to the time of year. On a national level, for instance, the peak federal budget appropriation season, from January through April, is an extremely busy time of year for Congress. A brief meeting might be forgotten if it is not immediately relevant to the budget cycle. Finally, Batten advises, "Keep your message simple, short, and to the point." [3]

Science policy and legal considerations are coming together to drive land use decision making in California. In Chapter 12, we mentioned California's innovative new climate change legislation, AB 32, which puts regulatory and market mechanisms in place to reduce greenhouse gases. AB 32 became law through the collaboration of state policy officials, scientists, environmental advocates, and stakeholders. A group of academics and sector leaders expert on the issues, California's Climate Action Team, also informed officials along the way toward crafting a law. At this point, state agencies also play a role as consumers of the information put forward by scientists and with responsibility for the implementation of AB 32. The lead state agency is the California Air Resources Board. But the innovation comes from the top-to-bottom integration of goals from the state house to the town house.

According to Chris Pyke, director of climate change services for Constructive Technologies Group, Inc., California Attorney General Jerry Brown has created a link between AB 32—a high-level state policy—and local planning and projects to ensure that these local plans are consistent with the state's land use goals. Specifi-cally, in California the attorney general requires (1) consideration of the impacts of the project on emissions and the impacts of changing conditions on the project, (2) quantification and disclosure of the emissions, (3) consideration of the alternatives, and (4) demonstration of consistency with state emissions reduction goals. [12] (See Table 13.1 for the desired regulatory process.) Naturally, imposing a great deal more public scrutiny of land use development will by itself be a significant public education process.

As Pyke's colleague Heather Rosenberg summarizes, the consequences of California's new standards mean buildings and land use projects need to: (1) contribute solutions to greenhouse gas emission reductions and climate vulnerability, (2) provide quantitative analysis of baseline and "alternative" designs, and (3) demonstrate both design and operational emissions reductions (by reduced emissions in 2020). [15] The expanded role for local governments is both central to the process and where a great deal of the public's interaction with the reducing greenhouse gas emissions will occur. The role of local government includes the following five elements: (1) develop local climate action plans, (2) update general and specific plans, (3) update zoning and ordinances, (4) review environmental documents, and most importantly, (5) develop an implementation process. [15] (See Table 13.1.)

As an example of more enlightened land use under AB 32, Pyke describes the plans for a community in San Diego, called Merriam

TABLE 13.2 California's Implicit Regulatory Process under AB 32

Implicit land use development requirements in California under recent statutes
1. Assess the implications of the project for climate change and the consequences of climate change for the project .
2. Quantify and disclose greenhouse gas emissions of the project.
3. Demonstrate a break from the business-as-usual land use development pattern.

Source: [15]

Mountains. The initial design for the project was utterly typical in its number of homes, area for retail, and amount of open space allotted. Pyke and his company worked with the project leader of the planned community to conduct a greenhouse gas inventory of the project and then to institute measures to decrease the environmental impact. The resulting plan for 2,700 homes on 2,327 acres will leave 74% of the land undeveloped and permanently preserved as open space. Approximately 1,192 acres of the open space will be maintained in its current natural state. By implementing seven measures, Merriam Mountains will decrease the future greenhouse gas emissions impact by 32%. The measures include project design features such as (1) residential energy upgrades that are 20% better than the state's Title 24 energy code, (2) offset construction-phase emissions with 5% additional energy efficiency, (3) on-site electricity generation with 25% of dwelling units to have rooftop solar panels, and (4) water use efficiency that represents a 50% reduction in overall projected water use over business-as-usual design. Whether these design changes are universally praised or not, the resulting emissions scenario is far better than if the environmental impact under AB 32 had not been considered. In other words, buildings and land use projects need to demonstrate emissions reductions in both their design and operation. [12, 15]

Planning can only take one so far. The process must keep repeating itself so we can continually develop better policies. Roger Pulwarty, the director of NOAA's National Integrated Drought Information System emphasizes the need to reconsider and reevaluate policies and scientific strategies as time goes by, because climate scenarios are likely to change over time. "We have to ask, how do we learn from event to event? How is learning evolving over time?" Although scientific consensus on issues is important to have, he cited a colleague of his in saying, "We can have consensus while the resource degrades." From Pulwarty's experience with drought in the western United States, he notes that a "focusing event" almost always precedes successful agreement over adapting to and managing a climate event. These key events could include the occurrence of a severe drought, a storm such as Hurricane Katrina, the new listing of a species under the Endangered Species Act, or the publication of an IPCC report. To react well to such events, agencies need to support each other across jurisdictional lines and embrace a strong collaboration between research and management. It is not enough to have this collaboration after an event; it must be in place before the event in order to work effectively, "or we're fighting the last war each time." Finally, Pulwarty underlines the need to account for uncertainty in climate variability and impacts by leaving a little bit of slack in predictions and adaptations. [11]

Let's Make "Glocal" News

Most of the public does not hear about climate events directly from government agencies. Local television is a very important source of information for the public. Local newscasters and weather forecasters are often well known and generally trusted in their communities. Climate coverage has been a challenge for the broadcast media for several reasons. First, the topic, climate, itself is often not an established category, unlike politics or business or sports. Second, climate is often slow-moving compared with the crises and catastrophes that are news each morning. Third, it may be beyond the grasp of the local news broadcasters to make the accurate connections between local and global events. Finally, the underlying physical and biological mechanisms can be complex scientifically, so simple metaphors are often lacking or not readily employed as they could be. Let's take a moment to look at these challenges as opportunities.

According to Joe Witte, meteorologist for WJLA-TV in the Washington, DC, broadcast area, television news departments "need help from scientists finding climate change stories." Efforts like the AAAS "glocal" strategy have been very helpful in promoting local engagement in global science. A glocal strategy seeks to promote local engagement with global science and technology-related issues. Glocal means providing a local context and relevance for global conditions.

Specifically, AAAS Science Insights and News Service staff have been successful by publishing regional op-ed articles, providing local experts and speakers, and working collaboratively with colleagues throughout AAAS. [1] This collaboration often partners AAAS member scientists with local opinion leaders, policymakers, school board members, clergy, the public, and the news media. The AAAS has been able to deploy this strategy in a broad range of topics, from climate change (e.g., "Time to Get Serious about Climate Change" on the *San Francisco Chronicle* editorial page, 2006) to evolution (e.g., "Teaching Evolution and Creationism" on the Diane Rehm Show, National Public Radio, WAMU FM, 2008) to biomedical research (e.g., "Standing in the Way of Stem Cell Research" on the *Washington Post* editorial page, 2007) and general science awareness (e.g., "Among Science-Debate Questions Put to Candidate" in the *St. Louis Post-Dispatch* editorial page, 2008). [2]

Of course, the interest for residents anywhere about what is happening or what has happened in their own community is an age-old curiosity, and one that can be abetted with new digital technologies. Between Google Earth users, who share photographs or stories by tagging them to specific locations, and the expanding group of geocache hobbyists, who use handheld global position satellite calculators, it becomes increasingly possible to portray change in the landscape—even subtle change—to many more fellow citizens. Locative journalism, or "lojo," is not specifically devoted to climate disruption news. It is one step beyond the "broadcast" and combines location-based technology and journalism. As NewsLab's Deborah Potter writes, "If you have a GPS-equipped cell phone, for example, your location could automatically trigger news and information developed specifically for that place." It is easy to imagine observed natural patterns or energy use patterns that could be reported at a neighborhood or even street-intersection level via Web-enabled phones and social-networking sites. [10]

National Public Radio (NPR) had planned a major series on climate change a few years ago. When Hurricane Katrina hit in 2005, the planned coverage was put on the back burner. Eventually, though, NPR returned to the idea of a climate project that would try to connect all the separate content areas. NPR science correspondent David Malakoff recalls how NPR began by looking at "how we change climate and how climate changes us, beyond the dying polar bear stories." On May 1, 2007, NPR and *National Geographic* started a partnership, Climate Connections, that sought to publish news stories on "how we are shaping climate and how climate is shaping us." In the first 7 months alone, 175 Climate Connections stories aired on NPR news programs. Every month, reporters traveled to a different part of the world to cover the story from a different human angle. [9]

Doyle Rice, weather editor at *USA Today*, explains that *USA Today* mainly covers climate as it relates to weather and thus only peripherally covers climate change—and then, only "when it's in the news." *USA Today* has only done a couple of specific packages on climate

INSIGHT 13: TAKE YOUR PICTURES DIRECTLY TO THE PUBLIC!

From the dying daily newspapers to the shrinking science desks at the broadcast networks, the traditional news outlets that science and public policy organizations have relied on are withering right before our very eyes. With YouTube, Facebook, and Twitter expanding daily, the public gathers its information in ways that are rapidly evolving. "That long and great press release that worked for you 10 years ago is no longer enough," comments Karl Leif Bates, director of research communications at Duke University. "The good news is you are now a publisher. Write and produce for your audience, for direct consumption." Tom Kennedy, managing editor for multimedia at Washingtonpost.Newsweek Interactive elaborates on the role of multimedia in online stories: "Seeing content visually allows me to understand it more readily. If the video is really well crafted, people are going to bond to that story emotionally, and they'll make an effort to really process it and internalize it and make the understandings that emerge from it more a part of their life."*

Scientists and those who prepare public information about science need to learn how to create bite-size doses of information that can be served up directly to the public via the Internet. The American Association for the Advancement of Science has been expanding its global news service, EurekAlert!, to accommodate short video clips and other multimedia in order to allow these to be viewed by the public. EurekAlert! is a central place through which universities, medical centers, journals, government agencies, corporations, and other organizations engaged in research can bring their news to more than 6,600 reporters worldwide who use the service and to the public. EurekAlert! features news and resources focused on all areas of science, medicine, and technology. "In a 90-second video, you can see the researcher and hear the passion in her voice, and see her in her setting in a way that I probably couldn't capture in a thousand words of prose," concurs Bates. "Journalism 101 is 'show, don't tell.' Web video does that to a really great degree." [4,7] Much as political campaigns now rely on social media tools, so too campaigns for disseminating scientific and policy information will increasingly incorporate the publishing tools of the Internet that greatly facilitate fact sharing.

*For more from the NCSE conference participants, see their complete talks at http://ncseonline.org/2008 conference/.

---

change. As of early 2008, climate change did not fit cleanly into the paper's four main categories—news, money, life, and sports. In 2008, *USA Today* employed only one full-time science writer. However, climate articles online do well, when such stories are run. Rice stated that the topic lends itself more to the Web, with its ability to house interactive features, than to printed newspapers. In a January 2007 poll *USA Today* ran online, people were asked What's your big-gest fear about global warming? The most common response: "I'm not worried about it." But more rigorous and more recent polling, like that examined in a prior chapter, indicates that 8 out of 10 Americans believe climate change is a serious problem.†

---

†For more, see www.usatoday.com/weather/resources/ ask-the-weather-guys.htm or Rice's complete talk at http://ncseonline.org/climatesolutions.

Why We Need "Cathedral Thinking"

Stephen Schneider, a professor at Stanford University, described a session at the 2008 National Council for Science and the Environment climate change conference itself as a metaphor for the problem facing climate change: "They changed the room, and they didn't tell us. And there were all of these people wandering aimlessly in the halls for a while. Slowly, eventually we trickled in. We delayed for a while, but pretty soon we all got here, and now we're rolling." And "metaphor, of course, is the whole point," he states. In this climate change trial, the judges and juries are the members of Congress and the public. The lawyers are the media. The case is being told in short snippets in newspapers and on television. This "sound bite" method is a very poor way to communicate something complex.

In sum, Professor Schneider finds that metaphors that convey both urgency and uncertainty are best—particularly for controversial cases like climate change. For example, says Schneider, "I often say climate is like a die: it has some hot faces, some wet faces, some dry faces, etc. I think our [inaction] on global warming is loading the climatic dice for more heat and intense drought and flood faces." Schneider shows that metaphors need not be exact but, rather, readily understood. For example, Schneider might ask an audience, If we put a pan full of water in the sun and another in the shade, which will evaporate first? The water pans metaphor is somewhat imprecise in its characterization of the effects of global warming on the hydrological cycle, but for Schneider, "it drives the point home well enough, and I can live with that metaphor for mass consumption and supplement that with longer articles and books for those who really want to know more about the real physical processes." [19]

We will close this chapter with a metaphor offered by a businessman for the needed approach to addressing climate disruption. Rather than worry about metaphors of the science itself, James Rogers, chief executive officer of the large utility Duke Energy, offers the following thought:

We need cathedral thinking in this country. It took 104 years to build Notre Dame. The architect never saw its completion. Those that worked on the stone foundations never really saw the stained glass windows. Those that worked on the walls never really saw the roof, because building a cathedral took three generations of work at that time. Why were they able to do that? Because they had a vision. They saw the possibilities, they had faith in their vision. They had confidence in what they hoped to achieve. We need that same kind of vision. We need that same kind of commitment, the same confidence, because the cathedral that we are building is the cathedral of tomorrow, a planet where we solved this problem. We can do it. We will do it with possibilities and with confidence. [13]

CONNECT THE DOTS

- Governmental decision makers desire strategies to lessen the impact of climate change on humans specifically at a local level.
- In working with policymakers, scientists should have a clear, definable goal for the interaction.
- Building and land use planning are activities in which good opportunities exist to have scientific knowledge help guide public policy decisions.
- A "glocal" strategy can promote local engagement with global science and the related technology by emphasizing local context and relevance for larger-scale events or research.
- Using metaphors for the complex aspects of climate science can be helpful for bringing the public up to pace.
- The science and policy communities will increasingly use social media tools to disseminate information directly to the general public.

Online Resources

American Association for the Advancement of Science, www.aaas.org/news/press_room/climate_change/

AAAS Communicating Science, http://communicating science.aaas.org

AAAS EurekAlert!, www.eurekalert.org

Center for American Progress, www.american progress.org

California Air Resources Board, www.arb.ca.gov

Climate Central, http://climatecenter.org

Coalition on the Public Understanding of Science, www.copusproject.org

Communication Partnership for Science and the Sea, http://compassonline.org

Dot Earth (Andrew Revkin), http://dotearth.blogs .nytimes.com

Earth Portal (NCSE), www.earthportal.org

E-print Network, www.osti.gov/eprints/

Global Warming Art, www.globalwarmingart.com

Internet Scout Project, http://scout.wisc.edu/

News Lab, http://newslab.org

North American Environmental Atlas, www.cec.org/ naatlas/

Online Access to Research in the Environment, www .oaresciences.org/en/

Open Education Resource (OER) Commons, www .oercommons.org

Real Climate, http://realclimate.org

World Changing, www.worldchanging.com

www.eoearth.org/article/Communicating_climate_ change_motivating_citizen_action

www.eoearth.org/article/Global_land_use_models

www.eoearth.org/article/Landscape_ecology%3A_ Its_role_as_the_scientific_underpinning_of_ land-use_planning

www.eoearth.org/article/Land-use_and_land-cover_ change

www.eoearth.org/by/topic/environmental%20policy

See also extra content for Chapter 13 online at http:// ncseonline.org/climatesolutions

Climate Solution Actions

Action 30: Should There Be a National Climate Service? If So, What Should It Do and Where Would It Be?

Action 31: Communicating Information for Decision Makers—Climate Change at the Regional Level

Works Cited and Consulted

[1]   AAAS (2008) Center for Public Engagement with Science and Technology: Science Insights and News Service. *American Association for the Advancement of Science* (read October 15, 2008). www.aaas.org/programs/centers/pe/news_svc/

[2]   AAAS (2008) Global Climate-Change Resources. *American Association for the Advancement of Science* (read October 4, 2008). www.aaas.org/news/ press_room/climate_change/

[3]   Batten K, Goldstein B, Hendricks B (2008) Investing in a Green Economy. *Center for American Progress* (read October 10, 2008). www.american progress.org

[4]   Eurakalert! (2008) An online, global news service operated by the American Association for the Advancement of Science, Washington, DC. www .eurekalert.org

[5]   Grantham Prize (2008) The Climate Policy Puzzle: Piecing Together the Solutions. *2008 Grantham Prize Seminar on the State of Environmental Journalism.* http://dl2.newmediamill.net/media/ metcalf/flash/080908d/080908d.html

[6]   Jones SA, Fischoff B, Lach D (1999) *Evaluating the science-policy interface for climate change research.* Climatic Change 43(3):581–599. www.springer-link.com/index/NG60835839642511.pdf

[7]   Lohwater T (2008) Great Expectations for Science Multimedia as Print News Shrinks. *American Association for the Advancement of Science* (read December 26, 2008). www.aaas.org/news/ releases/2008/1224pio_seminar.shtml

[8]   Milinski M, Semmann D, Krambeck HJ, Marotzke J (2006) *Stabilizing the Earth's climate is not a losing game: supporting evidence from public goods experiments.* Proceedings of the National Academy of Sciences 103(11):3994–3998. www.pnas.org/cgi/ content/abstract/103/11/3994

[9]   NPR (2007) Climate Connections. www.npr.org/ templates/story/story.php?storyId=9657621

[10] Potter D (2008) "Locative" News. *Advancing the Story: Broadcast Journalism in a Multimedia World.* May 16, 2008. http://advancingthestory.wordpress. com/2008/05/16/locative-news/

[11] Pulwarty R (2008) Symposia: Climate Scientists and Decision-makers: The Communication Interface. *National Conference on Science, Policy, and the Environment: Climate Science and Solutions.* http://ncseonline.org/2008conference/

[12] Pyke C (2008) Green Buildings & Climate Change: Best Practices to Demonstrated Performance. *CTG Energetics* (read September 30, 2008). http:// ctgenergetics.com

[13] Rogers J (2008) Keynote: Climate Change: Science to Solutions—The Case for Business Leadership. *National Conference on Science, Policy, and the Environment: Climate Science and Solutions.* http:// ncseonline.org/2008conference

[14] Romsdahl R (2008) *Addressing institutional challenges to adaptation planning for climate change impacts on the northern Great Plains: a case study of North Dakota.* Interdisciplinary Environmental Review (in press). http://www.ieaonline.org/ier.htm

[15] Rosenberg H (2008) Reducing GHG Emissions: Report from Projects in California. *Urban Land Institute 2008 Fall Meeting and Urban Land Expo.* http://ctgenergetics.com

[16] Rosenberg S, Cowman D, Vedlitz A, Zahran S (2007) *Climate Change: A Profile of US Climate Scientists' Perspectives.* Institute for Science, Technology, and Public Policy Working Paper. http://bush.tamu.edu/istpp/scholarship/

[17] Rosenberg S, Vedlitz A (2007) *Climate Scientists and Decision-Makers: Exploring the Communication Interface.* Institute for Science, Technology, and Public Policy Working Paper. http://bush.tamu.edu/istpp/scholarship/

[18] Rosenthal E (2007) "UN Report Describes Risks of Inaction on Climate Change." *New York Times,* November 17, 2007. www.nytimes.com/2007/11/17/science/earth/17climate.html

[19] Schneider SH (2005) Mediarology (read October 1, 2008). http://stephenschneider.stanford.edu/Mediarology/MediarologyFrameset.html

CHAPTER 14

Scaling Up Amidst the Curse of Knowledge

Each country, each state, each community may decide to deploy or not deploy certain technologies, but as technologists, our message has to be that we need to be working on everything. [9]

STEVEN SPECKER,
president, Electric Power Research Institute, 2008

On March 12, 2008, the price of a barrel of light crude oil exceeded $110 for the first time in history. The next day, more than 800 people gathered in Washington for the National Academies Summit on America's Energy Future. Global energy use has doubled over the last 30 years and could likely double again in the next 30 years. Most of this increase will come from developing nations, including China and India. But most of the current atmospheric emissions come from the developed world, led for decades now by the energy-intensive and emission-heavy United States.

As US Senator Jeff Bingaman of New Mexico chair of the Senate Energy Committee said at the summit, "Energy policy does not have a single goal. It is extremely complex and multifaceted. . . . We run a real risk of heading in the wrong direction in energy policy if we try to oversimplify the issues, if we try to overstate the potential of any single energy initiative, or if we try to understate the difficult nature of the energy problems that we face. [9]

Psychologists use the term *curse of knowledge* to describe a situation in which a group of experts in a field all think they know the answer to a problem and these answers are accepted as conventional wisdom and turn out to be wrong. Andy Grove, past chairman of the board of Intel and no stranger to technological challenges, has summarized this problem as follows, "when everybody knows that something is so, nobody knows nothing." When everybody knows something, that fact alone is insufficient for taking action. Such is the case with developing solutions to climate disruption. To create collaboration that enables learning and discovery, we need to leave conventional wisdom behind. Only then will true technological and policy innovation begin.

Overcoming the Curse of Knowledge

Lewis Milford, president of the Clean Energy Group, spends a lot of time thinking about how we will arrive at solutions for climate disrup-

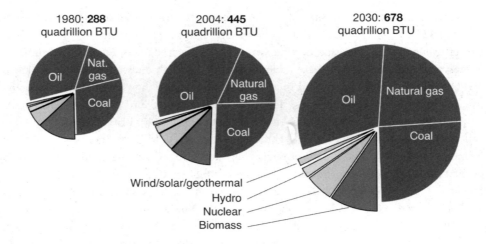

1980: **288**
quadrillion BTU

2004: **445**
quadrillion BTU

2030: **678**
quadrillion BTU

Wind/solar/geothermal
Hydro
Nuclear
Biomass

FIGURE 14.1 Global energy demand by 2030

The use of energy worldwide is projected to increase 235% between 1980 and 2030
to almost 680 quadrillion Btus, or 680 quads. A Btu is a British thermal unit and the
energy equivalent of 0.293 watt-hours. The US economy consumed 99.75 quads in
2005, just under a quarter of the world's entire energy consumption. Note that by
2030 natural gas and coal use will pick up the biggest share of the increase among
the fossil fuels, based on these IEA projections from 2006. Source: [9: p. 33]

tion. The scientific community has collaboratively demonstrated the problem, he states. It
was Nobel prize–winning material. However,
the scientific community has not dedicated anywhere near the same time and resources to the
solutions side of things. There has been research
into modeling climate scenarios, as well as the
pros and cons to various technologies. But no
coherent path to stabilization has yet emerged.

The scale of the challenge is enormous. To
stabilize atmospheric levels at 450 ppm—far
higher than today's 380 ppm—we will need to
create twice as much net "carbon-free" energy by
2050 as all the energy consumed today throughout the world. [11] Put another way, the increase
in worldwide energy demand by 2030 will need to
be met almost entirely with "carbon-free" energy
sources. Yet the International Energy Agency's
most recent projections show that the slice of the
energy pie that comes from fossil fuels in 2030 is
likely to be barely different from its share in the
current decade (see Figure 14.1). [9]

To put the numbers in Figure 14.1 into perspective, we should ask, How much is 1 quadrillion Btus (1 quad)? In terms of electricity,
1 quad (10 to the 15th power) is equal to 293
gigawatt-hours. This amount is equal to the
energy obtained from burning 45 million tons
of coal, which would be a pile 3.3 meters (10 feet)
tall, 1.6 kilometers (1 mile) wide, and about 5.3
kilometers (3.3 miles) long. At 60 mph, it would
take about 9 minutes to drive around the pile.
[25] Coal is typically shipped to power plants
by freight train. This much coal would require
315,000 rail freight cars to deliver.

Two main goals arise in energy policy for the
United States: (1) Reduce emissions of greenhouse gases, especially carbon dioxide from
fossil fuel combustion, and (2) reduce consumption of oil, especially imported oil. [24] Given
these two straightforward goals, why have we
not made more progress? We currently have no
plan to get there and many powerful interests
with competing perspectives on these goals.

Decreased energy demand through a combination of increased conservation and increased energy efficiency could decrease dependence on foreign sources of energy, increase national security, reduce the trade deficit, and increase innovation for efficient technologies. Therefore, a rapid scaling up of efficiency and energy reduction technology and policy is the first order of business. How do we find a path of energy innovation and efficiency that can scale up from lab to global implementation quickly enough?

Neither governments nor nongovernmental groups show consensus on what structure is needed to address the scaling up of technology and the transfer to a carbon-free energy strategy. "Except for that," Milford admits, "we're in good shape!"*

The answer may lie in a series of "parallel regimes," as was suggested by Ambassador Richard Benedick, President of NCSE, in which many different approaches develop simultaneously, largely from the bottom up, rather than the more typical top-down approach. A way forward can be found, through distributed innovation, collaborations between fields of study, new technology policies, and new finance strategies. [3]

To bring the energy sector into the 21st century, we need to look at how other fields, such as medicine and computing, have reached large-scale change. The energy sector is highly segmented by source. Fossil fuel specialists rarely interact with nuclear or renewable energy specialists. The energy sector is also highly stratified by delivery channel and end user. Electricity grid specialists rarely interact with transportation specialists. Too often, those working in the energy sector only talk to others in their field and remain isolated from the innovations of other related sectors. This segmentation exists in the academic energy research fields, too. This isolation must end. If we do not get moving toward creating 20 terawatts of carbon-free energy within the next decade or so, we will have missed the trajectory needed to be on course for this unprecedented, but extraordinarily necessary, goal.[†] A terawatt is 10 to the 12th power watts, or one thousand billion watts. One terawatt delivered for 1 hour is the equivalent of 3.4 to the 15th power British thermal units (Btus).

To start, Milford argues government and international organizations need to create a technology track different from cap and trade and other market-based pricing incentives. Market-based mechanisms generally and by design create motivation and investment in the least-cost technology. Therefore, we see carbon credits going to the most efficient coal plants in Germany, but not to the least efficient plants elsewhere that need the technology the most, such as highly inefficient plants in China. For those facilities, we need to prioritize ways to reduce costs of more-expensive technologies, such as carbon capture and sequestration. The emission trajectory will overtake us if we do not.

> Cap and trade alone will not stabilize carbon emissions. A new complementary technology innovation track must be initiated now to serve as a twin pillar of the post-2012 climate framework. This track must include complementary policies, innovation strategies, and finance mechanisms that support the rapid development and deployment of low carbon technologies, all within new forms of global infrastructure. [16]
>
> *Lewis Milford*

Two pivotal questions lie immediately ahead for the international community in constructing a climate stabilization framework with a greater focus on technology innovation:

---

*For more from the NCSE conference participants, see their complete talks at http://ncseonline.org/climatesolutions.

[†]To create a more unified energy science and technology field, the National Council for Science and the Environment has formed the Council of Energy Research and Education Leaders (CEREL). CEREL provides a means for leaders in different fields of energy research and education to work together to solve the energy challenge.

(1) What technology-based policies can be adopted to drive massive technology innovation? (2) How can a new technology innovation approach be structured—what is the most effective international architecture to advance global climate technology innovation and involve other players from the private and public sectors and civil society? [16]

As the Kyoto Protocol and its related agreements are due for revision in 2012, preparing now for a new framework under the United Nations Framework Convention on Climate Change (UNFCCC) has a great deal of urgency. The world's nations meet in Copenhagen in December 2009 to hammer out the concrete details of new emissions standards and paths to reaching them. The Copenhagen agreements will form the basis for individual nations to ratify and adopt the new standards and policies in the 2 years before the Kyoto agreements expire in 2012.

Organizations like the Bill & Melinda Gates Foundation use a process of distributed innovation to accelerate technology development, identify barriers, and scale up. By allowing grassroots innovation to rise from the bottom up, researcher and Harvard Business School faculty member Karim Lakhani writes, "Distributed innovation systems are an approach to organizing for innovation that seems to meet the challenge of accessing knowledge that resides outside the boundaries of any one organization." [13]

Climate change technology could borrow acceleration methods from the distributed innovation schemes increasingly used by businesses to speed up time-sensitive product development. Other market and policy failure areas, such as agricultural productivity, HIV drugs, and vaccine development, use these open source strategies to spark innovation. We could ask whether the ultimate policy failure is thinking that a market by itself can solve a societal problem.

Global problems such as poverty and HIV are increasingly supported by very large multi-billion-dollar collaborative efforts that connect governments of many nations, major organizations and corporations, and other key players. There is nothing similar at work yet on the problem of climate change.

No single organization, public or private, can count among its staff all the expertise needed for such a large challenge as inventing and implementing 20 new terawatts of carbon-free energy. Yet, according to researchers Karim Lakhani and Jill Panetta, the development of open source systems such as Linux operating system, Apache Web server, and Firefox Web browser show that traditional "closed models of proprietary innovation will have difficulty completing knowledge intensive tasks when most of the needed knowledge resides outside the organization."

Lakhani and Panetta also write, "Many practitioners and scholars of innovation did not anticipate the emergence of a distributed and open model for innovation that can aggressively compete with additionally closed and proprietary models. That complex software systems running mission critical applications can be designed, developed, maintained, and improved for 'free' by a virtual 'community' of mostly volunteer computer programmers has come as a great surprise to them." [13]

Software system development may seem an unlikely potential innovation model for developing climate protection systems. But consider the parallels. The needed knowledge in any given energy technology application may reside well outside the confines of any one specialist or any single proprietary firm or laboratory. Indeed, such a highly interdisciplinary sharing process has been instrumental in allowing the Intergovernmental Panel on Climate Change to amass its assessment reports in the first place.

People and organizations do not contribute to open source projects unless they see a benefit in doing so. Open source collaboration has caught on because participants see a mutual benefit to contributing to a project and thereby influencing how certain aspects of the project will behave for them. An added incentive may well be the realization that a huge pool of knowledge workers who might contribute will be able to create a better outcome more useful to more people.

A new collaborative group for climate innovation could and should be created to connect all players, bringing in scientists and teams of experts from around the world, with the support of government, to develop key technologies and create initiatives to research the breakthrough technologies needed to bring energy into the 21st century. [17]

The energy sector is monopoly regulated, and thus it is the last place that catches innovation. As long as energy research and development continues to be dominated by monopolies, there will not be the needed innovation. Indeed, the US government has already been successful with highly collaborative, ground-breaking basic research. The Advanced Research Projects Agency within the Department of Defense (DARPA), established in 1958, conducts basic research at taxpayer expense that has led to the foundation for the Internet and global positioning systems. Congressman Bart Gordon of Tennessee, chair of the House of Representatives' Science Committee, has proposed we establish a similar agency for energy research, an ARPA-E: "We can't have incremental change. We need a major , out-of-the-box breakthrough." As Gordon explains it, the private sector alone is not up to the challenge. "It's also a unique opportunity to bring together the public sector, the private sector, industry, the national labs, and the universities. By doing that, not only do you make breakthroughs, but you already have the community involved, so they can take it to the next step, to market." DARPA currently has an annual budget of just over $3 billion. Gordon would like to see ARPA-E funding reach $1 billion within 2 years: "We can take the approximately $20 billion over ten years in tax breaks [for oil companies] and shift them into alternative-energy research. That way we're not adding to the deficit, but rather we're shifting the incentives." [27] Funding to initiate ARPA-E was provided in early 2009 as part of the Ameri-can Recovery and Reinvestment Act (ARRA) approved by the US Congress.

Technology policies must also embrace a collaborative approach. Discussion of such policies often centers around the voluntary sharing of information and ideas. But to create the right environment for global adoption of non–greenhouse gas–emitting technologies, policies must be collaborative, setting multinational targets, time lines, and goals. Regulation will be a necessary part of this. "We already know how to regulate the energy industry and the transportation industry," asserts Clean Energy Group's Milford, "but we just haven't done it. We need climate protection policies to move these industries along, which will create incentives for further innovation and investment. But first, we need to create the infrastructure." [17]

Regarding the potential gridlocking obstacle of selecting research priorities, when a new technology issue comes up, says Milford, there is always the question of whether we should try to pick winners. But, he says, "Technologies are not selected because they are efficient, they are efficient because they are selected." Government policy has driven space exploration technology, for example. It seems as though we have had a bit of memory lapse about how we got to where we are today. With the support of new finance tools and opportunities available with organizations such as the World Bank, it seems there is a path to get there. More good news is that models for successful international collaboration of governments, researchers, nonprofits, and businesses exist. For example, the Consultative Group on International Agricultural Research (CGIAR) is a strategic partnership established in 1971. CGIAR's members include 21 developing and 26 industrialized countries, four cosponsors, and 13 other international organizations. More than 8,000 CGIAR scientists and staff are active in over 100 countries throughout the world, bringing research from the laboratory to the farm and marketplace. "The new crop varieties, knowledge and other products resulting from the CGIAR's collaborative research are made widely available to individuals and organizations working for sustainable agricultural development throughout the world." [4]

Why Kyoto Failed

Over 190 nations participate in the UN's climate talks. Yet, 85% of emissions come from just 25 of the wealthiest nations. And, arguably, the nations most vulnerable are among the least developed. So who should be at the table as the key negotiators? Should the many climate change issues be addressed as an aggregate all at once, from biodiversity threats to disease prevention to cap and trade schemes? Or one by one? Should climate protections be addressed at the global stage or in smaller, regional forums? Are short-term or long-term targets more important in addressing the problem successfully? Which policy tools will be most effective, cost-efficient, and politically feasible? What is the role of technology in an international agreement?

Currently, the model for international negotiation involves one big table with 190 chairs. This one-table-many-chairs approach has been in place for all key meetings on the Kyoto Protocol to reduce greenhouse gas emissions. After a decade, the results of this method of hammering out international agreement are mixed, at best. This lack of success has led diplomats to consider whether more could be accomplished with a fewer-chairs-at-each-table approach. Even if we look only at the overlapping international affiliations of Europe and the Organization for Economic Cooperation and Development (OECD) nations, as shown in Figure 14.2, we can see that trying to include every nation, with their own often opposing interests, leads to a dizzying array of allegiances and internal conflicts.

The Group of Eight (G-8) is an international forum of industrialized nations including: Canada, France, Germany, Italy, Japan, Russia, the United Kingdom, and the United States. A major challenge for climate protection is bringing the major emitters from both developed nations (the G-8) and developing nations (such as India and China) to the same table. The Global Leadership for Climate Action (GLCA) is a task force of world leaders committed to addressing climate change through international negotiations. The GLCA framework is the result of cooperation among former heads of state and other leaders in government, business, and civil society and international governmental organizations from 20 countries (10 developed and 10 developing). The framework includes 11 recommendations on greenhouse gas mitigation, for example, requiring that developed countries reduce emissions by 30% collectively from 2008 to 2020 and that all countries reduce emissions by 60% from 2008 to 2050. An element unique to this framework is a proposed target for rapidly industrializing countries to reduce energy intensity by 30% from 2008 to 2020. The framework also incorporates avoided deforestation mechanisms and considers forest degradation issues, elements that the Bali Action Plan of December 2007 also mentions. [12] Dilip Ahuja of the National Institute of Advanced Studies of India asserts that a plan that integrates these land use elements with strong emission reduction targets in developing countries is the only approach that makes sense.

How would we pay for this plan? The GLCA framework calls for increasing global research and development spending of US$20 billion per year and accelerating technology deployment through a climate fund. International technology cooperation is key. Finance drives the GLCA framework and must include provisions for both mitigation and adaptation, requiring a proposed US$10 billion per year, accumulated through an auction of emissions allowances and doubling of official development assistance. Ahuja concludes that Kyoto was a first step, not a solution, "To reach a new global agreement, we must build trust between the global North and South and create new modalities for cooperation." (See Table 14.1 for proposed North-South collaborations.)

Economist Scott Barrett of The Johns

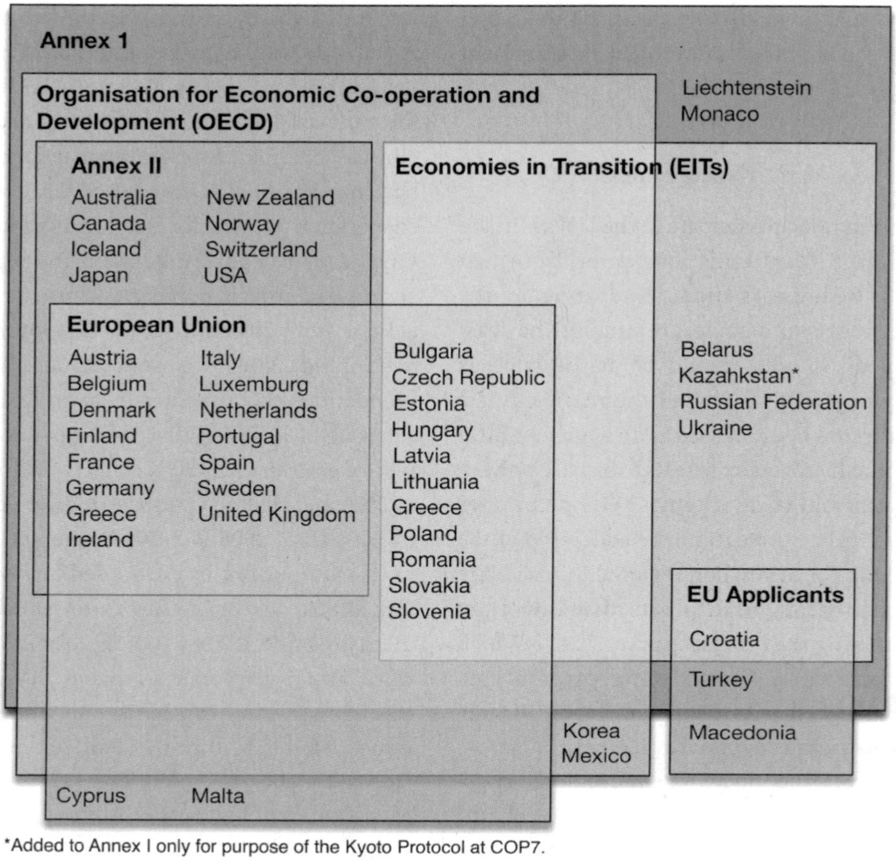

Annex 1

Organisation for Economic Co-operation and Development (OECD)

Liechtenstein
Monaco

Annex II

Australia	New Zealand
Canada	Norway
Iceland	Switzerland
Japan	USA

Economies in Transition (EITs)

European Union

Austria	Italy
Belgium	Luxemburg
Denmark	Netherlands
Finland	Portugal
France	Spain
Germany	Sweden
Greece	United Kingdom
Ireland	

Bulgaria
Czech Republic
Estonia
Hungary
Latvia
Lithuania
Greece
Poland
Romania
Slovakia
Slovenia

Belarus
Kazahkstan*
Russian Federation
Ukraine

EU Applicants

Croatia

Turkey

Korea
Mexico

Macedonia

Cyprus   Malta

*Added to Annex I only for purpose of the Kyoto Protocol at COP7.

FIGURE 14.2 Dizzying array of international organizations and national affiliations

This figure indicates the member nation overlap between the EU, OECD, and the UNFCCC. (UNFCCC Annex II indicates the developed nations, and UNFCCC Annex I indicates all those nations plus many of the economies in transition from the former Soviet sphere in Eastern Europe). Source: [10]

Hopkins University is known for his game-theoretical analysis of climate change treaties: "Global public goods include the prevention of nuclear proliferation, the suppression of killer pandemics, climate change mitigation, and fundamental scientific knowledge. Failure to supply these global public goods exposes the world to great dangers. Providing them expands human capabilities." [1: p. 1] Barrett finds that many proposals for an international climate change agreement make sense, but all choices have consequences. According to Barrett, the biggest challenge we face in formulating an international climate change agreement is getting industrialized countries to act and linking those actions.

First, climate change does not threaten the survival of the human species. . . . Even in a worse case scenario, however, global climate is not the equivalent of the Earth being hit by a mega-asteriod. Second, different countries will be affected in different ways by climate change. . . . Third, mitigating climate change on a significant scale will also have consequences . . . [such as] diverting resources from other good causes. . . . Finally, reducing the world's greenhouse gas

III: How We Work Together Now

TABLE 14.1 Stages of Technology Development and Proposed Funds for North–South Collaboration

Broad category	Stage	Where?	Financial incentives	Examples	Proposed fund and size
Invention	Research—pure or basic	Laboratory	Early stage—full-cost public funding	Second-generation biofuels, ultra super-critical coal plants, advanced solar thermal and photovoltaic energy, ocean energy, hydrogen, next-generation nuclear power	Consultative Group on Clean Energy Research (US$5 billion/year)
	Research—applied				
	Development				
Innovation	Pilot plants	Sheltered environments	Mostly public financing	Carbon capture and storage	Clean Energy technologies innovation (US$10 billion/year)
	Demonstration and first-of-a-kind commercial plants	Sheltered environments	Generally cost sharing		
Diffusion	Early deployment	Nascent or niche markets	Deployment incentives (guaranteed purchases, loan guarantees, tax credits, equity investments)	Hybrid vehicles, wind and other renewable energy sources, integrated gasification combined cycle, barrier removal for energy efficiency	Fund for the diffusion of clean energy technologies (US$20 billion/year)
	Dissemination (and scale-up)	"Tilted playing field" markets	Incremental cost financing, buy-down, learning by doing, concessional loans		
	Commercialization	Mature markets	No need for further incentives, technology performance standards	Natural gas combined cycle, large hydro plants, geothermal energy, super-critical coal plants	NA

Source: [12]

emissions depends on the aggregate effort of all countries. . . . It is really the combination of all the above four properties that makes climate change mitigation so hard to advance. [1: pp. 4–7]

The Kyoto Protocol is utterly unsatisfactory in this regard, particularly because the United States never agreed to be a ratified party to the agreement. By the time the Kyoto Protocol is renegotiated in 2012, it will have reduced greenhouse gas emissions very little. In fact, greenhouse gas emissions in many countries rose rapidly after Kyoto. The main reasons are incentives embedded in the Kyoto agreement that allow free riders to avoid making any real emission reductions.

The Montreal Protocol on Substances That Deplete the Ozone Layer is an international treaty that was first agreed to by just 24 nations in 1987. Subsequently ratified by over 180 governments, the Montreal Protocol is widely considered to be the most successful of the global environmental treaties. The Montreal Protocol to curb stratospheric ozone depletion has unintentionally done much more for climate change than Kyoto. Montreal achieved the elimination of chlorofluorocarbons, a powerful greenhouse gas, even though the goal of Montreal was to solve the ozone "hole" problem by protecting the ozone layer and not to address the climate problem by decreasing the accumulation of greenhouse gases in the atmosphere. Through Montreal, the international system showed it could act on big components of the climate puzzle by eliminating use of chlorofluorocarbons. The collective action that led to the effective Montreal accord was spurred by two stark new pieces of information. First, new science research showed the harmful effect of this human-made compound in the upper atmosphere, where it depleted the helpful ozone blanket. Second, new technology research showed that replacement compounds were readily available that were less harmful to the atmosphere and were economically competitive. The international community was ulti-

mately successful in its approach to defending the stratospheric ozone layer. This experience suggests several elements of the new diplomacy that is needed to address global ecological threats (see Table 14.2). As Ambassador Benedick, chief negotiator of the successful Montreal Protocol, writes, "there is no law that states that every aspect of complex scientific and environmental problems must be addressed by every nation at the same time and in the same forum, in an overheated atmosphere of public scrutiny." [2]

Yet, Benedick warns, "We took the wrong lessons from the successful Montreal Protocol ozone negotiations for the Kyoto Protocol." Reducing ozone-depleting chemicals was possible using a short-term timetable because limited numbers of facilities produced these harmful chlorofluorocarbons and immediately viable alternative chemicals could replace the harmful ones. The Montreal strategy of short-term timetables that worked for correcting ozone levels will not work for correcting greenhouse gas levels. Why not? So many of the likely greenhouse gas solutions are neither yet deployed in the market nor even developed beyond the research stage. Also, the scale of greenhouse gas producers is vastly larger and more diverse than the scale of CFC producers critical to ozone was. Yet the Kyoto Protocol was based on short-term strategies. Since the operational life of a coal-fired power plant may be 50 years, any short-term decision to build one with a certain emission output has a 50-year impact. Therefore, decisions based on short-term timetables lock in 50 years of plant infrastructure and undermine cash available for future long-term technology development.

Article 1 of the Kyoto Protocol states the following goals:

Implement . . . policies and measures . . . such as: (i) Enhancement of energy efficiency in relevant sectors of the national economy; (ii) Protection and enhancement of sinks and reservoirs of greenhouse gases not controlled by the Montreal Protocol . . . [and] promotion of sustainable forest management practices,

TABLE 14.2 Key Lessons from the Success of the Montreal Protocol

The Montreal Protocol was by no means inevitable. Knowledgeable observers had long believed it would be impossible to achieve. From its success, the following lessons are extracted.

1. Scientists must assume a critical new role in international negotiations.
2. Political leaders may need to act even while there are still scientific ambiguities, based on a responsible balancing of the risks and costs of delay.
3. A well-informed public opinion can generate pressure for action by hesitant politicians and private companies.
4. Strong leadership by major countries and/or institutions can be a significant force in mobilizing an international consensus.
5. A leading country or group of countries can take preemptive environmental protection measures, even in advance of a global agreement.
6. Both nongovernmental organizations (NGOs) and industry are major participants in the new diplomacy.
7. The effectiveness of a regulatory agreement is enhanced when it employs realistic market incentives to encourage technological innovation.
8. Economic and structural inequalities between North and South must be adequately reflected in an international regulatory regime.
9. The size and format of a negotiation may significantly influence the results.
10. Finally, the signing of a treaty is not necessarily the decisive event in a negotiation; the process before and after ratification is critical.

Source: [2]

TABLE 14.3 The Five Key Concepts of the Kyoto Protocol

Commitments	The Kyoto Protocol establishes commitments for the reduction of greenhouse gases that are legally binding for Annex I (industrialized) countries, as well as general commitments for all member countries.
Implementation	In order to meet the objectives of the Kyoto Protocol, Annex I countries are required to prepare policies and measures for the reduction of greenhouse gases in their respective countries. In addition, they are required to increase the absorption of these gases and utilize all mechanisms available, such as joint implementation, the Clean Development Mechanism, and emissions trading, in order to be rewarded with credits that would allow more greenhouse gas emissions at home.
Differentiation	The Kyoto Protocol sought to minimize the financial impacts on developing countries of lowering emissions, by establishing an adaptation fund for climate change.
Accountability	Accounting, reporting, and review standards would ensure the integrity of the Kyoto Protocol.
Compliance	The Kyoto Protocol established a Compliance Committee to enforce compliance with the commitments under the Protocol.

Source: [23]

afforestation and reforestation; (iii) Promotion of sustainable forms of agriculture in light of climate change considerations; (iv) Promotion, research, development and increased use of new and renewable forms of energy, of carbon dioxide sequestration technologies and of advanced and innovative environmentally sound technologies; . . . (v) Progressive reduction or phasing out of market imperfections, fiscal incentives, tax and duty exemptions and subsidies in all greenhouse gas emitting sectors that run counter to the objective of the Convention and apply market instruments; (vi) Encouragement of appropriate reforms in relevant sectors aimed at promoting policies and measures which

TABLE 14.4 Three Lessons from Kyoto

1. The first lesson is that a rigid system of targets and timetables for emissions reductions is difficult to negotiate.

    Developing countries have refused to participate in dividing up a fixed emissions budget. Disagreements occur on whether population rather than the historical emissions should be the basis of the Kyoto Protocol. Rigid targets and timetables force all players into a zero-sum game of winners and losers as countries must negotiate over shares of a fixed budget of future global emissions.

2. The second lesson is that it is difficult for countries to commit to specified emissions targets when the costs are large and uncertain.

    Countries facing potentially high costs, such as the United States, refused to ratify the Kyoto Protocol largely because the cost appeared too high, or countries simply failed to achieve their targets. Countries on track to meet their obligations were able to do so because of historical events largely unrelated to climate policy, such as German reunification, coal-mining reform in Britain, and the collapse of the Russian economy after independence was gained.

3. The third lesson is key: Even countries earnestly engaged may be unable to meet their targets because of unforeseen events.

    Two excellent examples are New Zealand and Canada. No one anticipated during the 1997 negotiations that a decade later New Zealand would be facing a dramatic rise in Asian demand for beef and dairy products that increased methane emissions in New Zealand so much that it has completely offset the earlier reductions there. No one predicted that Canada would find its tar sand deposits so valuable that extraction would be viable at the oil prices reached in 2006.

Source: Adapted from [15]

limit or reduce emissions of greenhouse gases not controlled by the Montreal Protocol; (vii) Measures to limit and/or reduce emissions of greenhouse gases not controlled by the Montreal Protocol in the transport sector; (viii) Limitation and/or reduction of methane through recovery and use in waste management, as well as in the production, transport and distribution of energy. [23]

In addition to a long-term strategy, we need three things for a successful international climate agreement. It should promote participation, promote compliance, and get countries to do something significant that they would not without an agreement. Though there would be different relative costs and benefits for individual countries, the benefits would be substantial for all countries together. Disaggregating the climate problem and addressing it by individual sectors (e.g., power generation, buildings, vehicles) may be one approach toward an agreement, but that technique alone is not sufficient. [3]

Although the [Kyoto] protocol has not been effective at reducing emissions, it has been very effective at demonstrating a few important lessons about the form future international climate agreements should take. [14: p. 1]

William McKibben and Peter Wilcoxen

We need the new technology to address climate change, which requires much more knowledge than exists today. Technology must be transformed worldwide. An agreement must include major R&D expenditures coupled with methods for diffusing technology. Barrett suggests an ambitious R&D target of $30 billion each year by the United States alone. [1] Others come to similar sums needed. For example, in 2008, Club of Madrid president Ricardo Lagos and UN Foundation president Timothy Wirth suggested as much as US$15–20 billion should be used for diffusion of clean energy technologies, by which they mean, the implementation of these technologies at wide scale (see Table 14.5) [11]. Additionally, adaptation plays a key role. Increased climate change will occur at some level no matter what actions people take. Developed countries are much better able to adapt than developing nations and so could leave develop-

TABLE 14.5 Potential Uses of Financial Resources for Global Climate Protection

Category	Range (US$ billion/year)
Avoided deforestation	5–10
Adaptation	10–15
Human and institutional capacity building	1–2
Collaborative research and development	4–5
Pilot and demonstration plants	5–10
Diffusion of clean energy technologies	15–20
Total	40–62

Source: [12]

ing nations high and dry (or rather low and wet), further increasing the global development gap. While reducing emissions seems to be the most plausible and lowest-cost mechanism to include in an international climate change agreement, if negotiations fail at this endeavor, incentives will be needed for other high-tech measures; however, these high-tech options will bring to light additional questions, such as Who decides the optimal temperature of Earth?

In late 2007, a significant international meeting took place in Bali, Indonesia. [21] Did this meeting make progress among the key nations on climate change? The jury is still out. Diplomats in Bali did agree that a long-term agreement should address the issue of a shared vision for long-term cooperative action, including a long-term global goal for emission reductions.§

The goals of the EU at the recent Conference of the Parties (COP) in Bali were to bring together all possible nations to determine post-Kyoto actions and avoid stretching negotiations out over a long time period. The United States went into Bali with objectives similar to those of the EU but also wanted to ensure that developing countries were fully included under the obligations of a future agreement. The resulting Bali Action Plan included a 2009 end date

for negotiations, the promise of US participation, and the beginnings of negotiations. The EU would have liked a US statement on potential long-term targets for 2050, but this did not occur. The EU suggested a long-term target of a 50% reduction in worldwide global emissions by 2050 with the goal of not exceeding a 2°C global temperature increase. The United States was not comfortable with the EU targets and raised the question, Why have a target that you might not meet? One important consideration is whether a target that can be met easily is really very demanding.

The EU's position is that the Kyoto Protocol was an essential first step to get action underway. It doesn't deny that the Montreal Protocol plays a significant role in reducing greenhouse gas emissions, but it states that even full implementation of Montreal will not solve the climate change problem. The EU sees Bali as the beginning of the international climate change agreement process, with many more meetings, conferences, negotiations, debates, and amendments to follow.

Richard Moss of the World Wildlife Fund (WWF), and previously of the US State Department, finds the most important objective is promoting early action. Moss brings a conservation perspective to the discussion. From this viewpoint, nations can adjust targets later in the process when science is more advanced and additional knowledge is available. Moss asks, "How many currently available solution tools, for example, renewable energy, does the US bring into international negotiations?"‡ This determination requires careful consideration of what would be most effective. Focusing on specific tools may define how the United States will participate in the international process more than it will define the US stance on the international policy itself. An international framework must take into account that a single tool is not appropriate for all countries that participate. It

§For more on Bali, see http://unfccc.int/meetings/cop_13/items/4049.php.

‡Source is unpublished symposia transcripts from http://ncseonline.org/climatesolutions.

TABLE 14.6 Snapshot of Post-Kyoto Emission Trends 1998–2008

Countries	Emissions trend 1998–2008	Positions 2008
United States	Emissions still rising	Most stringent proposed policies do not meet this target; reject target levels.
European Union	Emissions flat	Political proposals in key countries of similar stringency are already proposed/adopted; support the target.
Key developing countries	Emissions rising rapidly	Domestic policies do constrain growth; legal obligation to the target is not acceptable, but requirement to take action is acceptable.

Source: Adapted from [18]

follows that the most important element of an international agreement is comparing actions of different countries and verifying the successful completion of these actions; monitoring and enforcement are key.

The WWF is actively working on many potential solution tools, including ways to help communities avoid deforestation and degradation, enhance adaptation, achieve energy efficiency, and employ renewable energies. It sees these efforts being accomplished through building the capacity of local organizations to undertake changes in behavior. Regarding avoided deforestation, WWF is putting together a set of reports on impacts on places with unique collections of biodiversity that will be affected by climate change and must be preserved. Developed nations like the United States can help these valuable places by sharing the lessons of their own adaptation and demonstrating how to protect ecosystems and make them more resilient. Moss suggested, "We are not considering energy efficiency seriously enough; there is potential to build global trust between developed and developing nations on this topic. In terms of renewables, we need to push ahead while not exacerbating existing problems, like deforestation from bioenergy expansion." The major challenge with all of these tools is working through market and technological barriers in an aggressive way. Eventually we have to be willing to pay a price to solve this problem. The main idea should be to move beyond an exclusive focus on the target and think about what we must do to accelerate early action. [14]

How Are We Doing?

In 1997, the international community agreed to add the Kyoto Protocol to the UN Framework Convention on Climate Change. The Kyoto Protocol includes binding emission reduction targets for developed countries for the period 2008–2012. Jonathan Pershing of the World Resources Institute emphasizes the different positions of potential negotiating parties. For example, countries in the global North average four times the per capita emissions of countries in the global South. In terms of the Kyoto Protocol, Annex I (developed and industrialized) countries emit a much higher proportion of carbon dioxide over all greenhouse gases than non–Annex I (developing and industrializing) countries, which emit a higher proportion of methane because of their dependence on agriculture. These trends are summarized in Table 14.6.

As we noted in an earlier chapter, US energy intensity has headed in the wrong direction, increasing rather than decreasing in the most recent years. Proposed policies within the United States, such as the Lieberman-Warner act, failed to be adopted by 2009. Even if these had been adopted, these would not likely have reduced emissions. So the challenge remains to create political will in the United States for enacting targets and complying with those goals. The good news is that the United States is already part of a successful international effort, the Commission for Environmental Cooperation (CEC), created by Canada, Mexico, and the United States, as a complement to the environ-

TABLE 14.7 China's Goals from Bali, 2007

Energy efficiency: 20% reduction in energy per unit GDP by 2010

Renewable energy: 10% of total primary energy supply by 2010

Industry: stabilized nitrous oxide and increased use of coal bed methane recovery for methane reduction

Forestry: increased forest cover by 20% above 2005 levels by 2010 and sequestration of 50 million tons of carbon

Adaptation: forests, grassland, water use, coastal protection

Source: [21]

TABLE 14.8 Technology Wedges for US Emission Reduction

Today's technology	Action that provides 1 gigaton/year of greenhouse gas mitigation (1 stabilization wedge)	Major implementation issues
Coal plants	Replace 1,000 conventional 500 MW plants with "zero emission" power plants.	Technical, social, and economic viability
Geological sequestration of carbon dioxide	Install 3,500 Sleipners at 1 million metric tons of CO_2 per year.	Technical, social, and economic viability
Nuclear power	Build 500 1 GW plants.	Economics, safety, and nonproliferation
Energy efficiency	Deploy 1 billion cars at 40 mpg instead of 20 mpg.	Distributed opportunity that is hard to capture
Wind power	Install 750 times the current US wind generation capacity.	Geographic limitations, storage
Solar photovoltaics	Install 4,500 times the current US solar generation capacity.	Geographic limitations, storage

Source: [20] as adapted by [18]

mental provisions of the North American Free Trade Agreement (NAFTA) in 1994. To address public concerns about NAFTA's environmental impact, the Clinton administration created a side agreement to establish the CEC, whose specific role would be analyzing environmental impacts of liberalizing trade within North America. The CEC conducts ongoing environmental assessment of NAFTA trade in a host of impact areas, from forests to fisheries and from transborder hazardous waste shipments to water quality in shared water bodies. [7]

In the key developing countries, however, emissions are now rising rapidly. China is the single most significant developing nation whose emissions have grown rapidly. India also has a rapidly growing internal demand for automobiles and electrical power. These nations have policies in place to constrain emissions growth and are open to accepting a requirement to take further action on greenhouse gas reduction. See for example, China's goals in Table 14.7.

If Europe can reduce emissions while gross domestic product expands, what strategies could the United States adopt to accomplish similar reductions? The good news is that plenty of off-the-shelf technology exists to yield big greenhouse gas savings. Using Pacala and Socolow's stabilization wedges that we examined in Chapter 6, we can see at least 5 areas for immediate action, as set out in Table 14.8.

The Road Ahead

It can be hard to conceive of rational solutions when we face global-scale data (as we see in Table 15.1 in the next chapter). Ralph Cicerone, the

FIGURE 14.3 online at ncse.org/climate solutions

Chart shows trends and projections for greenhouse gas emissions for the EU, the US, and Japan.

INSIGHT 14: WHERE IS EMISSION REDUCTION SUCCEEDING?

In Europe, overall emissions have been curbed to a point where they are flat. The EU has committed itself to limiting the global average temperature increase to less than 2°C above preindustrial levels. As a unified group of 27 nations, the EU supports the Kyoto emission reduction targets. In 2007 the European Council endorsed the EU's independent commitment to reduce greenhouse gas emissions by 2020 to at least 20% less than 1990 levels—even if no international agreement is reached by the UN Framework Convention on Climate Change in Copenhagen in 2009. This commitment is to be honored until a new agreement is concluded, and without prejudice to its position in international negotiations. In other words, the EU members are adjusting key national-level policies in line with Kyoto emission reduction targets. Contrary to the prediction of doomsayers that emission reductions would reduce economic growth, in 2005 EU-15 emissions decreased by 0.8% compared with 2004, while the EU-15 economies grew by 1.6% (as measured by gross domestic product). In other words, economic growth occurred even as emissions were falling. Projections indicate that the EU is moving closer to achieving its Kyoto target, but additional initiatives need to be adopted and implemented swiftly to ensure success. [8]

president of the National Academy of Sciences, recommends that studies of climate change be scaled down to regional and local scales to create better understanding of the implications of climate change, to allow regions to better prepare for changes, and to create a framework for evaluating mitigation options. [6]

> There is no current belief that humans can control such [climate] changes once they are forced. [5]
>
> *Ralph Cicerone, Proceedings of the*
> *National Academy of Sciences, 2000*

Thomas Schelling, a Nobel laureate economist from the University of Maryland, makes the point that populations in the developing world will be affected the most by climate change. In 50 to 60 years, most of the world's population will be living in the nations we call the third world today. According to Schelling, the main problems of climate disruption for developing nations include decreasing agricul-

tural production, increasing vector-borne diseases, and decreasing production of forests and fisheries. For the foreseeable future, developing nations lack the resources to adapt as quickly as the disruption is happening. For these nations, Schelling notes, "Their best defense to climate change is their own development." While developed nations would prefer developing nations to use energy-efficient practices and renewable energy, we can not ask them to do this at the risk of their own development. [19]

Adrian Vazquez, executive director of the CEC, describes the primary goals of CEC as promoting conservation through cooperation, building the capacity to address environmental issues, and collecting information for policymakers. One of the accomplishments of the CEC is the *North American Power Plant Air Emissions* report. The report was the first look at power plant emissions in all three countries included under NAFTA. The process of creating the emissions report allowed the three governments to harmonize their data to create a useful document for decision makers. [7]

In the context of an international climate change agreement, Jonathan Pershing, director of the climate, energy, and pollution program at the World Resources Institute, recommends setting a target of a 25% to 40% reduction of carbon dioxide equivalent. But Pershing cautions that the notion of a single climate protection policy for all the nations on the globe is faulty, given the differences in demography and economics among nations. The solution will take a portfolio approach including market mechanisms, government regulation, technology research, development and deployment, public engagement, and behavioral adaptation. All of these elements should be designed to shift technology and behavior at an environmentally adequate rate while promoting sustainable development. The US Senate is considering several proposals that employ various combinations of these tools. Charting the projected effects of these proposals suggests that the United States may be able to nearly reach IPCC's suggested targets by 2050, but not by 2020. Financing these mechanisms cannot be overlooked. Approximately $100 billion per year may be required to address climate change in the United States alone; foreign direct investment should be an encouraged component of the financing of such efforts. [18]

As climate diplomats left the 2008 climate change talks in Bonn, Germany, Yvo de Boer, the UN's top climate change official, summed up a critical issue, financial engineering: "how to generate sufficient financial resources that will drive the technology into the market that allows developing countries to act, both to limit their emissions and to adapt to the impacts of climate change." [22]

In June 2008, the World Bank laid the groundwork for funding clean development initiatives after 2012 under a renegotiated Kyoto Protocol by approving Climate Investment Funds (CIFs). CIFs are a collaborative effort among the Multilateral Development Banks of the World Bank and individual countries to bridge the financing and learning gaps between now and a post-2012 global climate change agreement. CIFs are two distinct funds: the Clean Technology Fund and the Strategic Climate Fund. [26]

In December 2008, the Climate Change Conference parties met again in Poznań, Poland, for their 14th annual conference. The Poznań meetings, spread out over 12 days, provided the opportunity to draw together the advances made in 2008 and move from discussion to negotiation mode in 2009. In Poznań, the parties were expected to (1) agree on a plan of action and programs of work for the final year of negotiations on crucial issues relating to future commitments, actions, and cooperation; (2) make significant progress on issues required to enhance implementation of the convention and the Kyoto Protocol, including capacity building for developing countries, reducing emissions from deforestation, technology transfer, and adaptation; (3) strengthen commitment to the process and the agreed time line; and (4) advance a "shared vision" for a new climate change regime. Such an outcome at Poznań would build momentum toward an agreed upon outcome at Copenhagen in December 2009. In many ways, the final shared vision of a new climate regime will be the most interesting detail, including the decision on how many chairs should be placed at each table—to turn back to the metaphor used earlier in this chapter.

In late 2009, the parties to the UN Climate Change Conference will meet in Copenhagen with the goal of concluding two sets of critical negotiations long under discussion. The first set of agreements will outline an enhanced long-term response by the world's nations to climate change. The second set of agreements will seek to finalize commitments by the participants in the Kyoto Protocol for emission targets beyond 2012. "The world is expecting a Copenhagen deal to reach the goal set by science without harming the economy," Mr. de Boer said. "Parties will need to make real progress towards this goal." [22]

- We need energy technology revolution that brings breakthroughs of new technologies and new markets.

- Various policies and financing mechanisms can speed the energy-climate technology revolution.

- Achieving a global scale of adaptation and mitigation responses requires more flexible organizations and learning that can span borders and cross industry sectors.

- The Kyoto Protocol has been less effective in reducing global greenhouse gas emissions than the earlier Montreal Protocol was in reducing chlorofluorocarbons to protect the atmospheric ozone. The Kyoto Protocol negotiations involve over 180 nations. The original Montreal Protocol involved just 24 nations.

- Smaller-scale international environmental negotiation has produced good results, such as the EU's track record and commitment to greenhouse gas reductions and the CEC's *North American Power Plant Air Emissions* report, which standardized data from Canada, Mexico, and the United States.

- Should the 25 nations that account for 85% of emissions be the key negotiators for the Kyoto Protocol revisions? Would that leave out the nations that may be disproportionately negatively affected by climate disruption?

- The Kyoto Protocol is under negotiation for renewal with new standards and procedures taking effect after 2012.

Online Resources

www.eoearth.org/article/Global_Environmental_ Governance:_A_Reform_Agenda_%28e-book%29

www.eoearth.org/article/United_Nations_Framework_ Convention_on_Climate_Change_%28full_text%29

www.eoearth.org/article/ Convention_on_Biological_Diversity

www.eoearth.org/article/Lessons_from_the_ Montreal_Protocol

www.eoearth.org/article/Montreal_Protocol_on_ Substances_that_Deplete_the_Ozone_Layer

www.eoearth.org/article/Montreal_Protocol_in_ transition

www.eoearth.org/article/Kyoto_Protocol

Clean Energy Group, www.cleanegroup.org

Council of Energy Research and Education Leaders (CEREL), http://ncseonline.org/CEREL/

European Environment Agency, www.eea.europa.eu

Global Leadership for Climate Action, www.global climateaction.com

United Nations Framework Convention on Climate Change, http://unfccc.int

Commission for Environmental Cooperation, www .cec.org

World Bank, worldbank.org

G-8 2008 Summit Hokkaido, www.g8summit.go.jp/ eng/index.html

US Department of Energy calculator, www.eia.doe.gov/ kids/energyfacts/science/energy_calculator.html

See also extra content for Chapter 14 online at http:// ncseonline.org/climatesolutions

Climate Solution Actions

Action 22: Availability of Technology to Mitigate Climate Change

Action 33: Diverse Perspectives on Climate Change Education—Integrating Across Boundaries

Action 34: Building People's Capacities for Implementing Mitigation and Adaptation Actions

Works Cited and Consulted

[1] Barrett S (2007) *Why Cooperate? The Incentive to Supply Global Public Goods.* (Oxford University Press, Oxford). http://apps.sais-jhu.edu/faculty/ faculty_bio1.php?ID=2

[2] Benedick RE (2007) "Lessons from the Montreal Protocol," (topic ed) Cleveland CJ, in *Encyclopedia of Earth,* (ed) Cleveland CJ. (Environmental Information Coalition, National Council for Science and the Environment, Washington, DC). www.eoearth.org/article/ Lessons_from_the_Montreal_Protocol

[3] Benedick RE (2007) Perspectives: Avoiding Gridlock on Climate Change. *Issues in Science and Technology.* http://www.issues.org/23.2/p_ benedick.html

[4] CGIAR (2008) Consultative Group on International Agricultural Research (read October 8, 2008). www.cgiar.org

[5] Cicerone RJ (2000) *Human forcing of climate change: easing up on the gas pedal.* Proceedings of the National Academy of Sciences 97(19):10304– 10306. http://www.pnas.org/cgi/content/extract/ 97/19/10304

[6] Cicerone RJ (2006) *Finding Climate Change and Being Useful.* (National Council for Science and the

Environment, Washington, DC). http://ncseonline
.org/

[7] Commission for Environmental Cooperation (CEC) (2008) North American Mosaic. *Commission for Environmental Cooperation* (read November 12, 2008). www.cec.org

[8] EU Environment Agency (2008) Environmental Indicators. http://ec.europa.eu/environment/indicators/pdf/leaflet_env_indic_2008.pdf

[9] Fri RW (2008) *America's Energy Future: Summary of a Meeting*, in *The National Academies Summit on America's Energy Future*, eds Committee for The National Academies Summit on America's Energy Future, National Research Council. (The National Academies Press, Washington, DC). http://www.nap.edu/catalog/12450.html

[10] Gupta S, Tirpak DA, Burger N, Gupta J, Höhne N, Boncheva AI, Kanoan GM, Kolstad C, Kruger JA, Michaelowa A (2007) Policies, Instruments and Co-operative Arrangements (in *Climate Change 2007: Mitigation. Contribution of Working Group III to the Fourth Assessment Report of the Intergovernmental Panel on Climate Change*, eds Metz B, Davidson OR, Bosch PR, Dave R, Meyer LA) ar4-wg3-chapter13.pdf: www.ipcc.ch

[11] Hoffert MI, Caldeira K, Jain AK, Haites EF (1998) *Energy implications of future stabilization of atmospheric CO_2 content*. Nature 395:881–884. www.nature.com/nature/journal/v395/n6705/abs/395881a0.html

[12] Lagos R, Wirth TE, El-Ashry M (2008) Framework for a Post-2012 Agreement on Climate Change. *Global Leadership for Climate Action*. www.globalclimateaction.com

[13] Lakhani KR, Panetta JA (2007) *The principles of distributed innovation*. The Berkman Center for Internet and Society Research Paper No. 2007-7. http://ssrn.com/abstract=1021034

[14] Mallon K, Bourne G, Mott R (2007) in *Climate Solutions: WWF's Vision for 2050*. (World Wildlife Fund). www.panda.org/climate

[15] McKibben W, Wilcoxen PJ (2008) Building on Kyoto: Towards A Realistic Global Climate Agreement. *Brookings High Level Workshop on Climate Change*. www.brookings.edu/experts/mckibbinw.aspx

[16] Milford L (2007) Consultative Group on Climate Innovation: A Proposed Complementary Technology Track for the Post-2012 Period. *Road*

to Copenhagen 2009: Conference on Leadership, Sustainable Development and Climate Change.* www.cleanegroup.org

[17] Milford L, Dutcher D, Barker T (2008) Climate Choreography: How Distributed and Open Innovation Could Accelerate Technology Development and Deployment. www.cleanegroup.org

[18] Pershing J (2008) Thoughts on Bali (UNFCCC 13th Session; Kyoto Protocol Third Session). *National Conference on Science, Policy, and the Environment*. http://ncseonline.org/climatesolutions

[19] Schelling TC (2007) Developing Countries Will Suffer Most from Global Warming. *Resources for the Future*. www.rff.org/Publications/Resources/Documents/164/RFF-Resources-164_Thomas%20Schelling.pdf

[20] Socolow R, Pacala S (2004) *Stabilization wedges: solving the climate problem for the next 50 years with current technologies*. Science 304(5686):968–972. www.sciencemag.org

[21] UNFCCC (2007) The United Nations Climate Change Conference in Bali. COP 13. United Nations Climate Change Conference Bali, Indonesia *UNFCC* (read October 10, 2008). http://unfccc.int/meetings/cop_13/items/4049.php

[22] UNFCCC (2008) Bonn Climate Change Talks. *UNFCC* (read October 6, 2008). http://unfccc.int/meetings/sb28/items/4328.php

[23] UNFCCC (2008) Kyoto Protocol *UNFCC* (read August 24, 2008). http://unfccc.int/kyoto_protocol/items/2830.php

[24] Wilcoxen PJ (2008) Energy and Climate Policy for the Next Administration. Presentation to OASIS, October 27, 2008. *Maxwell*. http://wilcoxen.maxwell.insightworks.com/

[25] Wilcoxen PJ (2008) How Large Is a Quadrillion BTU? In Externalities and Public Goods: Facts about Energy. *Maxwell* (read November 11, 2008). http://wilcoxen.maxwell.insightworks.com/pages/index/

[26] World Bank (2008) World Bank Board Approves Climate Investment Funds: Targeting $5 Billion over Next Three Years to Support Developing Countries. *Press Release No:2009/001/SDN* (July 1, 2008). www.worldbank.org/cif

[27] Zakaria F (2008) "The Future of Energy: Best and the Brightest." *Newsweek*, October 13, 2008. http://www.fareedzakaria.com/articles/archive.html

CHAPTER 15

All of the Above!
Solutions in Perspective

Adapting to climate change and reducing vulnerability is best
done within the broader context of making development more
sustainable. [14]

MOHAN MUNASINGHE,
Vice-Chair, IPCC 2008

arbon dioxide molecules persist in our atmosphere for thousands of years. The atmospheric emissions driving global warming today are largely due to the activity of industrialized, wealthier nations from 1950 to today. In other words, the greenhouse gases we have emitted during the lifetime of the middle-aged decision makers alive today have already locked in about 2°C of warming. The climate disruption that our children and their children will suffer for decades to come will be determined by how much we continue to emit, precisely because carbon is so durable.

Before we examine climate solutions in the next, and final, chapter of this book, we will examine the interactions between climate disruption and the development of economies and civil societies all over the globe. The question is simple: How can the pattern of development around the world sustain opportunities for health, education, and economic security for all the Earth's residents? However, the answer is complicated. The fate of current and future generations depends on how well we answer the question today.

Billions and Billions:
A Tale of Two Nations

With a population of about 82 million, Germany is a heavily industrialized nation with an active political movement around issues of sustainability. As the 14th most populous country in the world, Germany has experienced no population growth for about a decade. This zero-growth pattern is also found elsewhere, especially in the newly democratic economies of Eastern Europe, where the average annual fertility has fallen below the replacement level. [10]

Germany's total emissions have actually fallen in recent years, as it makes significant changes in energy use and efficiency. The main sectoral contributions to greenhouse gas reductions between 2004 and 2005 in Germany came from public electricity and heat production, households and

FIGURE 15.1 online at ncse.org/climate solutions

Map amd population growth curves for developing versus developed countries

services, and road transport. Germany switched a significant amount of electricity generation from coal to natural gas. As a result, Germany is making progress toward meeting the lowered emissions it agreed to in the Kyoto Protocol. In truth, the European Union (EU) candidly admits that "most emission reductions (in the industrial sector) had already been achieved by 1993, mainly due to efficiency improvements and structural change in Germany after reunification, and to the relatively small economic growth in the EU-15." [6] However, Germany is certainly setting a good, if not perfect, example of emission reduction. Twelve EU nations—the United Kingdom, Sweden, Germany, the Netherlands, Portugal, France, Finland, Belgium, Ireland, Austria, Greece, and Luxembourg—project that they will meet their 2010 targets through a combination of current and future domestic policies and measures, as well as the use of carbon sinks and flexible Kyoto mechanisms.

By contrast, India is a rapidly industrializing and developing nation with a steep population growth. Currently, the nation is home to 1.15 billion people who have a rising life expectancy, currently at 68.6 years. Life expectancy in 1980 in India was 54 years, so this 14-year gain in less than three decades is a very positive indicator of improved quality of life. [4] Between 2000 and 2020, India's population is expected to grow by about 300 million. That is the equivalent of adding the entire population of the United States today in just 20 years. [12]

Unlike Germany, India as a developing nation was not bound to meeting lowered carbon emissions standards in the Kyoto agreement. India's per capita greenhouse gas emissions are far smaller than those of Germany. However, in the past few decades, India's huge population has gained access to motorized transport, and the country's electricity generation has accelerated to reach millions of new households.

According to the US Department of Energy's Carbon Dioxide Information Analysis Center at Oak Ridge National Laboratory, the source of emissions has shifted dramatically to developing countries like China and India. As a source of atmospheric carbon, India surpassed Germany in the past decade. By 2005, India's emissions were almost double those of Germany—1.8 times, or 170,000 metric tons, of additional carbon [3]

In terms of emission intensity, Germany's per capita emissions—while high relative to the rest of the world—had fallen by 2005 to the levels equal to those of 50 years earlier, just as the country struggled to rebuild its economy after the devastation of World War II. India's per capita emissions, while very low, have risen steadily each year for 50 years, largely due to India's steady increase in economic activity and prosperity (see Figure 15.2).

Even if industrialized and developed nations become more successful in stemming emissions growth—and the key nations have not been as successful at reduction as Germany—the future growth of emissions clearly hinges on how nations with not-yet-mature economies manage their growth. As this century progresses, the world will need to accommodate the growth in energy demand from 6 billion people today to 9 billion people within the lifetime of today's youth. Most of that demand will come from the developing world's nations, such as already well-populated India, China, Indonesia, and the nations in sub-Saharan Africa.

Yet, climate disruption is indiscriminate. In the Prologue, we learned how Bangladesh will be a huge loser when sea level rises. Many nations least responsible for the warming taking place now will be most negatively affected by the impacts of the climate disruptions. Those who are poor, elderly, young, or otherwise marginalized in the developing nations—or in impoverished pockets of industrialized nations—will bear the brunt of the negative impact of climate disruption, from food supply shocks to shortages in potable water and energy. Those who are not yet born will enter into a world compromised by the lifestyles of their ancestors.

For the first time in history, more than half [the world's] human population, 3.3 billion

15. All of the Above! Solutions in Perspective

231

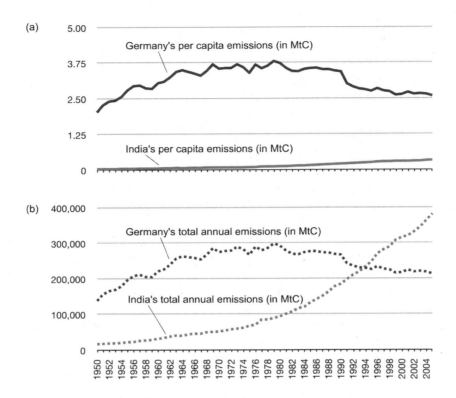

(a)

5.00

Germany's per capita emissions (in MtC)

3.75

2.50

1.25

India's per capita emissions (in MtC)

0

(b)

400,000

Germany's total annual emissions (in MtC)

300,000

200,000

India's total annual emissions (in MtC)

100,000

0

1950 1952 1954 1956 1958 1960 1962 1964 1966 1968 1970 1972 1974 1976 1978 1980 1982 1984 1986 1988 1990 1992 1994 1996 1998 2000 2002 2004

FIGURE 15.2 A tale of two nations' emissions: Germany and India, 1950–2005

Notice how closely the curve for per capita emissions for Germany (a) tracks with the curve for Germany's total emissions (b). By contrast, notice how a very small per capita rise in more-populous India (a) causes a steep rise in total emissions in India (b). In India, as energy demand rises with a growing population, emissions rise dramatically. (All emission estimates are expressed in thousand metric tons of carbon, MtC.) Source: [3]

people, will be living in urban areas. By 2030, this is expected to swell to almost 5 billion. Many of the new urbanites will be poor. [21]

United Nations Population Fund, 2007

Climate Change and Sustainable Development

What is the interaction between the developing world and climate change? As the poorer nations develop higher-technology economies, certain patterns in their emissions emerge. Methane, 20 times more powerful than carbon as a greenhouse gas, tends to be produced in higher volumes than carbon dioxide. Heavy methane emissions are largely due to agriculture. Livestock husbandry employs 1.3 billion people and creates livelihoods for 1 billion of the world's poor. [7] As a nation industrializes, changes in its land use, such as urbanization, lead to decreases in forest cover. This loss of green space lowers the land's natural capacity for storing carbon. Brazilian inroads in Amazonia are prime examples of deforestation. Diseases tend to spread under urbanizing conditions as well. As millions of poor flock to cities, they often reside in areas with little or no sanitary water supply or waste management. Warmer or wetter climate patterns also bring expansions of the ranges of insects and other animals that spread diseases.

III: How We Work Together Now

INSIGHT 15: TWO DEGREES OF SEPARATION

Our decision on how much warming constitutes an acceptable risk to human health is a political choice. The European Union has made such a political decision by choosing 2°C as the upper limit of tolerable global warming. To limit overall warming to 2°C will require a reduction in carbon dioxide emissions of between 50% and 85% of the 2000 emission levels. To reach that lower emissions trajectory, given the 50-year residence time of carbon once airborne, global emissions must peak no later than 2015. The laws of nature—not of human organization—govern the consequences of our emissions decisions.

This projection is based on science and is not disputed by those most knowledgeable about the data. In other words, we have less than a decade to get this right, not only in the developed world, such as in Germany, the United States, and Russia, but also in the developing world, such as in China, India, Indonesia, South Africa, and Latin America.

TABLE 15.1 Three Target Scenarios: IPCC Projections 2007

Scenarios (in increasing level of climate disruption)	CO_2 concentration at stabilization*	Year in which global emissions peak	Global average temperature above preequilibrium	Global average sea level rise above preindustrial at equilibrium[†]	Change in global CO_2 emissions in 2050 (% of 2000 emissions)
I	350–400 ppm	2000–2015	2.0–2.4 °C	0.4–1.4 meters	−85 to −50
III	440–485 ppm	2010–2030	2.8–3.2 °C	0.6–1.9 meters	−30 to +5
V	570–660 ppm	2050–2080	4.0–4.9 °C	0.8–2.9 meters	+25 to +85

* CO_2 concentration in 2005 was measured at 379 ppm.
[†] Anticipated changes resulting exclusively from thermal expansion.
Source: [2]

The Intergovernmental Panel on Climate Change (IPCC) has projected six different scenarios in its *Climate Change 2007 Synthesis Report.* We reproduce only three of these in Table 15.1 (scenarios I, III, and V). Long-term thermal expansion is projected to result in a sea level rise of 0.2 to 0.6 meter per 1°C of global average warming above preindustrial levels. The sea level rise shown in this table reflects only that caused by the expansion of warmer seawater and does not reflect the additional rise caused by melting polar caps or glaciers. So the sea level conditions projected here likely understate the actual rise.

Our success in reducing the ozone-depleting chlorofluorocarbon gases (CFCs) under the Montreal Protocol offers some hope that we can change course in time. In Table 6.2 we examined the relative global warming potentials of different greenhouse gases. If we can reduce emissions of methane and other gases that have a more powerful short-term warming effect than carbon dioxide, such action could provide some cushion to deal with CO_2. Table 15.1 sums up just three of the numerous scenarios the Intergovernmental Panel on Climate Change (IPCC) scientists have updated in *Climate Change 2007*: Scenario I, in which we succeed in capping emissions by 2015; Scenario III, in which we don't cap emissions until 2030; and Scenario V, in which we do not cap emissions until four to eight decades from now.

Scenario I is a virtual certainty, given the current level of emissions. Our fossil fuel and

15. All of the Above! Solutions in Perspective

233

land use appetite since World War II have locked in a warming of 2 degrees Celsius (°C). If we fail to reduce emission levels by 2015, then the planet's strict adherence to the laws of physics takes us automatically toward scenarios II and III. If we manage to reduce global emissions within the next 20 years (by 2030 or sooner) and we manage to keep emissions from rising above the levels of 2000, then we may see warming of around 3°C and a sea level rise approaching the range of 2 meters (m), or 6 feet (ft), from the thermal expansion of warm ocean water—as we examined in Chapter 4.

If we fail to alter our current emissions growth path by carrying on with business-as-usual behavior, then the world faces the likelihood of scenario V or VI, in which emissions may not be capped before 2050, and the warming impact may reach 5°C, and the sea level impact may approach 3 m (9 ft) and much, much more from melting of glaciers.

We will never have complete certainty in the precise levels of future warming due to specific increases or decreases of greenhouse gas (GHG) emissions. Yet, we have a very high level of confidence that business-as-usual emission growth will lead to scenario V (in Table 15.1), with 4°C or more of warming. And we have a very high degree of confidence that such warming will lower human life expectancy and erode human welfare as both direct and indirect consequences of global climate disruption.

Anthropogenic warming and sea level rise would continue for centuries even if GHG emissions were to be reduced sufficiently for GHG concentrations to stabilise, due to the time scales associated with climate processes and feedbacks. [2]

IPCC, 2007

Risk Equals Probability Times Consequence

So how much risk do we want to take? Risk equals probability times consequence. Not long ago, we faced the cold war standoff between the nuclear power of the Soviet Union and that of the United States. The probability of a nuclear holocaust, much deliberated, turned out to be relatively low as the years since 1960 elapsed. Yet the consequences of mutually assured destruction—preemptive missile launches by both sides against each other—were so unacceptable that the risk (or perceived risk) remained high for 30 years until the collapse of the Soviet Union into its successor states in the early 1990s.

We never have 100% certainty. If you wait until you have 100% certainty, something bad is going to happen on the battlefield. That's something we know. [5]

General Gordon R Sullivan,
Chief of Staff, US Army, 2007

Unlike the cold war's potential for nuclear war as an all-or-nothing event, climate change is already underway. Hence the probability for a relatively low but significant impact (at 2°C warming) is virtually certain. The probability for a higher impact is low now, but it increases more the longer we delay meaningful remedies. Additionally, as we have learned, climate consequences are not linear. Thresholds and tipping points can lead to even more catastrophic impacts. The consequences of climate change for humans—something we examined in earlier chapters—are dire. Hence the risk lurking behind climate disruption is huge. In Figure 15.3 we can compare the risk versus consequence of nuclear war to climate change disruption.

"Climate change can act as a threat multiplier for instability in some of the most volatile regions of the world, and it presents significant national security challenges for the United States. . . . The increasing risks from climate change should be addressed now because they will almost certainly get worse if we delay." [5] In these sobering terms, a blue-ribbon panel of retired admirals and generals from the Army, Navy, Air Force, and Marines introduces a new report. They composed the Military Advisory Board at the CNA Corporation, a nonprofit research group, and

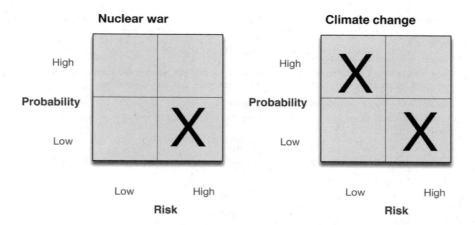

FIGURE 15.3 Risk equals probability times consequence

During the cold war, the United States and Soviet Union were locked in a three-decade game of brinkmanship with low probability of nuclear destruction that carried with it a high negative impact. Global climate change presents a new and very different type of national security challenge. The specter of global climate change comes with both high probability and high impact. Source: [5]

they studied how climate change could affect our nation's security over the next 30 to 40 years—the time frame for developing new military capabilities. Global climate change presents a serious national security threat that could impact Americans at home, impact US military operations, and heighten global tensions, according to the Military Advisory Board. Climate change, national security, and energy dependence are a related set of global challenges. In many locations today already, civil strife and warfare can be traced in part to environmental causes. Darfur, Ethiopia, Eritrea, Somalia, Angola, Nigeria, Cameroon, and Western Sahara all have been hit hard by tensions that drought, flood, famine, and disease have heightened. [5]

We can see some of the security risks from climate disruption in Table 15.2.

In addition, if we examine only weather-related disasters, the data point to both increasing frequency of disaster events and rising numbers of people who are victims of these events, as shown in Figure 15.4.

Sustainomics

Mohan Munasinghe of Sri Lanka is an optimist. He is vice chair of the IPCC and has been active for decades as a scholar and educator in environmental economics and sustainable development. Despite the daunting challenges, Dr. Munasinghe notes, "Although the problems are serious, an effective response can be mounted to make development more sustainable, provided it is initiated immediately." Munasinghe defines sustainable development as "a process for improving the range of opportunities that will enable individual human beings and communities to achieve their aspirations and full potential over a sustained period of time, while maintaining the resilience of economic, social and environmental systems." [14]

Can humans sustain a civilization for many centuries and make durable use of natural resources? The answer is Yes! The pharaonic system lasted over 4,000 years in the Nile River basin

FIGURE 15.4 online at ncse.org/climate solutions

Number of weather-related disasters, and number of victims, 1975–2007

15. All of the Above! Solutions in Perspective

235

TABLE 15.2 The Security Risks of Climate Disruption

Examples in Africa

- Unstable governments and terrorist havens

  When the conditions for failed states increase—as they most likely will over the coming decades—the chaos that results can be an incubator of civil strife, genocide, and the growth of terrorism.

- Less effective governance and more potential migrations

  In the past decade, severe food shortages affected 25 African countries and as many as 200 million people.

- Land loss and weather disasters

  The Niger delta is home to 20 million people who will be displaced as sea level rises.

- Escalating human health challenges

  Excessive flooding is conducive to the spread of cholera.

Examples in Asia

- Sea level rise

  The 130,000 miles of ocean coast in southern Asia (along the coasts of Pakistan, India, Sri Lanka, Bangladesh, and Burma) and in Southeast Asia (along the coasts between Thailand and Vietnam, including Indonesia and the Philippines) are especially vulnerable.

- Water scarcity combined with flooding

  Snow melting in the high Himalayas and increased precipitation across northern India are likely to produce flooding, while substantial declines in agricultural productivity will result from higher temperatures and more-variable rainfall patterns.

- Rising spread of infectious diseases

  Much of the Asia/Pacific region is exposed to malaria and dengue or has conditions suitable for their spread.

Examples in Europe

- Hotter temperatures and rising seas

  The heat wave of 2003 killed over 30,000 people, and acute water shortages are projected in the Mediterranean area, especially in the summer.

- Internal migration

  Some northern migration within Europe might be expected, as is already happening from Albania to Italy.

- Impact of climate change elsewhere

  Migration of people from across the Mediterranean from the Maghreb, the Middle East, and sub-Saharan Africa will increase.

Examples in the Middle East

- Water shortages

  Roughly two-thirds of the Arab world depends on sources outside their borders for water.

- Inflaming political instability

  Precipitation may decline by as much as 60% in some areas, which will cause high tension for the region's fragile governments and weak infrastructure.

Examples in the Western Hemisphere

- Increasing water scarcity and glacial melt

  The Peruvian plains, northeast Brazil, and Mexico, already subject to drought, will find that droughts in the future will last longer. In the United States, three of the top grain-producing states—Texas, Kansas, and Nebraska—each get 70% to 90% of their irrigation water from the Ogallala aquifer, which is under duress.

- Storms and sea level rise

  Warming seas and their link to storm energy are especially worrisome for Central American and small Caribbean island nations.

- Refugee migration

  An increased flow of migrants northward into the United States is likely, especially from the more impoverished nations of Latin America and the Caribbean.

Source: [5]

in Egypt with sustainable resource use and reasonable quality of life for vast numbers of inhabitants. In the Yellow River basin of China, the imperial system was stable for many millennia and supported a flourishing society. Similarly, in India, the Saraswati River region hosted a flourishing civilization for 4,000 years. Why did it end? The Saraswati eventually dried up as a result of tectonic activity (earthquakes), climate change, desertification, and water piracy. But today, we certainly do not have 4,000 years to work with!

Climate change undermines sustainable development and unfairly penalizes the poor, whether they live in developed or developing nations. Dr. Munasinghe points out that the problem of climate change for sustainable development can be approached from one of two perspectives: We can assume that the cost of mitigating climate change would be high and, thus, the cost-effectiveness of mitigating climate change is unknown. Or conversely, we can look at the level of mitigation that we can afford. As John Holdren has said, "The only choices are mitigation, adaptation, or suffering." [11]

In order to mitigate the effects of climate change, global carbon dioxide emissions would have to stabilize at below 500 ppm, a decrease that Munasinghe deems economically feasible. Many groups, including the IPCC, have estimated the net cost of this mitigation as an estimated 0.1% of global GDP annually. [8, 9]

Climate change would severely impact vulnerable populations, especially the poor. To describe the global distribution of income, we can picture the "champagne glass" metaphor. At the top of the champagne glass, 82.7% of the world's income is concentrated for use by 20% of the world's population. At the bottom, 1.4% of the world's wealth is shared by an equally numerous 20% of the population. Put another way, in at least 25 of the world's nations—including the 2 most populous, India and China—15 citizens or more out of 100 live on less than US$1 per day (see Figure 15.5). But the news is not all grim. A recent update from the UN reports that on the number-one goal of

eradicating extreme poverty, steady progress is underway. The 2007 *Millennium Development Goals Report* states, "In particular, impressive results have been achieved in sub-Saharan Africa in areas such as raising agricultural productivity (in Malawi, for example), boosting primary school enrollment (as in Ghana, Kenya, Uganda and the United Republic of Tanzania), controlling malaria (as in Niger, Togo, Zambia, Zanzibar), widening access to basic rural health services (Zambia), reforesting areas on a large scale (Niger), and increasing access to water and sanitation (Senegal and Uganda). These practical successes now need to be replicated and scaled-up." [20]

Munasinghe stressed the importance of development within the nations where much of the world's poor are concentrated in order to improve their ability to adapt to climate change, but he also stressed that these developing nations need to take an alternate route to development rather than the route that developed nations took. Sustainable economics, or "sustainomics"—as Munasinghe coins the phrase—aims for a more holistic and practical synthesis that would help to make development more sustainable. The core sustainomics framework draws on three basic principles: first, and most important, making development more sustainable; second, balancing the social, economic, and environmental dimensions of development; and third, ensuring that any discussions transcend traditional boundaries, across space and time and among academic disciplines or interest groups. In Munasinghe's view, sustainomics applies a full cycle of practical and innovative analytical tools. Sustainomics also seeks to balance people-oriented southern hemisphere priorities, including promotion of development, consumption and growth, poverty alleviation, and equity, with

FIGURE 15.5 online at ncse.org/climate solutions

Developing countries where people live on less than $1 a day

TABLE 15.3 Climate Change and Sustainable Development: Mutual Challenges

Sustainable development challenges due to climate change, especially in developing countries
• Alleviate poverty for about 1 billion people who live on less than $1 per day (including more than 40% of India) and 3 billion people who live on less than $2 per day.
• Provide adequate food, especially for the 800 million people who are malnourished today. This will require us to double food production in the next 35 years without causing further environmental degradation, such as deforestation.
• Provide clean water for the 1.3 billion people who live without clean water now, and provide sanitation for the 2 billion people who live without sanitation.
• Provide energy for the 2 billion people who live without electricity.
• Provide a healthy environment for the 1.4 billion people who are exposed to dangerous levels of outdoor pollution and the even larger number exposed to dangerous levels of indoor air pollution and vector-borne diseases.
• Provide safe shelter for those who live in areas susceptible to civil strife due to environmental degradation and for those vulnerable to natural disasters.

Source: [14]

environment-oriented northern concerns about issues like natural resource depletion, pollution, unsustainable growth, and population increase. [13] (See Figure 15.6 for a depiction of a model climate decision-making cycle.)

Munasinghe's essential point is that the poorer developing nations should invest in reducing their vulnerability by adapting to climate disruption, especially to safeguard the health and welfare of their impoverished populations. Meanwhile, the richer industrialized nations should lead the effort to reduce greenhouse gas emissions, while assisting the poorer nations. The richer nations have a great deal to gain from a more sustainable development path in the poorer nations, not the least of which is avoiding widespread conflict and migration. [15] (See Table 15.3.)

All of the Above!

The emissions data the IPCC assembled for the 2007 Assessment Report show that the devel-oped world ("Annex I nations" in IPCC parlance) represents about 1.5 billion people, or just under 20% of the world total, with an average per capita annual emission of a staggering 16 metric tons of carbon dioxide equivalents (tCO_2e). In other words, less than 20% of the world population in the richest nations produces almost half the total emissions (45.7%). The developing world ("non–Annex I nations" in IPCC parlance) represents about three times as many people (4.5 billion people) with an average per capita annual emission one-fourth that of the developed world (4 tCO_2e). Some portion of those emissions comes from producing goods or extracting natural resources in the developing world to sell to the developed world. (See Figure 15.7 for relative GDP and emission data by nation.)

Clearly, the biggest current mitigation efforts will be the responsibility of the emitters who have been at it the longest and who also have better access to resources necessary to decarbonize their economy, namely, the financial, educational, research, and material capital.

According to Munasinghe, there are four

FIGURE 15.6 online at ncse.org/climate solutions

Caricature shows the climate policy process as a decision cycle that includes decisions, reduction of uncertainty, range of decisions, and outcomes.

FIGURE 15.7 online at ncse.org/climate solutions

Charts show the CO_2e emitted per capita and the relative GDP by various groups of similar nations.

TABLE 15.4 Response Options for a National Climate Change Strategy

1. Grow fast (to reduce vulnerability to climate change by gaining wealth).
2. Improve adaptive capacity (to reduce negative impacts of climate change).
3. Mitigate (financial incentives are needed to offset costs).
4. Integrate climate change and sustainable development strategies by combining options 1, 2, and 3.

Source: [14]

types of responses to the problem of climate change: increased development of all nations, improved adaptive capacity, improved mitigation activity, or a combination of all of these strategies. (See Table 15.4 for a summary of response options.) The last option—"all of the above"—is the best option of the four. The problems of climate change and development should be solved together. We know enough about the physical climate and social impacts to move forward, but we still need to collect more information and learn more.

Adaptation is the first priority of developing countries that are most vulnerable to climate change. Why? Climate change is likely to impact disproportionately the poorest countries and the poorest persons within all countries, exacerbating inequities in health status and access to adequate food, clean water, and other resources. Net economic effects will be negative in most developing countries. Impacts will be worse in developing countries. Many areas are already flood and drought prone, and these economic sectors are climate sensitive. Poorer nations have a lower capacity to adapt because they lack financial, institutional, and technological capacity and ready access to knowledge.

In sum, conventional economic approaches to development focus on optimality, that is, on maximizing growth. It is possible to grow one's way toward higher health and welfare, but such a path, as shown by India, usually results in higher emissions—similar to overdeveloped nations.

Environmental and social approaches to development rely instead on durability, that is, on maintaining the system's health. Poorer nations face the steep challenges to adapt to ongoing climate change and mitigate their own contributions to future climate. The chief assets needed for sustainable development are social capital, natural capital, and manufactured capital. Balancing the contributions of each is the key for attaining a sustainable path for all the world's nations.

In closing, a few words from Buckminster Fuller, who was an early environmental activist, educator, designer, and popularizer of terms such as *spaceship Earth*, seem worth contemplating:

> All of humanity now has the option to "make it" successfully and sustainably, by virtue of our having minds, discovering principles and being able to employ these principles to do more with less.
>
> *R. Buckminster Fuller (1895–1983)*

As we think about doing more with less, let us turn to the solutions described in the next and final chapter. Let us make the world a better place by acting—starting today.

FIGURE 15.8 online at ncse.org/climate solutions

Three approaches toward sustainable development

FIGURE 15.9 online at ncse.org/climate solutions

Asset triangle available for climate protection

FIGURE 15.10 online at ncse.org/climate solutions

Demographic transition to sustainability

- As the 21st century progresses, we need to somehow accommodate the growth from 6 billion to 9 billion people worldwide in terms of energy demands.

- Regions and individuals that are the least responsible for climate disruption will suffer the most.

- What is the connection between the developing world and climate change? Higher methane is a bigger factor than carbon dioxide; decreased agricultural, forest, and fishery production; increased vector-borne diseases.

- Four responses to climate change are possible beyond business as usual: increase development, improve adaptation, improve mitigation, or a combination of all of the above.

- Sustainable development is the intersection of economic, social, and environmental assets.

Online Resources

www.eoearth.org/article/Making_Development_More_Sustainable~_Sustainomics_Framework_and_Applications_(e-book)

www.eoearth.org/article/Adaptations_to_climate_change

www.eoearth.org/article/Business_strategy_and_climate_change

www.eoearth.org/article/Tools_and_methods_for_integrated_analysis_and_assessment_of_sustainable_development

www.eoearth.org/article/Measuring_sustainable_economic_growth_and_development

Online Atlas of Millennium Development Goals, http://devdata.worldbank.org/atlas-mdg/

United Nations Millennium Development Goals, www.un.org/millenniumgoals

World Bank Millennium Development Goals, www.developmentgoals.org

EU Energy Commission Citizen's Corner, http://ec.europa.eu/energy/citizen/index_en.htm

See also extra content for Chapter 15 online at http://ncseonline.org/climatesolutions

Climate Solution Actions

Action 14: Engaging China on a Pathway to Carbon Neutrality

Action 17: Climate Change Adaptation for the Developing World—Expanding Africa's Climate Change Resilience

Action 33: Diverse Perspectives on Climate Change Education—Integrating Across Boundaries

Action 34: Building People's Capacities for Implementing Mitigation and Adaptation Actions

Action 35: Climate Change and Human Health—Engaging the Public Health Community

Works Cited and Consulted

[1] Adams WM (2006) The Future of Sustainability: Re-thinking Environment and Development in the Twenty-first Century. *IUCN Renowned Thinkers Meeting*. iucn_future_of_sustanability.pdf: www.iucn.org

[2] Bernstein L, Bosch P, Canziani O, Chen Z, Christ R, Davidson O, Hare W, Huq S, Karoly D, Kattsov V, et al. (2007) Synthesis Report (in *Climate Change 2007: Fourth Assessment Report of the Intergovernmental Panel on Climate Change*, 74 pp, eds Allali A, Bojariu R, Diaz S, Elgizouli I, Griggs D, Hawkins D, Hohmeyer O, Pateh Jallow BP, Kajfež-Bogataj L, Leary N, Lee H, Wratt D) ar4_syr.pdf: www.ipcc.ch

[3] Carbon Dioxide Information Analysis Center (CDIAC) (2008) Fossil-Fuel CO_2 Emissions. *Carbon Dioxide Information Analysis Center*. Oak Ridge National Laborartory, US Department of Energy, Global Change Data and Information System, Oak Ridge, TN (read October 15, 2008). http://cdiac.ornl.gov/trends/emis/meth_reg.html

[4] CIA (2007) *The World Factbook*. (Central Intelligence Agency, Washington, DC). www.cia.gov/library/publications/download/index.html

[5] CNA (2007) National Security and the Threat of Climate Change. http://SecurityAndClimate.cna.org

[6] European Environment Agency (2007) *Greenhouse gas emission trends and projections in Europe 2007: tracking progress towards Kyoto targets*. EEA Report No 5/2007, 108. http://eea.europa.eu

[7] FAO (2006) Livestock's Long Shadow. www.fao.org/docrep/010/a0701e/a0701e00.htm

[8] Fisher BS, Nakicenovic N, Alfsen K, Corfee Morlot J, de la Chesnaye F, Hourcade J-C, K Jiang K, Kainuma M, La Rovere E, Matysek A, et al. (2007) Issues Related to Mitigation in the Long Term Context (in *Climate Change 2007: Mitigation. Contribution of Working Group III to the Fourth Assessment Report of the Intergovernmental Panel on Climate Change*, eds Metz B, Davidson OR, Bosch PR, Dave R, Meyer LA) ar4-wg3-chapter3.pdf: www.ipcc.ch

[9] Gupta S, Tirpak DA, Burger N, Gupta J, Höhne N, Boncheva AI, Kanoan GM, Kolstad C, Kruger

JA, Michaelowa A (2007) Policies, Instruments and Co-operative Arrangements (in *Climate Change 2007: Mitigation. Contribution of Working Group III to the Fourth Assessment Report of the Intergovernmental Panel on Climate Change*, eds Metz B, Davidson OR, Bosch PR, Dave R, Meyer LA) ar4-wg3-chapter13.pdf: www.ipcc.ch

[10] Haupt A, Kane TT (2004) *Population Handbook.* (Population Reference Bureau, Washington, DC). www.prb.org/pdf/PopHandbook_Eng.pdf

[11] Holdren J (2008) Meeting the Climate Change Challenge: Eighth Annual John H. Chafee Memorial Lecture on Science and the Environment. *Climate Change Science and Solutions: Eighth National Conference on Science, Policy, and the Environment.* http://ncseonline.org/climatesolutions

[12] Mari Bhat PN (2001) Indian Demographic Scenario 2025. www.iegindia.org/

[13] Munasinghe M (1995) *Making economic growth more sustainable.* Ecological Economics 15(2):121–124. http://linkinghub.elsevier.com/retrieve/pii/0921800995000666

[14] Munasinghe M (2007) *Making Development More Sustainable: Sustainomics Framework and Practical Applications.* www.mindlanka.org/sustainomic .htm

[15] Munasinghe M (2008) *Warming signs.* Nature 456:28–29:doi:10.1038/twas08.28a; Published online 30 October 2008. http://www.nature.com/nature/journal/v456/n1s/full/twas08.28a.html

[16] PRB (2008) Presentation Graphics. *Population Reference Bureau* http://prb.org/Publications/GraphicsBank.aspx

[17] Sachs JD (2008) *Common Wealth: Economics for a Crowded Planet.* (The Penguin Press, New York). www.sachs.earth.columbia.edu/commonwealth/

[18] Scheuren J-M, le Polain de Waroux O, Below R, Guha-Sapir D, Ponserre S (2008) Annual Disaster Statistical Review. *EM-DAT Emergency Events Database.* www.emdat.be

[19] Sen A (2004) *How Does Culture Matter?* in *Culture and Public Action,* (eds) Rao R, Walton M. (Stanford University Press, Palo Alto, CA). www .sup.org

[20] UN MDG (2008) The Millennium Development Goals Report 2008. www.un.org/millenniumgoals

[21] UNFPA (2007) *State of the World Population.* www .unfpa.org/swp/swpmain.htm

[22] WHO (2008) World Health Day: Protecting Health from Climate Change. (April 2008). *World Health Organization.* www.who.int/globalchange/en/

PART IV

Thirty-Five Immediate Climate Actions

This section presents our national "to do list"—the Climate Solutions Consensus agenda as prepared by the 1300 scientists, engineers, educators, managers, policy-makers, and citizens who participated in the 8th National Conference on Science, Policy and the Environment in January 2008. This agenda presents some 300 tasks organized under 35 Actions that can be carried out by individuals, organizations, businesses, universities, government agencies and others to reduce the threat and impacts of global climate disruption.

The Actions represent key areas of need including Strategies for Stabilization, Mitigation, and Adaptation; Multidisciplinary Research; and Expanding Understanding. The Actions and the Tasks within Part IV follow this pattern.

The process that led to this Action Agenda involved experts in climate science and solutions who organized sessions at the NCSE Climate Solutions conference. A list of the organizers and discussants in these sessions along with background material can be found at http://ncseonline.org/climatesolutions. Each of the sessions involved a diverse set of 15–50 individuals who deliberated and developed a set of top priority tasks within the topic area. The tasks are generally in the form of recommendations to a specific organization or other potential implementers.

We invite every reader of this book to become involved in this Action Agenda. There is more than enough work for all of us. To learn more and get involved go to: http://ncseonline.org/climatesolutions/.

Strategies for Stabilization, Mitigation, and Adaptation

Green Buildings and Building Design

Building construction and operations account for about half of the national energy budget and a disproportionate amount of carbon emissions, because electrical power that is used to light, heat, and cool buildings is fueled principally by coal. Numerous design and construction practices, technologies, and standards are currently available under the rubric of green building. They could reduce building energy use dramatically. However, most projections for building energy efficiency and greenhouse gas (GHG) reductions show only modest improvements over the next 20+ years.

What barriers stand between the current trend and more rapid achievement of building energy efficiency and greenhouse gas emission reductions?

What can be done to speed the deployment of existing green building practices and technologies into the marketplace?

What emerging technologies offer the most promise to reduce building energy use and greenhouse emissions?

Policy

Task 1 The building community should develop strategies to incorporate energy efficiency and green practices into both existing and historical buildings. One specific action to accomplish this goal is to reduce energy use and urban "heat islands" through the use of cool-roof efforts, as advocated by the One Degree Less campaign. This practical action uses simple and relatively inexpensive techniques and technologies, such as painting roofs white or using other reflective or insulative roof materials.

Task 2 Federal, state, and local government organizations and lenders and builders should collaborate to make green buildings accessible to all income groups.

Task 3 Private insurance and government codes should be modified to facilitate green building measures.

Research

Task 4 Measuring, verifying, and modifying systems should reflect increasingly stringent energy standards and improved technologies.

Task 5 The federal government, including the Environmental Protection Agency (EPA), Department of Energy (DOE), Department of Housing and Urban Development (HUD), Department of Agriculture (USDA), and National Science Foundation (NSF), should support research into green buildings. Social science can contribute to research and promulgation of green building practices.

Education

Task 6 The building and construction industries should collaborate to create a green building wiki; a free Web-based encyclopedia built collaboratively.

Task 7 Government, industry, and civic organizations should advance education and public awareness on the importance of green buildings among multiple stakeholders.

Task 8 Teachers should use green building practices—especially during early education—as education tools.

ACTION 2:
Moving Forward—Transportation and Emissions Reduction

According to the US Greenhouse Gas Inventory for transportation, US GHG emissions from transportation sources grew by about 30% between 1990 and 2007. As of 2005, transportation sources were nearly 28% of US GHG emissions overall. The three biggest segments of the transportation sector in terms of GHG emissions are light-duty passenger vehicles (cars and sport-utility vehicles), freight trucks, and aviation. Each of these segments has increased in overall emissions in the past 10 years.

There is a need to better understand emissions trends and underlying driving forces, as well as current strategies and technology and policy options to reduce emissions. Relevant societal trends, such as land use patterns and changes in manufacturing, must be taken into consideration, along with transportation system priorities that affect GHG emissions, such as congestion reduction and safety. These considerations will be useful in determining possible future scenarios for transportation with respect to GHG emissions and opportunities for reducing emissions.

http://epa.gov/climatechange/emissions; www.energy.gov/energyefficiency/transportation.htm; www.trb.org

Policy

Task 1 Policymakers should understand transportation market forces, to inform pricing policy or a carbon tax, and answer questions such as

- When and where will there be a rebound effect?

- Under what conditions are travel behavior and resulting GHG emissions changed or not changed for multimodal freight and passenger travel?

- What are the implications of various policy scenarios for social justice?

Task 2 Policymakers should understand implications of federal transportation infrastructure investment on climate change, to inform reauthorization formulas and discretionary programs. Issues include

- relative value of investment for new projects versus enhancements to existing infrastructure

- life cycle analysis (LCA) including trade-offs for different approaches to achieving mobility objectives

- public versus private transit vehicles

IV: Thirty-Five Immediate Climate Actions

- optimizing multimodal travel in metropolitan areas

Task 3 Federal, state, and business decision makers should understand the impact of information on consumer behavior and resulting GHG emissions and provide information where and when it is most useful for reducing emissions. Needs include

- GHG implications of shipping options
- better public education on GHG emissions and trade-offs for vehicles
- instantaneous miles per gallon (mpg) information for drivers

Task 4 Public and private fleet managers across all transportation modes should understand the best pace for advanced technology investment and adaptation from a perspective of life cycle GHG emissions and cost. This will inform strategies and policies to encourage faster turnover and incentives for acceleration of better technologies to increase the pace of environmental benefits.

Task 5 Policymakers should draw on standardized wells-to-wheels/wings analyses of environmental emissions, land use, and water use when supporting the use of advanced fuels in vehicles.

Task 6 Freight shippers should package goods for more efficient shipping.

Research

Task 7 The Transportation Research Board of the National Academy of Sciences (NAS) should conduct a study of minimum potential energy intensity with trade-offs for environment, economics, and travel time for each transportation mode so that inspirational benchmarks can be set to drive technology innovation and implementation of solutions.

Task 8 Researchers and policy analysts across all transportation modes should model decision-making tools and analyses after the cross-governmental Next Generation Air Transportation System (NextGen) and develop planning

tools that integrate strategies, measures, and visualization of trade-offs between GHGs and other environmental impacts at a system level.

Animal Agriculture and Climate Change

Despite the findings reported in the Food and Agriculture Organization's *Livestock's Long Shadow: Environmental Issues and Options,* much of the recent discussion about climate change has focused on personal and business energy use while failing to account for the gross contributions by the meat, egg, and dairy industries and supporting sectors or the significance of intensive animal agricultural practices that have become the norm in Western nations and increasingly are exported into lesser-developed countries.*

The direct connection between farm animal production and climate change is not as well known as the linkages between climate change and other industries, such as transportation. It is essential to identify the ways in which energy use in confinement production facilities, deforestation, production of nitrogen fertilizers to grow feed crops, and farm animal waste management systems contribute to greenhouse gas emissions. Moving toward solutions, attention should be paid to agribusiness industries' existing mitigation techniques, as well as the impacts of converting to more-sustainable production systems.

www.fao.org; www.usda.gov

Policy

Task 1 Congress should encourage the US animal agriculture sector to participate in carbon markets and consider soil carbon sequestration (primarily emphasizing the use of pastures).

Task 2 USDA and Congress should revisit animal product labeling laws so that labels allow

---

*FAO (2006) Livestock's Long Shadow. Food and Agriculture Organization of the United Nations www.fao.org.

for identification of the carbon footprint of the product.

Task 3   Congress should consider how existing infrastructure makes it more difficult for smaller-scale producers to reduce transportation associated with slaughtering and processing.

Research

Task 4   USDA should set a research priority for comparing methane and other GHG emissions (in a life cycle analysis) from pastured animals as compared with animals raised on grain in confinement.

Task 5   USDA should review and analyze the impact of subsidies for various crops on climate change. (This analysis has never been done and is necessary for any redirection of subsidies.)

Task 6   Researchers should develop sample policies and modeling analyses for local land use organizations so they can actively preserve land for management-intensive grazing of animals in peri-urban areas.

Task 7   To better assist communities implementing GHG inventories, researchers should evaluate how to best measure and quantify emissions from production of meat, eggs, and dairy products.

Task 8   The NAS should conduct a study that leads to a national science-based dialogue about how meat consumption, processing, packaging, and waste impact GHGs.

Education

Task 9   There should be a public communications campaign to educate the public about the issue of animal agriculture and climate change, in order to impact individual consumption patterns (similar to calling attention to how our driving habits impact GHGs).

Task 10   Environmental and other organizations (including public health professionals) should bridge work on food/agriculture issues with work on climate change.

Task 11   Institutions, including universities, should identify the sources of animal products they use in a way that considers the GHGs/carbon footprint, including increasing funding for existing farm-to-institution programs.

ACTION 4:

Minimizing Agricultural Impacts on Climate; Minimizing Climate Impacts on Agriculture

Agriculture is subject to climate change, both directly (i.e., via temperature and precipitation effects) and indirectly (e.g., through changing pest and weed ranges). At the same time, agricultural management contributes to the atmospheric greenhouse gas concentrations responsible for climate change. That this is occurring over an already complex landscape of regional geographic considerations, changing land use patterns, innovations in adaptation, and a multifaceted socioeconomic environment suggests that multiple possibilities may exist for addressing the challenges that agriculture faces in maintaining widespread food security while preserving environmental integrity. However, the scale of information needed does not necessarily match the scale at which information is available, and the application of that information can face challenges related to specific production types, finances, and social acceptance of climate change as a fundamental management consideration.

http://dels.nas.edu; www.ars.usda.gov; www.nationalacademies.org/agriculture

Policy

Task 1   USDA should provide monetary incentives for creative technical approaches to coping with climate change impacts on plants and livestock.

Task 2   Agricultural producers should rethink agriculture and energy systems so that energy and agricultural waste streams can be utilized, for example, high-value agricultural production coupled with urban waste energy.

Task 3 State agriculture departments should include climate change considerations in nutrition management programs.

Research

Task 4 The NAS should assess regionally appropriate management recommendations on mitigation and adaptation to protect agricultural production in conjunction with producers.

Task 5 USDA should fund development of stress-resistant varieties and management practices to cope with climate stresses for agriculture and forestry systems.

Task 6 USDA should develop new approaches to spread out producers' risk over time and space.

Task 7 The USDA Agricultural Research Service (ARS) should develop long-term data sets to quantify and understand the impacts of climate on agriculture.

Task 8 USDA should perform a life cycle GHG analysis on all production systems, including controlled-environment production systems in northern latitudes.

Task 9 USDA should study the effects of climate change on pests and invasive species.

Task 10 The NAS should conduct a comprehensive assessment of the impacts of climate stress on livestock production and identify potential management practices to alleviate stress.

Education

Task 11 The USDA Cooperative Extension Service should make climate change a priority in educational and engagement efforts.

ACTION 5:
Mitigating Greenhouse Gases Other Than CO_2

Reducing emissions of non-CO_2 gases can help minimize global climate change and yield broader economic and environmental benefits. Recent analysis by the Massachusetts Institute of Technology indicates that feasible reductions in emissions of methane and other non-CO_2 gases over the next 50 years could make a contribution to slowing global warming that is as large as, or even larger than, similar reductions in CO_2. Mitigation costs for non-CO_2 gases are lower than for energy-related CO_2. Because sources of "other gases" are much more diverse, not just energy and land use, there is a large portfolio of mitigation options and the potential for reduced costs for a given climate policy objective.

http://epa.gov/climatechange; www.global change.gov; www.globalreporting.org

Policy

Task 1 Policymakers should recognize the substantial opportunities and benefits of mitigation of non-CO_2 GHGs.

Task 2 Mitigation technologies and best management practices exist for many of the non-CO_2 gases and their sources. Industry and others should incorporate and implement these practices as aggressively as possible.

Task 3 Getting to near-zero emissions for some of the non-CO_2 sources is not currently technically feasible (e.g., methane from ruminant livestock), and/or to do so may require large-scale societal changes.

Task 4 Researchers and decision makers should use a systems approach for integrated thinking across gases and sectors in order to make sure not to create a new problem by addressing another.

Research

Task 5 EPA should improve research and understanding regarding co-benefits and the range of environmental impacts associated with the interrelationships between air quality and climate change.

Task 6 EPA and other researchers should work to better articulate the relationship between emissions, concentrations, and radiative forcing for all GHGs, not only carbon dioxide. This will help us better understand the role of non-CO_2 greenhouse gases in climate stabilization.

Education

Task 7 EPA should initiate or support targeted education campaigns to inform relevant sectors and decision makers of opportunities for reductions of non-CO_2 GHGs.

ACTION 6:
Energy Efficiency and Conservation

The cheapest carbon is what we don't emit (see Chapter 6). Energy efficiency is the fastest and least expensive first step in tackling carbon emissions. Analytical and policy issues must be addressed if energy efficiency is to realize its very large potential contribution (perhaps 25% or greater) to the climate challenge.

www.eere.energy.gov; www.energystar.gov; www.energysavers.gov

Policy

Task 1 Decision makers should utilize price signals, to create incentives to achieve increased energy efficiency and in the context of cap and trade systems.

Task 2 Researchers and regulators should focus on plug loads, electrical devices that receive power from AC wall outlets, such as cell phones and small appliances, in developing strategies for energy efficiency.

Task 3 Policymakers should integrate energy efficiency into other related policy arenas (e.g., health care, criminology, education).

Task 4 Policymakers should support and use social science research to understand and influence consumer behavior in energy markets.

Research

Task 5 Federal and state agencies, in partnership with industry, should increase research on use of heat and energy capture technology.

Task 6 Federal and state agencies and utilities should increase research on maximum achievable energy cuts to provide a more conclusive projection of the role of energy efficiency in mitigating climate change.

Task 7 DOE should increase research on energy storage technology in order to increase efficiency.

Education

Task 8 EPA and DOE should initiate and fund major public education campaigns to encourage substantially increased energy efficiency and conservation.

ACTION 7:
Biofuel Industry and CO_2 Emissions— Implications for Policy Development

Biofuels are gaining in popularity as replacements for fossil fuels. Biofuels result from the conversion of biomass into liquid fuels, which are then burned for energy. Various biofuel conversion options are being researched. These include biological processes and thermal chemical processes, with different processes more appropriate for different crops, conversion plant size and location, ecosystem service efficiencies, and logistic options. Collectively these components will impact energy efficiencies and the carbon footprint of biofuels. Certain biofuels have great potential to reduce greenhouse gases; however, current biofuels (particularly corn kernels) may be neutral, at best, in terms of net energy production.

http://bioenergy.ornl.gov; www.biodiesel .org; www.nrel.gov/biomass

Policy

Task 1 Policymakers should include biofuels in a comprehensive energy policy, including energy conservation and efficiency.

Task 2 Policymakers should use the results of a life cycle analysis of biofuel systems when developing policy options.

Task 3 Researchers and policymakers should account for the carbon and energy footprint in research and policy on climate implications of large-scale biofuel production and sequestration strategies.

Task 4 Policymakers should develop incentives for outcomes, not technologies.

Task 5 Policymakers should base incentives such as the blenders credit for biofuel use on energy balance in the fuel of consideration. All biofuels are not created equal.

Task 6 Efforts should be made to maximize compatibility of new biofuels systems with existing fuel infrastructure.

Task 7 The federal government should increase support for research and curricular development (kindergarten through 12th grade and up) on current technologies for biofuels.

ACTION 8:
Solar Energy Scaling Up—Science and Policy Needs

Solar energy is an important, but currently tiny, component of a low-carbon economy. Barriers to expansion of solar energy are economic, scientific and technological, and educational. Technologies include large-scale concentrating solar power (CSP) plants, new solar and fuel cell manufacturing, solar thermal energy, nonsilicon thermal energy, and related technologies for commercial and residential use in the United States, other industrialized nations, and in the developing world.

www.ases.org; www.nrel.gov/solar; www1 .eere.energy.gov/solar; www.dsireusa.org

Policy

Task 1 Policymakers should emphasize implementation rather than developing new technologies.

Task 2 Legislators should prevent utilities from passing the risk of volatile energy resource costs onto the consumers.

Task 3 Policymakers and financers should develop strategies to alleviate the financial risk of commercial solar power.

Education

Task 4 The solar industry and its allies (such as community colleges, colleges, and universities) should organize to deal with problems unrelated to the technology, such as lack of work force to expand.

Task 5 College students should serve dual roles—as the solar workforce and also as a community of solar proponents.

ACTION 9:
How to Ensure Wind Energy Is Green Energy

Wind energy has become an increasingly important and the fastest-growing sector of the electrical power industry, largely because it has been promoted as being emission free and is supported by government subsidies and tax credits. However, large numbers of birds and bats are killed at utility-scale wind energy facilities, especially along forested ridgetops in the eastern United States. These fatalities raise important concerns about cumulative impacts of proposed wind energy development on bird and bat populations. Research and information are needed to better inform researchers, developers, decision makers, and other stakeholders and to help minimize adverse effects of wind energy development.

www.awea.org; www.windpoweringamerica .gov; www.windustry.org; www.dsireusa.org

Policy

Task 1 State and federal regulatory agencies should improve the consistency of requirements and regulation and discourage policies that reduce research and environmental review prior to granting permits for new facilities.

Task 2 Decision makers should ensure that all positive and negative impacts of wind energy are analyzed in their proper contexts in relation to other sources of energy generation.

Task 3 An independent body should explore the development of a process to certify wind projects that adequately minimize or mitigate impacts on wildlife and habitat.

Task 4 All stakeholders must increase funding for priority monitoring and research, and federal and state agencies should increase funding and staffing to address wind permitting issues.

Task 5 Permitting agencies and public utility commissions should account for monitoring, research, and mitigation in up-front planning and permitting of wind projects to improve cost certainty.

Research

Task 6 Federal and state guidelines should define and identify high risk areas that may warrant additional research, mitigation, or avoidance.

ACTION 10:
Nuclear Energy—Using Science to Make Hard Choices

The future of nuclear energy is most often set forth in absolutist terms: either "nuclear energy is necessary to combat climate change" or "nuclear energy is an unacceptable option." A more fruitful debate might follow from a conversation that begins by establishing the set of characteristics that are important for future energy sources and then evaluates nuclear energy in the context of these characteristics. Guidelines are needed for appropriate norms for discussing nuclear energy in the context of climate change. These include, but are not limited to, roles for economics, ethics, expertise, technical information, government funding, health and safety, and uncertainty. Exploration of what role nuclear energy might have if subsidies on all energy sources are made transparent, and if the external costs associated with carbon emissions are internalized, is also

needed. The availability of qualified expertise and educational programs in nuclear power generation must also be considered.

www.iaea.org; www.ne.doe.gov; www.key stone.org/spp/energy07_nuclear.html

Research

Task 1 An independent, respected organization such as the NAS should develop a set of appropriate and transparent life cycle comparison metrics for all energy technologies, as well as conservation and efficiency.

Task 2 An independent, respected organization should conduct a complete analysis of subsidies, mandates, and market directives associated with all electricity generation options.

Task 3 NSF should issue a program announcement to fund further research in perception and communication of nuclear and climate issues.

Education

Task 4 An independent, respected organization should further develop broadly acceptable communication materials about the advantages and disadvantages of nuclear energy (e.g., The Keystone Project, www.keystone.org/spp/energy07_nuclear.html).

Task 5 The federal government should increase funding for nuclear engineering and science education at the undergraduate and graduate levels.

ACTION 11:
Economics—Setting the Price for Carbon

There is growing political momentum in the United States to set a price for carbon via a cap and trade mechanism. However, many substantive questions remain concerning the design of cap and trade and the role of complementary policies. Political questions remain on how to coalesce the political forces necessary to enact national legislation (as well as ratify new international agreements applicable in the post-2012 period). Key topics include (1) design issues such as stringency, timing, and "cost containment" provisions (banking and borrowing mechanisms, price caps and floors, and the use of offsets); (2) whether to create a GHG emission standard for new power plants to accelerate deployment of carbon capture and sequestration and to complement cap and trade; and (3) how the role of the coal industry in the public policy debate is likely to evolve and how to get it more actively involved in finding solutions. Other issues include the likely interplay of various ways to allocate and/or auction allowances, and the vastly different state regulatory systems for electrical utilities. This interplay impacts both the economics and politics of cap and trade.

www.ceres.org; www.chicagoclimatex.com; www.globalreporting.org; www.rggi.org; www.us-cap.org; www.westernclimateinitiate.org

Research

Task 1 There should be more economic and policy research on merits or demerits of government oversight, regulation, and management of the allowance market. Research should examine both price ceilings and price floors.

Task 2 State-level regulation of electric utilities varies widely between traditional cost-of-service regulation and varying degrees of deregulation at the electricity generation level (coupled with continued regulation at the distribution level). There should be more economic and policy research on the complex interactions between state-level utility regulation and state and national climate change policies that are likely to occur. State legislators, public utility commission officials, and stakeholders should have a credible and accessible set of research findings to guide them in future regulatory decisions that interact with climate policy.

Task 3 There should be more research into the optimal combination of "carrots and sticks" that can accelerate the commercialization of carbon capture and sequestration. Issues of liability should also be addressed along with other legal/regulatory issues.

Task 4 There should be more economic and policy research into how nations could make "border adjustments" to account for imports from countries that do not control GHG emissions. This is relatively easy in the case of carbon taxes but problematic in the case of cap and trade. Topics would include how World Trade Organization policy should treat such border adjustments.

Forests and Markets for Ecosystem Services (ES)

Land managers, owners, and users continue to explore new and innovative ways to accomplish land management objectives. Markets for ecosystem services provide opportunities and challenges for forest/land stewardship; yet, a number of questions remain about the challenges to implementing most ES markets. It is useful to examine the markets for specific ecosystem services—conservation banking, water quality trading, wetlands banking, and carbon markets—to advance the research behind and implementation of ES markets.

www.fs.fed.us/ecosystemservices; www.millenniumassessment.org; www.unep.org/ecosystemmanagement

Research

Research to better understand ecosystem services and to provide proper valuation of these benefits of nature should include the following:

1. Scientists should develop better methods to measure, map, model, and value ecosystem services at multiple spatial and temporal scales:

   - Technology/tools needed
   - International global land use observatory (built with the following tools)
   - Landsat Data Continuity Mission
   - LIDAR systems

- Streamlined clearinghouse for remote sensing data with broad international access
- Science needed
- Methods for modeling carbon storage and approaches to modify existing models
- Dealing with scaling factors and different measurement techniques to relate data sets, models, and projections across regions

2. Valuation science (i.e., monetizing ecosystem services) should emphasize the following:
   - How to improve spatial targeting mechanisms for identifying and valuing ES
   - How to balance demand and supply of ES
   - Explicit models for the demand side of ES (a set of models and tools with standardized measures of ecosystem service demand and value)

3. How do forest management activities (e.g., research should consider logging, thinning, burning) affect provision of ecosystem services individually and bundled?
   - How does the area, type, and condition of a forested area affect the quantity and quality of water provided?
   - How well do models predict carbon storage, and what modifications are needed?
   - How can "leakage" of forest ecosystem services be detected and prevented?

4. Researchers should standardize language regarding ecosystem services.

5. The government should develop registries of ecosystem services.

6. Professional associations should develop verification standards across regions.

7. Funders should support research to improve understanding of ecosystem bundling:
   - How do we best add value to carbon sequestration?
   - Under what circumstances is it best to bundle multiple ecosystem services (climate regulation, water provision, bio-

diversity, etc.), and what are the implications and the trade-offs of doing so?
   - How do we market and price bundled services; we need to understand the relationship between resilience and bundled ecosystem services (ecosystem functioning)?
   - Research should develop a systems approach to understanding ecosystem service bundling and processes.
   - Solve conflicts between bundled services.

8. Research should improve the connections between social and natural sciences within ecosystem services research.

Policy Challenges of GHG Rule Making—Where the Rubber Meets the Road

A new law to reduce GHG emissions will be a major milestone, yet much of the "fine print" requirements will be addressed later by agencies through detailed agency rules based on a public process. Complex, contentious rules, especially those affecting major swaths of the US economy, can take 5 years or more to implement. Given the need to "get it right the first time," can federal agencies expeditiously issue numerous rules before the rules become obsolete? If cap and trade regimes, along with offsets, are enacted, decisions must be made about the applicability (e.g., what gases/sectors) and design (e.g., trading/offsets/agency discretion of future GHG regulation). How can expediting rule making occur without sparing analytic rigor, given the need to provide incentives to foster data sharing among key parties; the culture clash between science and policy making; the potential for increased use of dispute resolution; and the role of states and the impacts of a regulatory "patchwork"? Presidential leadership and new legislation are necessary for action.

www.ec.gc.ca/cc; www.eea.europa.eu; www.globalcarbonproject.org; www.globalchange.gov

Research

Task 1 Lack of knowledge about climate change mitigation is a barrier to a comprehensive GHG reduction program.

Task 2 The president should include funding for research and development (R&D) and pilot projects in the budget. Congress should appropriate money for R&D and pilot projects.

Policy

Task 3 Strong presidential leadership is essential. The presidential message should identify what we need to do now and where we need to be headed. This will

- decrease the chaos and make the regulatory process more linear
- give clear marching orders to the agencies
- create a firm political position from which to work with Congress

Task 4 Congress must draft legislation to avoid a patchwork of state approaches and should incorporate deadlines.

Task 5 Leadership, both in Congress and by the president, can be stimulated by national industry and business leaders demanding government action to

- ensure that Congress clarifies the linkages between a GHG reduction scheme and the Clean Air Act (CAA)
- enact GHG legislation that is comprehensive and obviates the need to regulate GHGs under the CAA and that preempts regulation of GHGs under the CAA to address climate change
- maintain the CAA to address traditional pollution (e.g., ozone) while enacting new legislation to minimize climate change

Task 6 Constituent pressure on both Congress and the White House is important to enable enactment of strong legislation.

Task 7 Compromise will speed action.

Task 8 Congress should specify the nexus among climate change, air, and water regulation, perhaps beginning with existing regulations.

Task 9 Policymakers should utilize information and reports from existing advisory groups to inform policy and regulation.

Task 10 In the short term, a patchwork approach may be inevitable.

Engaging China on a Pathway to Carbon Neutrality

China will play a key role in the development of any global effort to address climate change. According to the International Energy Agency, China is now the world's largest emitter of carbon dioxide, and its emissions are likely to continue growing strongly in the decades ahead. China's position on climate change will also influence the action or inaction that other countries consider. While the United States has emitted roughly twice the cumulative greenhouse gas emissions as China over the past century, it has noted China's potential to overwhelm other global mitigation efforts as at least one reason for not ratifying the Kyoto Protocol. China's success or failure in curbing emissions will also be a powerful example for other developing countries to follow or avoid.

www.ccchina.gov.cn/en; www.state.gov/p/eap/ci/ch; www.usmayors.org/climatechange protection

Policy

Task 1 Congress and the US administration should play a leadership role in limiting GHG emissions domestically and reengaging in international negotiations. This will help to encourage China to reduce its own GHG emissions.

Task 2 Governments should continue and expand work to reduce trade barriers for green technology.

Task 3 Associations of mayors and governors should establish climate sister city/province/state relations with counterparts in China that face similar energy and climate issues.

Task 4 The United States should provide financial and technical assistance to China to reduce GHG emissions from existing and planned coal-fired power plants.

Research

Task 5 China and the United States should support a joint study on the energy and carbon embedded in goods and products traded between the countries (Department of Commerce and Ministry of Commerce).

Task 6 A joint Chinese-US governmental task force should be established to assess climate security in Pacific Rim countries.

Task 7 The United States should increase assistance to build Chinese capacity for emission inventories and monitoring and for identification of carbon sinks.

Task 8 The United States should support capacity building to help bring locally appropriate technology to scale in China.

Education

Task 9 There should be a massive cultural and educational exchange program to build a base of mutual understanding between the two countries.

Task 10 US universities should offer green MBAs in China that include sustainable development as a core component of the curriculum.

ACTION 15:
Human Population and Demographics—Can Stabilizing Population Help Stabilize Climate?

Population growth is one of several drivers of climate change. Programs designed to improve access to reproductive health care and slow the

future growth of the world's population can also serve as long-term mitigation strategies at the global level and adaptation strategies at the community level.

www.developmentgoals.org; www.prb.org

Policy

Task 1 The US Congress and other policymakers should

- allocate funding to achieve universal access to voluntary family planning as a means to slow the growth in greenhouse gas emissions and reduce human vulnerability to climate change impacts
- promote human rights – based strategies to reduce population growth, increase resilience, and build capacity for adaptation in regions most vulnerable to climate change impacts

Research

Task 2 Climate science should fully integrate demography and population dynamics—including fertility, mortality, migration, geographic distribution, and age structure—into climate change research and models in order to better understand how these factors can contribute to optimum reductions in greenhouse gas emissions globally.

Task 3 Climate modelers, including the Intergovernmental Panel on Climate Change (IPCC), should work with demographers to clarify the effect and feasibility of slowing the growth in greenhouse gas emissions and reducing human vulnerability to climate change impacts by achieving the UN's low population growth projection of 7.8 billion people in 2050; likewise, they should clarify the climate change outcomes that would be likely to result from the UN's high population growth projection of 10.8 billion people in 2050.

Task 4 The US government should support research initiatives that facilitate the integration of demographic factors into climate change research and modeling, as outlined above. In addition, the US government should support research initiatives that

- quantify both the costs of providing universal access to voluntary family planning and reproductive health services and the benefits of such universal access, in terms of reducing future greenhouse gas emissions and human vulnerability to climate change impacts

- highlight links among demographics, household income, consumption, and other socioeconomic factors as they relate to climate change

- examine the connections among food security, biofuel development, and population dynamics in the context of climate change

- explore the role of migration, both international and internal, in greenhouse gas emissions growth and human vulnerability to climate change impacts

ACTION 16:
Urban Responses to Climate Change in Coastal Cities

All urban decision makers and planners must recognize the urgency of climate change on the local level. Communities must address the ongoing and escalating threat beginning *now*. In order to effect this kind of social change, educators, researchers, policymakers, and planners must recognize that local, individual, and institutional perception and response to climate change is both culturally dependent and culturally specific. Policymakers and advocates for change must engage the local culture; the deep change in thinking required to address climate change will come from within it.

www.usmayors.org/climateprotection; www.icleiusa.org

Policy

Task 1 The National Flood Insurance Program should take into account the risks posed by climate change in urban areas.

Task 2 Project and program review criteria at federal, state, and local levels should include climate change impacts and vulnerabilities.

Task 3 The appropriate agencies should establish climate change–triggered threshold levels for existing critical infrastructure.

Task 4 Elected officials who make land use decisions should base these decisions on a long-term land use plan, design standards, and building codes that include vulnerability analysis, certified by a planner.

Research

Task 5 The IPCC needs to develop user-friendly tools to improve access to information in the Program for Climate Model Diagnosis and Intercomparison (PCMDI) Web site so planners can incorporate climate scenario information into their decision-making tools.

Task 6 Funding agencies should support the scientific community in the incorporation of the socioeconomic side of local impacts into adaptation issues associated with climate change.

Education

Task 7 University accreditation boards and professional accreditation boards in planning, architecture, and civil engineering should include an understanding of climate change mitigation and adaptation in their criteria for accreditation. This will require the development of education programs for professionals.

Task 8 Climate change scientists, professionals, and advocates must improve the way they communicate climate change and its urgency in order to make it locally relevant to schools, engineers, planners, and communities.

Task 9 City officials, planners, and decision makers should meet together regularly in informal social settings to exchange information and opinions on climate change as related to their responsibilities.

Climate Change Adaptation for the Developing World—Expanding Africa's Climate Change Resilience

The Fourth Assessment Report of the IPCC states that Africa is one of the most vulnerable continents to climate change and climate variability. Representatives from African and US-based research institutions, development institutions, and nongovernmental organizations (NGOs) need to identify the most salient research questions to improve Africa's ability to cope with the projected impacts of climate change and develop the most practical solutions based on what we know to date.

www.developmentgoals.org; www.undp.org; www.worldbank.org

Research

Task 1 The appropriate international scientific and donor organizations should develop an international scientific research program to which governments, private entities, NGOs, and academics both in and out of Africa can contribute to develop fundamental natural science understanding for sustainable development (surficial geology, soil science, mineral resources, geochemistry, surface water and groundwater, land cover, ecology, biodiversity conservation, etc.). They should also develop greater understanding of climate change at the regional to local scale, including observations, models, and verification of models.

Task 2 International agencies and governments should support clean energy research and development—specifically solar, geothermal, and biofuel generation—at the regional level to expand energy access in Africa.

Task 3 Donors should support research on climate change impacts on water resources and infrastructure for water systems.

Task 4 Scientists should conduct research on relationships among population growth, demographic movements, urbanization and available agricultural land base, and carrying capacity, with multiple climate change scenarios, to advance climate change adaptation and technology.

Task 5 Funders should support outcome-directed research to enable climate change adaptation, and they should improve efficiency of projects based on African priorities.

Task 6 Researchers and policymakers should explore policy mechanisms to bridge the competition between short-term relief of food crises and longer-term rural development assistance in drought-prone countries.

Task 7 Agricultural development agencies should support research into the spread of nonnative agricultural crops.

Task 8 The World Bank and other development organizations should conduct research to understand how to implement microcredit programs and other credit vehicles in Africa with large and growing informal economies.

Education

Task 9 International agencies and governments should expand training of African climate change scientists.

Task 10 Governments should support national educational programs to promote understanding of climate change and its impacts on natural and human systems at multiple levels and to promote career opportunities in solutions and sustainability.

Coastal Management and Climate Change

State coastal management programs are on the front lines dealing with the impacts of climate change—sea level rise, dropping water levels in the Great Lakes, ocean acidification, and changes in temperature and precipitation patterns. The National Coastal Zone Management Program requires states to balance competing uses of the coastal zone and to address the full range of coastal issues, including managing development in high-hazard areas, protecting natural resources, providing public access, redeveloping urban waterfronts and ports, siting energy facilities, protecting coastal water quality, and ensuring that the public and local governments have a role in coastal decision making. This voluntary federal-state partnership was authorized under the Coastal Zone Management Act (CZMA) of 1972. Currently 34 states and territories participate in the program. In 1996, the CZMA was reauthorized, and Congress called for coastal states to anticipate and plan for global warming, which may result in a substantial sea level rise and fluctuating water level in the Great Lakes.

www.coastalmanagement.noaa.gov; www.coastalstates.org

Policy

Task 1 Coastal management agencies should translate climate scenarios into best management practices for planning, regulation, and engineering.

Task 2 Congress and state governments should increase funding for coastal habitat restoration and address the long-term sustainability of restoration projects.

Task 3 The Coastal States Organization and state coastal management programs should initiate planning for regional adaptation to climate change.

Research

Task 4 States and federal agencies should collaborate regionally to conduct a data inventory to identify data gaps relating to climate change and coastal environments and communities. They should also develop strategies to fill data gaps and disseminate data and information through a clearinghouse mechanism (e.g., a portal).

Task 5 The federal US Global Change Research Program (USGCRP) and the Union of Con-

cerned Scientists should synthesize IPCC information into more-relevant, regionally focused formats.

Task 6 The National Oceanographic and Atmospheric Administration (NOAA), US Geological Survey (USGS), Army Corps of Engineers, Federal Emergency Management Agency, and other federal agencies should develop integrated models that link climate to ocean and coastal processes and impacts.

Task 7 Academia should assist coastal managers in determining scenarios for land use planning, infrastructure, and habitat impacts.

Education

Task 8 Home buyers, homeowners, and renters should be given information through printed and online resources about adverse effects and consequences of sea level rise and natural hazards.

Task 9 Congress should fund education programs supporting integrated natural science and public policy to develop and acquire curricula specific to regional climate impacts at kindergarten through university levels.

Task 10 States should set up speed-dating-like interfaces for scientists and managers to facilitate communication of needs and sharing of research results.

ACTION 19:
Forest Management and Climate Change

Forests in the United States are managed for many goals under diverse ownerships. Goals range from long-term environmental protection and biodiversity sustainability with new possibilities for carbon sequestration to short-term production of fiber and biomass with new possibilities of biomass energy. Adapting to climate change impacts and mitigating anthropogenic drivers of climate change will require new practices for the full range of forest management goals. The changing climate complicates forest management because sequestration and emissions goals are added to the more traditional goals of protection and production. Increased uncertainty about how forests are responding to climate change complicates management.

www.fs.fed.us/ccrc; http://cfs.nrcan.gc.ca

Policy

Task 1 Coordinate landowners and land management agencies on joint decision making about adaptation actions to address fragmentation of habitats and management.

Task 2 Federal agencies should incorporate field and monitoring data about all forest management into publicly available, Web-accessible databases.

Research

Task 3 Economists should incorporate linkages among energy supply, demand, and policy into forest-sector models for carbon management.

Task 4 Government and academia should develop predictive tools and models for land managers that are designed to predict

- climate change at regional and local scales
- species shifts at regional and local scales
- the GHG implications of alternative forest management activities and strategies, including "no active management"

Note that the different groups will need different types of tactical and strategic information.

Task 5 Researchers should develop and evaluate options for facilitated adaptations to enhance ecosystem resilience to climate change.

Task 6 Climate impact modelers should develop "hot spot" analyses to help decision makers and stakeholders prioritize adaptation opportunities.

Task 7 Scientists should work across disciplines and with local communities to assess vulnerabilities and impacts, and they should

develop integrated adaptation and mitigation strategies within local communities.

Education

Task 8 Local and regional stakeholders should develop and implement a communications strategy and an educational strategy to engage all stakeholders in dialogue and action about the consequences of current land management and societal behaviors in the context of climate change.

ACTION 20:
Climate Change, Wildlife Populations, and Disease Dynamics

Most scientific evidence related to climate change and its effects on biological organisms is about plant species and vertebrate animals. Changes are occurring in terrestrial, aquatic, and marine ecosystems because of changes in climate. These changes will lead to further changes in wildlife diseases, vectors of disease, intermediate host alterations, and susceptibility to disease.

www.audubon.org; www.millennium assessment.org; www.who.int

Research

Task 1 Priority should be placed on filling these critical knowledge gaps in our understanding of climate change and wildlife diseases:

- effects of invasive species
- vector-borne diseases
- rapid evolution of pathogens
- host-species movement patterns
- ecosystem fragmentation
- seasonality of wildlife disease events
- ecosystem dynamics

Task 2 These critical knowledge gaps could be filled through development and implementation of standardized data collection systems, to detect ecosystem changes, and development of models, to explain observed data trends and forecast future events. These data and models would then allow for development of risk assessment models. Existing scientific expertise is probably sufficient for organizing and analyzing data and for defining data needs, but dramatically more capacity is needed for data collection. This capacity could be expanded by training citizen-scientists, engaging citizen-based organizations, and engaging traditional and other local communities.

Task 3 It is critically necessary to expand resources to support research in and management of wildlife in the face of climate change, with increased funding from

- federal and state governments
- private foundations
- private industry and financial institutions like the World Bank and World Health Organization (WHO)

Task 4 Economic metrics are needed for the values of ecosystems and habitats.

Task 5 There should be a global ecosystem assessment based on these economic metrics as indicators of environmental health and resources. This information could be used by national policymakers around the world.

Education

Task 6 There should be a concerted effort to enhance the education and awareness of the public about wildlife and climate change by

- providing information for curriculum development in the nation's schools
- developing curricula based on information from professional societies (e.g., The Wildlife Society, Society for Conservation Biology, and the Wildlife Disease Association)
- engaging state and tribal wildlife agencies to provide public programs on wildlife and climate change
- engaging local media and organizations to provide a format for increased awareness

Guiding and Fostering Multidisciplinary Research

ACTION 21:

The US Global Change Research Program (USGCRP)—What Do We Want from the Next Administration?

The USGCRP was created during the latter part of President Reagan's administration when the scientific community, other expert observers, and the public policy communities noted that there were trends and changes, often on global scales, that exceeded historic patterns. For example, marked changes in weather and climate, a rush of historically rural societies to more urban regions and other demographic shifts, changes in tropical rain forests and other accelerating alterations in land use, and disruptions to the structure and biodiversity of ecological systems were being observed and reported in the scientific literature and the media.

www.globalchange.gov; www.global change .gov; www.nap.edu/catalog.php?record_id =12595

Research

Task 1 The federal budgetary process should more effectively represent the needs of the nation to address the issues of climate and global change.

Task 2 The USGCRP should be reframed to better address the 21st century opportunities and challenges:

- Enhance focus on adaptation research and response strategies.
- Enhance focus on mitigation research and response strategies.
- Enhance support for international, national, and regional-scale climate and global change assessments and related analyses.
- Enhance support for observations and monitoring of essential climate and global change variables.
- Enhance support for capacity building within schools, universities, and the general concerned public.
- Enhance effectiveness of decision-support and communication activities.

Task 3 Implement the recommendations of the National Research Council of the National Academy of Sciences, as decribed in "Restructuring Federal Climate Research to Meet the Challenges of Climate Change."

Task 4 Enhance research, assessment, and communication activities at regional to local scales.

Task 5 Enhance and broaden the social science research agenda.

Task 6 Enhance implementation of the statutory mandate for the USGCRP.

Task 7 Invest in and amplify the use of the collaborative capabilities of Web-based systems for real-time data and monitoring.

Task 8 Reform the management of the USGCRP.

Availability of Technology to Mitigate Climate Change

Global emissions of GHGs are increasing at an unsustainable rate. The current driving forces for CO_2 emission growth are economic and population growth, which are powerful and not likely to change. It will be necessary to counteract these vectors by moving as quickly as possible toward technologies that generate fewer GHG emissions per economic activity and per capita. This would need to be accomplished in all the key sectors: power generation, transportation, building, and industrial. The following issues should be considered:

- Which are the most important sectors for which technology has the greatest potential for mitigating GHG emissions?

- What are the most promising technologies by sector, what is the state of their development, and is the research community focusing on these most promising technologies?

- For these key technologies, what are the remaining technical, economic, and environmental challenges?

- What has been the history of funding for such technologies, and is it deemed adequate to the challenge?

- What should be the relative roles for government, industry, and academia in developing and deploying key technologies?

- If additional resources were made available

to accelerate technology development in a timescale consistent with the challenge, where should they be invested?

- How important is fundamental research versus pilot and full-scale research/ development/demonstration activities?

www.climatetechnology.gov; www.eea .europa.eu/themes/technology; www.energy .gov/sciencetech; www.globalchange.gov; www .netl.doe.gov; http://iea.org

Policy

Task 1 Energy efficiency is the low-hanging fruit that can lead to the greatest reduction in CO_2 at least cost, particularly in transportation, appliances, and buildings. The federal and state governments should develop new incentives (and remove disincentives), promulgate new regulations, and foster changes in public behavior to decrease emissions from energy use by 1% to 2% per year.

Task 2 The administration and Congress should greatly increase funding, by at least doubling it, to promote implementation via development and demonstration of technologies that are commercial or near commercial to reduce carbon emissions at the fastest possible rate.

Task 3 The US Climate Change Technology Program (CCTP) should lead in setting priorities for the implementation of such technologies. Priority should be given to projects that address more than one issue; for example, carbon-free power production supports carbon-free transportation and simultaneous production of biomass energy with carbon capture.

Task 4 The federal government should address two widely acknowledged problems slowing progress toward attainment of stabilized GHGs: lack of a price in the marketplace on GHG emissions and the underinvestment in GHG-reducing solutions, including plant, equipment, best practices and services, and advanced GHG-reducing technology and related R&D.

Task 5 The administration and Congress should explore a new investment-stimulating mechanism that might address both problems. The new mechanism would be privately held environmental security accounts, or ESAs, modeled loosely after individual retirement accounts. Each participating entity or individual would pay into its ESA a fee (not a tax) based on the amount of its GHG emissions. Accrued funds would be made available for investment by the ESA account holder to grow the ESA tax free, or they could be withdrawn, provided that the funds were applied in ways that furthered the goals of environmental security. A fee schedule would be set by national legislation, which would authorize the ESAs and establish the criteria for the withdrawal of the funds.

Task 6 The ESA mechanism would create a price in the marketplace on GHG emissions, setting into motion private creativity to reduce such emissions. It would also provide a source of funds (or collateral for third-party financing) for payer-directed investments in GHG-reducing solutions.

Task 7 Compared with a tax (or equivalent cap and trade mechanisms), where revenues are collected by government and redistributed politically, the ESA mechanism could prove to be more environmentally effective and, perhaps, less objectionable to payers. The adverse effects of the higher near-term costs might be offset intra-entity by the stimulating longer-term benefits of new investment. Analysis would be needed to estimate macroeconomic and sectoral effects on the economy and international competitiveness. Pilot programs could be carried out to test the concept, work out administrative procedure, and identify costs.

Research

Task 8 The administration and Congress need to triple the level of funding for strategic research to develop the next generation of end use and production energy technologies with efficiencies to meet the 2100 goal at a low enough

cost that they can be adopted by lesser-developed economies. Emphasis should be on renewable energy and energy enablers such as energy storage (batteries, capacitors), plug-in hybrids, hydrogen, and gas separation membranes. The CCTP should provide the road map for ensuring the balance of funding technologies at the fundamental, strategic, and demonstration levels and for providing the correct mix of participation by government, industry, and academia.

Task 9 Federal agencies such as the DOE, EPA, USDA, and Department of Defense should be involved in prioritization and management of these technologies. On the research side, it is important to utilize all relevant federal capabilities, such as the EPA Office of Reserach and Development's technology assessment and environmental characterization expertise. Opportunities for synergy between the climate change mitigation programs of different agencies might be promoted by providing resources for rotating positions at CCTP for key researchers from the participating agencies.

Task 10 The DOE should develop regional climate change commercialization centers that can adapt mitigation technologies to local climates, topographies, vegetations, and demographics. The role of the regional centers would be to apply modeling to determine the effects of climate change on the local climates to assist in

- setting standards for efficiency that take into account future changes

- mapping of wind, solar, and biomass resources in order to promote the optimum utilization of renewables

- planning adaptation strategies

ACTION 23:

CO_2 Capture and Storage (CCS)— How Can It Play a Major Role in Mitigating Climate Change?

There is rapidly growing national and international interest in the use of carbon capture and storage (CCS) as part of a climate change mitigation strategy for controlling CO_2 emissions from coal-fired power plants and other large industrial sources. All three components of the CCS system—CO_2 capture, pipeline transport, and geological storage (sequestration)—are found in industrial operations today, and there are now several projects worldwide that each capture and sequester a million tons of CO_2 or more per year. However, CCS technologies have not yet been applied to a large-scale power plant, nor has the integration of capture, transport, and storage at a commercial scale yet been demonstrated in the United States. Current CCS technologies also incur significant costs and energy penalties. A number of important technical, economic, legal, regulatory, and public acceptance issues therefore must be resolved before CCS can be widely deployed as a part of a climate change strategy.

www.netl.doe.gov/technologies/carbon _seq; http://iea.org; www.eea.europa.eu/ themes/tecnnology

Research

Task 1 The private sector, in collaboration with the federal government, should conduct multiple commercial-scale demonstrations of integrated CCS systems at power plants with geological sequestration to validate large-scale performance and reliability. Projects should span a range of power plant and CO_2 capture types (e.g., combustion, gasification systems), new and retrofit applications, and a range of geological formations (e.g., deep saline formations, depleted oil and gas fields).

Task 2 A financing mechanism should be developed by government and industry to fund these projects.

Task 3 The federal government and the private sector should collaborate to conduct risk assessments of geologic sequestration to identify data needed by the insurance industry and regulatory agencies concerned with site approval and risk management.

Task 4 Congress should significantly increase funding for basic and applied R&D to develop new and advanced (lower-cost) CO_2 capture and storage technologies.

Task 5 The EPA and DOE should develop life cycle assessment tools for CCS projects covering all aspects from resource requirements through geological storage, including impacts of capture and storage systems involving mixtures of CO_2 and other acid gases.

Education

Task 6 The public and private sectors, including universities and environmental organizations, should jointly undertake an initiative of education and dialogue to facilitate public awareness or acceptance of potential future deployment of CCS technologies.

Task 7 Scientists and engineers should increase their efforts to work with and educate policymakers and regulators about CCS (including the risks, benefits, and additional needs).

ACTION 24:

Counting Carbon—Tracking and Communicating Emitted and Embodied Greenhouse Gases in Products, Services, Corporations, and Consumers

As corporations, countries, consumers, and communities attempt to measure and report their greenhouse gas footprints, they face many daunting challenges, particularly in the United States, where awareness lags and emissions (sometimes embedded in products imported from abroad) soar. There are many challenges and opportunities in measuring and conveying to stakeholders the quantities of greenhouse gases emitted into the atmosphere, sometimes hidden in the life cycles of various products or services. Such information can assist individuals, companies, communities, and nations in meeting specific goals and fostering more energy-efficient and climatically savvy societies.

Policy

Task 1 Those working to track and reduce emissions should keep an eye on the big picture and high-magnitude solutions and not be distracted by noise and minutia. The variety and scale of factors can be overwhelming.

Research

Task 2 The measurement of emitted and embodied GHGs should be standardized, and transparency should be ensured, using techniques and strategies such as those developed by the World Resources Institute (WRI) and the Carbon Disclosure Project.

Task 3 Standardize and simplify data gathering without sacrificing quality and integrity, perhaps by following WRI's approach, which is already taking the lead at developing standards and protocols for data.

Task 4 More data must be produced and assured for quality. Existing and emerging tools and technology should be employed. These include "smart metering" in homes and the BEES calculator tool (see the National Institute of Standards and Technology's Building for Environmental and Economic Sustainability software, http://www.bfrl.nist.gov/oae/software/bees/). Case studies of particular products to build should be highlighted.

Education

Task 5 Awareness should be raised by training and education of corporations and consumers, supported by coordinated, multidisciplinary efforts to convey the service life cycles of specific products, including their contributions to greenhouse gas emissions.

Task 6 Training programs should be developed to increase expertise in tracking, reporting, and reducing greenhouse gas emissions.

Task 7 Opportunities should be provided to individuals and corporations for sharing information, including a clearinghouse of information and resources. www.nist.gov

ACTION 25:
Ocean Fertilization for Carbon Sequestration

Ocean iron fertilization is the process by which iron is deposited onto the surface of the ocean to stimulate a large bloom of phytoplankton in order to remove CO_2 from the atmosphere by photosynthesis. This mimics a natural process that happens via dust storms, coastal interaction, and deep water upwelling. Iron is a necessary trace nutrient used in photosynthesis and is the primary limiting factor to plankton growth in much of the world's open oceans far from land. Carbon sequestration occurs as dead phytoplankton or fecal pellets from zooplankton sink into the deep ocean. This process of sequestration is known as the "biological pump," and it has been the Earth's primary atmospheric carbon removal mechanism since photosynthesis first began over 1 billion years ago—contributing to the storage of nearly 86% of the world's mobile carbon in the deep ocean.

Like all plants, phytoplankton require various nutrients to grow. In the central ocean basins, the scarcest of those nutrients is iron, only episodically supplied by large wind-driven dust events. Ocean fertilization involves the use of ships to apply trace amounts of iron to these iron-limited regions of the ocean. This process has been demonstrated in 12 publicly funded experiments since 1993 to effectively trigger large bloom events, which may accelerate the transfer of CO_2 to ocean depths.

Recently, several commercial entities have proposed the use of iron fertilization to sequester CO_2 and to generate carbon offsets for sale in the voluntary carbon market and/or eventually the regulated market. What combination of scientific research and public dialogue is needed for informed decision making about iron fertilization?

http://esd.lbl.gov/climate/ocean/fertilization.html; www.whoi.edu

Research

Task 1 It is essential that both iron fertilization experiments and any potential commercial fertilization in the ocean be regulated internationally to assure that the environmental impacts of the activity are understood and, in the case of commercial fertilization, that offsets for emitted carbon are legitimate.

Task 2 Fertilization activities should be monitored for compliance with regulations.

Task 3 The scientific community should evaluate ocean areas to determine whether any are inappropriate for fertilization because of negative environmental impact (e.g., marine protected areas) or for oceanographic reasons (e.g., areas with upwelling that would prevent sequestration).

Task 4 Research on the environmental impacts of ocean iron fertilization should include the entire water column and the open ocean food web.

Task 5 Biological monitoring of iron fertilization of the ocean should include genomic approaches that provide better evidence of impact on organisms than only sampling and standard identification.

Task 6 The scientific community should evaluate the long-term impacts of fertilization, even if it becomes accepted for carbon credits.

Task 7 The scientific community should identify the parameters and metrics that are necessary to demonstrate sequestration and to identify environmental impacts.

Education

Task 8 There should be a dialogue about concerns over ocean iron fertilization with international scientific, conservation, government, and business communities.

ACTION 26:
Geoengineering as Part of a Climate Change Response Portfolio

Geoengineering refers to the deliberate modification of the environment. It has been suggested that, in order to reduce the magnitude of future anthropogenic (largely CO_2-induced) warming, humans might deliberately reduce the net amount of incoming solar radiation received by the Earth by putting reflectors in orbit around the planet, by injecting aerosols or aerosol precursors into the stratosphere, or by changing the albedo (reflectivity) of marine clouds by using artificially produced cloud condensation nuclei. While these ideas have been around for many decades, they have recently received renewed attention because of the rapidity of current climate change and the increased confidence in projections of substantial future change. Geoengineering must be viewed, therefore, as a possible complement to mitigation. It may either be held in reserve as a means to ward off major changes should the climate system be judged to be heading for an otherwise irreversible "melt down" or, if the technological challenges of timely mitigation be judged too difficult, be employed as a way to gain time to develop and implement appropriate new climate-neutral technologies.

www.eea.europa.eu/themes/technology

Policy

Task 1 Geoengineering (solar radiation management) is not now well enough understood to be considered as an option that is complementary to mitigation and adaptation for dealing with global warming.

Research

Task 2 More research on the efficacy, effects, and ethical considerations of geoengineering is needed.

- A well-managed, multiagency program focused on geoengineering should be established.

- This research should be multidisciplinary, including the climate system, biological, and ecosystem aspects.
- The research program should study governance questions and ethical issues.

Task 3 A geoengineering research program should not be at the expense of a much larger increase in research about mitigation and adaptation. Geoengineering should only be considered in emergencies if those larger programs are inadequate.

Task 4 To be accepted and monitored by the people of the world, the research needs to be published in the peer-reviewed, open literature, and the research program should be internationally sponsored.

Task 5 Large-scale field experiments of geoengineering measures should not be carried out until detailed theoretical assessments are conducted of how they would work and their possible consequences .

Task 6 The capability for long-term monitoring of the climate system, particularly by satellites, needs to be maintained and enhanced so that climate change, and the effects of any geoengineering approaches, can be measured and detected in an accurate and robust manner.

Task 7 Beyond solar radiation management, other novel approaches to counterbalancing climate change and its impacts should be explored.

ACTION 27:
Looking into the Past to Understand Future Climate Change

The Earth's climate history is invaluable to understanding future change and guiding policy. Paleoclimatology is a multidisciplinary field that uses past geologic records to understand changes in climate; this understanding can help guide decisions about adaptation and mitigation. For example, for sea level rise, the geological record provides information that

enables scientists to determine realistic levels of risk, in both time scale and magnitude. The past also reveals links between sea level rise and "rapid ice melt," between climate change and ocean circulation, and between atmospheric greenhouse gas concentrations and global and regional climate change. Changes in droughts and floods and their impacts on past societies can indicate ecosystems' abilities to adapt to climate change. Understanding the sensitivity of Earth's climate to greenhouse gases will help policymakers to determine the levels of mitigation that will be required.

www.giss.nana.gov/research/paleo; www.ipcc.ch; www.ncdc.noaa.gov/paleo

Research

Task 1 The scientific community should develop integrated land-based (e.g., ice cores and lake cores) and ocean-based (e.g., sediments from scientific ocean drilling, corals) paleoclimatic data sets.

Task 2 The federal government should create funding mechanisms and institutional arrangements to encourage researchers of climate/ocean dynamics and those making paleoclimate observations to collaborate to improve climate models.

Task 3 Research is needed to understand the sensitivity of ice sheets to climate change and their impact on sea level.

Education

Task 4 Scientific professional organizations should train and encourage scientists to communicate paleoclimate research to policymakers, educators, and the public at large.

Task 5 Publishers should incorporate paleoclimate research into environmental science textbooks.

Task 6 Scientific societies should increase the number of Congressional Science and Technology Fellowships.

A National Strategy for Wildlife Adaptation to Climate Change— What Should It Include?

Proposed climate change legislation calls for development of a "national strategy" for assisting wildlife and ecosystems, both terrestrial and marine, to adapt to the impacts of climate change. Such legislation would provide significant new funding for conservation activities, land acquisition, and other actions to implement such a strategy. Efforts to define a national strategy raise challenging scientific and policy questions. What should such a national strategy include? What should be its goals, and how should it measure progress toward achieving them? What does "adaptation to the impacts of climate change" mean? What actions and approaches should such a strategy include to help wildlife and ecosystems faced with disruption by a changing climate, and who should implement such actions? What scientific research is needed to help define such a national strategy and to refine it as it is implemented over the course of decades? A national strategy for wildlife adaptation should include the goals and provisions outlined below.

www.wwf.org; www.milleniumecosystem .org

Policy

Task 1 Protect biodiversity and the ecological and evolutionary processes that produce and maintain it.

Task 2 Employ a transparent process that is iterative and adaptive, supported by research and monitoring of ecosystem structure and functioning.

Task 3 Ensure early action to invest in habitat conservation, including buying land.

Task 4 Include a quick response mechanism for ecological catastrophes and other episodes.

Task 5 Ensure coordination with all stakeholders, including Mexico, Canada, and other countries, and integrate with strategies to address the impacts of climate change on public health and the built environment.

Task 6 Focus on a broad range of stresses on wildlife (non-climate as well) to promote resilience.

Task 7 Integrate with and implement the strategy through planning and management for federal lands.

Task 8 Consider international biological diversity and opportunities to provide assistance to other countries.

Research

Task 9 Create an unbiased, IPCC-like commission to identify the best available science.

Education

Task 10 Develop a strategy for education, communication, and public outreach.

Expanding Climate Information, Education, and Communication

Mass Action—How Scientists Can Engage the Public in Global Dialogue Toward Shared Policy and Behavior Change Solutions for Global Climate Change

Can solutions for global climate change come from nontechnical, democratic movements? Is climate science shared in such a way that the public can share both pain and hope in climate change actions? Scientists need to present statistical, economic, and technical materials in plain language, and across cultures, through the media and through changes to educational curricula and materials. Is simply educating the public enough? Do scientists have a role in encouraging social action on energy and climate change?

www.eurekalert.org; http://communicating science.aaas.org; www.nasonline.org; www.nae.edu

Education

Task 1 A broad, cross-sectional partnership including government foundations, philan-

thropic organizations, corporations, educators, students, NGOs, and scientists should be formed to take collaborative action on climate change. The partnership should be built across audiences and issues.

Task 2 The partnership should establish a working group to communicate with scientists about how to share data with the lay public.

- The working group should create a protocol for communication (climate literacy) and common language, including terminology.

- The working group should design standardized training for scientists that encompasses the communication protocol, as well as media techniques that improve basic communication.

Task 3 The partnership should work with youth organizations and media organizations to reach youth and popular segments through "new media."

Task 4 The partnership should mobilize community leaders for community action, using appropriate messengers.

Task 5 Scientists working with the National Council for Science and the Environment (NCSE) should select articles that are most important for public awareness. These articles should be edited in such a way that the lay public can comprehend them. (Note that NCSE's online Encylopedia of the Earth, www.eoearth. net, does this and welcomes authors.) NCSE should work through journalists such as science writers to communicate key information in popular journals and other venues.

Task 6 The subject of climate change should be included in college and other educational curricula.

ACTION 30:
Should There Be a National Climate Service? If So, What Should It Do and Where Would It Be?

As the nation advances its policymaking and scientific activities related to global climate change, the federal government must ensure that agency programs are administered and organized effectively. Edward Miles and colleagues at the University of Washington recently proposed the establishment of a national climate service "to connect climate science to decision-relevant questions and support building capacity to anticipate, plan for, and adapt to climate fluctuations." (See *An Approach to Designing a National Climate Service* in Proceedings of the National Academy of Sciences 103 (52):19616-19623, available at www.pnas.org in pdf.)

Establishment of a national climate service is part of the American Clean Energy and Security Act of 2009, introduced by Congressmen Waxman and Markey and approved by the House of Representatives in June 2009. The national climate service would define the activities to be undertaken by the National Oceanic and Atmospheric Administration to fulfill three primary functions: "advance understanding of climate variability and change at the global, national, regional, and local levels; provide forecasts,

warnings, and other information to the public on variability and change in weather and climate that affect geographic areas, natural resources, infrastructure, economic sectors, and communities; and support development of adaptation and response plans by Federal agencies, State, local, and tribal governments, the private sector, and the public" (HR 2454).

www.ostp.gov; www.omb.ogv

Policy

Task 1 The Office of Science and Technology Policy (OSTP), the Office of Management and Budget (OMB), and the Department of Commerce, with authorization by Congress, should move quickly to establish a national climate service and in parallel establish an advisory committee of nonfederal representatives (information providers and users) to define a mission and responsibilities, identify priorities, estimate required resources, and propose an organizational structure.

Task 2 OSTP and OMB should undertake a federal interagency initiative to mobilize the nation's vast resources to better understand, mitigate, and adapt to the changing climate.

Task 3 The national climate service should bring together the best and brightest from government, industry, academia, and the nongovernmental sector to tackle the urgent and unprecedented information challenges associated with climate change.

Research

Task 4 The national climate service should specify scientific and technical needs and requirements and should work with the science and technology community to deliver improved products and services.

Education

Task 5 To ensure an informed citizenry, the national climate service should be the federal focal point for climate change communications and education.

Task 6 The national climate service should work in an ongoing, close partnership with the broad user community—within and outside government—to define needs and continually develop products to meet them.

Task 7 To ensure continued public awareness, the national climate service should effectively communicate to society the risks and adverse consequences of climate change.

Task 8 The national climate service should ensure the scientific integrity, transparency, and accuracy of its products and services.

ACTION 31:
Communicating Information for Decision Makers—Climate Change at the Regional Scale

Societal impacts of climate change and climate variability are experienced most acutely at regional (subcontinental), state, and local levels. Likewise, planning for adaptation to climate change and climate variability over the next 30 years most likely will be done by decision makers focusing on these scales. Many regions of the United States have shown trends in climate variables over the last 30 years that are likely to be related to global climate change. Much of this analysis has been done by the North American Regional Climate Change Assessment Program, an interagency regional-climate modeling program for creating future-scenario climates at regional scales for impacts assessment. Also, NOAA's Regional Integrated Sciences and Assessments program and regional climate centers communicate climate science to decision makers.

www.eurekalert.org; http://communicating science.aaas.org; www.nasonline.org; www .nae.edu

Task 1 There should be proactive, early and frequent, meaningful, and purposeful dialogue between the scientific community and the intended audiences to increase climate science

and technical literacy among decision makers and the community at large.

Task 2 Regional climate-change-impacts projects should be facilitated by professionals with expertise in communications, decision making, and conflict resolution.

Task 3 Institutional structures, such as inter-disciplinary teams and extension services, need to be encouraged and embedded in projects.

Task 4 Analysts should use several climate models to create ensembles and hence characterize probabilities of future climate conditions.

Task 5 Communicators should frame and direct information for specific intended audiences.

ACTION 32:
Adaptation and Ecosystems— What Information Do Managers and Decision Makers Need?

In the coming decades, environmental change driven by climate disruption and complicated by other factors promises to be both significant and surprising and will place new and complex demands on decision makers working in the areas of ecosystem conservation and natural resource management. Providing timely and relevant information to this community is a crucial, national infrastructure need. Unfortunately, there is an emerging consensus that existing environmental observational and reporting systems are inadequate and that the gap is most acute at the local to regional level where many adaptation decisions will be made. Overarching observations include the following:

- There must be increased communication with, and education of, managers, decision makers, and policymakers concerning effects of climatic disruption on flows of ecosystem goods and services to society.

- The effect of climate disruption on ecosystems and natural resources is a dynamic problem; to facilitate adaptation, managers and decision makers need flexible policies.

- Adaptation requires that managers and decision makers receive near real-time delivery of customized data and decision-support products.

Policy

Task 1 Policymakers should understand the urgent need for a national-scale, comprehensive assessment of the needs for specific data, information, and "tools" of all types of decision makers (including private landowners) to better enable adaptation to ecosystem changes.

Task 2 Integration and synthesis capability must be improved. Decisions about ecosystem adaptation require integrated analysis from many sectors and monitoring programs, with special efforts needed to make analyses relevant to regional and local-level decision makers. There should be increased emphasis on translating and communicating scientific information to decision makers at all levels, including funding for science integrators and translators, as a vital component in the information system.

Task 3 Adaptation to climate disruption should be addressed through an overall ecosystem sustainability framework of "whole systems" thinking that also incorporates social, economic, and political considerations. Policymakers should facilitate flexible policies.

Task 4 The proposed national climate service should include a component capable of providing timely and relevant information needed by the ecosystem adaptation community—a national climate effects network. Operating in a manner comparable to an "integrated threat center," this component would serve as a one-stop source of science, data, information, and modeling from all branches of the federal government and provide national-scale oversight and management to coordinate between agencies. Operationally, this component might be based at networked centers distributed around the nation.

Research

Task 5 There is a great need for strategic design and long-term maintenance of ecosystem monitoring and reporting programs that deliver tailored and optimized products and tools. New systems should be designed to enhance and build on the value of existing monitoring and reporting programs.

Task 6 To inform decisions about ecosystem adaptation and to support active adaptive management, regionally and locally relevant data, projections, and other information are needed, both in near real time and from downscaled global General Circulation Model forecasts.

Task 7 High-intensity monitoring of selected ecosystems or watersheds may provide "early warning" signals of climate disruption, threats to species, and ecosystem thresholds.

ACTION 33:
Diverse Perspectives on Climate Change Education — Integrating Across Boundaries

The need to integrate climate change education, both formal and informal, into existing initiatives, businesses, programs, and curricula is increasingly recognized, as the urgency and seriousness of climate change grows. Different organizations involved in climate change education have different target audiences and face a diversity of challenges, and they need to work together to ensure that climate education is coordinated enough to be broadly effective. Opportunities for cross-sectoral collaboration aimed at improving climate education strategies are rare.

www.aess.info; www.ncseonline.org

Policy

Task 1 President Obama should deliver and support a clear, compelling national call for citizens and leaders in all sectors to take well-informed action in response to current climate change science in the workplace and in home life.

Education

Task 2 The US Global Change Research Program (USGCRP) should coordinate with NSF (multiple directorates), NOAA, EPA, Department of the Interior (USGS, National Park Service), NASA, DOE, Department of Education, and USDA (Forest Service and CSREES) to support and guide the development of a national-level strategic plan for climate change education that includes specific mechanisms for working with a wide range of nongovernmental partners (including formal, informal, and nonformal education, corporations, foundations, and NGOs).

Task 3 The US Global Change Research Program and its federal partners should co-convene a workshop with climate-education networking organizations, such as the Climate Literacy Network, and other nongovernmental partners to ensure multisectoral and diverse stakeholder input into the design of a climate change education strategic plan.

Task 4 This coordinated effort and strategic plan should include outreach to and collaboration with governmental and nongovernmental funding sources (i.e., private foundations) to initiate and support multidisciplinary research to benchmark and assess the effectiveness of existing climate change education programs and to identify and evaluate promising integrative approaches ("best practices").

Task 5 To infuse popular culture with accurate and appropriate climate change science, educational NGOs and their university, community, and business partners should facilitate opportunities for scientists and engineers to partner with artists, fashion designers, novelists, game designers, and other conduits to the public.

Task 6 Textbook publishers should integrate climate change into the long-term development of textbooks across the range of academic disciplines. In the short term, these publishers should provide multidisciplinary climate change information to supplement existing publications.

Task 7 All climate change educators (instructors and curriculum developers for kindergarten through college, nonformal, and informal education) should

- utilize pedagogical strategies that are responsive to target audiences, (e.g., use localized examples of climate change impacts and incorporate financial implications of climate change)

- adopt an adaptive approach to education (recognize that climate change science will continue to evolve)

- use IPCC and other reliable sources as a framework to build trust among the public (particularly important for informal educators)

- incorporate existing and emerging social science research on how to most effectively frame messages based on scientific evidence

Task 8 Organizations like the Climate Literacy Network should cross-link key high-traffic Web resources (e.g., Windows to the Universe, Encyclopedia of the Earth, Keystone Center) to make them visible and accessible and to connect relevant groups. All organizations could include links on their Web sites to other resources.

Task 9 With the assistance of federal, educational, business, and NGO coalitions, the American Association for Retired Persons (AARP) should develop and implement a climate change education campaign for their constituency (adults over 50).

ACTION 34:
Building People's Capacities for Implementing Mitigation and Adaptation Actions

Research-based strategies are needed for efficient climate change education and communication strategies. How do we educate people of all ages about climate change? How do we design messages about climate change? Which barriers limit people's intentions to get involved in mitigation and adaptation actions? Which people's capacities should be reinforced in order to help them in proposing and implementing adaptation measures?

Task 1 Communicators should tailor messages to the audience (appropriately research the target audience, and use appropriate messengers with appropriate target audiences).

- In communicating with individuals, connect climate change solutions to everyday actions/decisions; explain them to people using examples they understand in their everyday lives; gradually build social norms and trends toward climate change solutions; utilize status symbols to drive initial changes; use funny messages and cartoons to convey information; make it personal; focus on one message at a time; use analogies; focus on addressing people's motivation to act, skill sets, and permission to act.

- In educating the community, provide and encourage community activities and involvement.

- In mass education, use multiple media pathways (school curricula, wikis, commercials, gyms, and collections of personal stories of "Why I Care about Climate Change" like PostSecret), and train teachers and other educators.

Task 2 Measure the effectiveness of communication strategies, and continually improve.

Task 3 Address paradoxical issues and false beliefs.

Task 4 Provide empowerment and opportunity to get involved.

Task 5 Tap into positive motivators; focus on immediate benefits of climate mitigation.

Task 6 Frame issues in a generational context.

Task 7 Encourage innovative behavior.

Task 8 Reduce barriers to make it easy for people to do something.

Task 9 Use the school system to improve science literacy and spread climate change solutions.

Task 10 Create youth-produced advertisements to publicize awareness and solutions.

ACTION 35:
Climate Change and Human Health—Engaging the Public Health Community

The immense implications of climate change for health and well-being are still not sufficiently recognized. The American Public-Health Association featured Climate Change: Our Health in the Balance as the theme for their National Public Health Week in April 2008. Building on NCSE's 2007 National Conference: Integrating Environment and Human Health, these recommendations are intended to help make the case for increased attention to the need to protect the environment and to protect human health. www.globalchange.gov; www.apha.org

Policy

Task 1 Policymakers should include life cycle analysis of the potential health, environmental, economic, and social consequences and co-benefits when considering proposed technologies or practices to mitigate or adapt to climate change. This analysis is especially important for energy technologies and practices.

Task 2 Congress should support an increase in surveillance, monitoring, and response capacity for climate change–related health impacts in local, state, and federal public health agencies, with an emphasis on defining and protecting vulnerable populations.

Task 3 Congress should include, in relevant legislation, assessment of health impacts, positive and negative, of all technologies and policies related to climate change adaptation and mitigation.

Task 4 Congress should support the planning and implementation of climate change adaptation plans as part of public health preparedness strategies on the national, state, and local levels.

Task 5 Local, state, and federal public health departments should institutionalize collaborative relationships with a broader array of other governmental and nongovernmental organizations responsible for policies and projects about climate change and health.

Task 6 The United States should collaborate with international organizations to help the poorest and most vulnerable countries and populations.

Task 7 The recommendations on emerging infectious diseases and other health implications of climate change published in the 2007 report on NCSE's Seventh National Conference, Integrating Environment and Human Health, should be implemented.

Task 8 Congress should substantially increase funding for research on health impacts associated with climate change.

Research

Task 9 The USGCRP should create a working group to review and coordinate all federal research related to health impacts of climate change.

Education

Task 10 A national campaign should be initiated to educate the general public about the local and broader health implications of climate change.

Task 11 Curricula on climate change and health should be incorporated at all levels of education, with a special emphasis on programs in health, medicine, and environmental sciences and studies.

APPENDIX 1

Climate Change Time Line

LATE 19TH CENTURY   The level of carbon dioxide gas (CO_2) in the atmosphere is about 290 parts per million (ppm). The mean global temperature from 1850 to 1870 is about 13.6 degrees Celsius (°C).

The Industrial Era begins: Coal, railroads, and land clearing lead to increased greenhouse gas (GHG) emissions, while population growth is aided by improved agricultural and sanitation practices. During the "second industrial revolution" (1870–1910), electricity as well as fertilizers and other chemicals combined with improved public health lead to further growth.

1824   Joseph Fourier (French mathematician and physicist; 1768–1830) calculates that the Earth's temperature would be approximately −18°C if there were no atmosphere. The ability of gases in the atmosphere to increase the surface temperature of the Earth will one day be named the greenhouse effect.

1859   John Tyndall (British physicist; 1820–1893) discovers that some gases block infrared radiation. He constructs a spectrophotometer to measure the absorption of light by carbon dioxide, nitrogen, oxygen, ozone, water vapor, hydrocarbons, and other gases. He concludes that water vapor is the atmospheric constituent that absorbs heat most strongly.

1896   Svante Arrhenius (Swedish chemist; 1859–1927) proposes that anthropogenic CO_2 could increase the Earth's temperature. He calculates that a doubling of atmospheric CO_2 could increase the surface temperature of the Earth by 5°C to 6°C. (This estimate was remarkably close to the most recent value, in 2007, from the Intergovernmental Panel on Climate Change: 2°C to 4.5°C.)

1897   Thomas Chamberlin (American geologist; 1843–1928) models the way in which global carbon cycles through the various reservoirs on Earth, including the atmosphere, sea water, minerals, and living matter.

1924   Alfred Lotka (American chemist; 1880–1949) estimates that atmospheric CO_2 will double in 500 years due to industrial activity. He bases his assumptions on coal use in 1920.

1930   Milutin Milankovic (Serbian geophysicist; 1879–1958) proposes his theory on the relationship of the motion of the Earth's poles and glacial periods.

1940   Guy Callendar (British engineer; 1898–1964) proposes a link between warming observed in Europe and North America beginning in the 1880s and the increase of atmospheric CO_2 between 1850 and 1940.

1957   Roger Revelle and Hans Suess, from the Scripps Institution of
       Oceanography, report that atmospheric CO_2 produced by humans will not
       be readily absorbed by the oceans, as previously argued.

1958   Charles Keeling, also a Scripps scientist, begins the first accurate and
       continuous measurements of atmospheric CO_2 at Mauna Loa and detects an
       annual rise. The level when he begins his measurements is 315 ppm. The
       mean global temperature (5-year average) is 13.9°C.

1967   The first reliable climate simulation estimates that doubling of atmospheric
       CO_2 would raise world temperatures by 1°C to 2°C.

1970   Aerosol particles from human activity are shown to be increasing. Some
       scientists estimate that atmospheric cooling due to particulates may
       outweigh warming due to GHGs. There is too much uncertainty to
       determine which effect will dominate.

1972   Ice cores and other evidence of past climates show that large shifts between
       relatively stable modes can occur within about 1,000 years.

1976   Scientists find that chlorofluorocarbons, methane, nitrous oxide, and ozone
       are GHGs.

1977   Scientists estimate that cooling due to particulate matter has a small effect
       when averaged globally. Scientific consensus begins to form that there will
       be net warming, and not cooling, during the next century.

1979   A report by the National Academy of Sciences finds that doubling CO_2 will
       increase global temperatures by 1.5°C to 4.5°C. The panel states that "a wait-
       and-see policy may mean waiting until it is too late."

1983   Another National Academy of Sciences report predicts increased
       temperatures. A study by the US Environmental Protection Agency states
       that "agricultural conditions will be significantly altered, environmental and
       economic systems potentially disrupted, and political institutions stressed"
       as a result of climate change.

1985   Measurements of atmospheric CO_2 at Mauna Loa have increased steadily by
       about 1 ppm per year since Keeling began measurements in 1958.

1987   Analysis of an Antarctic ice core shows a very close correlation between
       atmospheric CO_2 and temperature. The record goes back approximately
       100,000 years.

1988   The IPCC is established. Delegates to the Toronto Conference on the
       Changing Atmosphere call for a reduction in CO_2 emissions.

1990   The IPCC's First Assessment Report states that the Earth has been
       warming and that continued warming is likely.

1992   The Earth Summit in Rio de Janeiro results in the UN Framework
       Convention on Climate Change (UNFCCC). The aim of the UNFCCC is to
       "prevent dangerous interference with the climate system."

1993   Analysis of Greenland ice cores suggest that large changes in regional
       climate can occur within a single decade.

1995   The first Conference of the Parties (COP 1) to the UNFCCC is held in
       Berlin. (The Conference of the Parties, or COP, is the "supreme body" of
       the UNFCCC and meets annually.)
       The IPCC's Second Assessment Report states that there is a discernible
       human influence on global climate.

1996   The second Conference of the Parties (COP 2) to the UNFCCC is held in Geneva and supports the findings of IPCC's Second Assessment Report.

1997   COP 3 is held in Kyoto and results in the Kyoto Protocol, which sets targets to reduce GHG emissions.

1998   COP 4 is held in Buenos Aires, and the delegates agree to an action plan to set the guidelines necessary to implement the Kyoto Protocol.

1999   COP 5 is held in Bonn, and the rules for achieving Kyoto Protocol targets are further discussed.

2000   COP 6 in The Hague again address issues surrounding Kyoto Protocol targets.

2001   The IPCC's Third Assessment Report (TAR) states that net warming of the atmosphere is very likely, with possibly severe consequences.
The legal text for the Kyoto Protocol is produced during COP 7, held in Marrakech.

2005   The Kyoto Protocol treaty goes into effect, and nations begin efforts to reduce emissions.

2006   The *Stern Review on the Economics of Climate Change* concludes that 1% of global GDP per year should be invested in order to avoid the negative effects of climate change.

2007   The IPCC's Fourth Assessment Report (AR4) states that the effects of warming have become apparent and the cost of reducing emissions would be far less than the damage they will cause.
Atmospheric CO_2 reaches 382 ppm. The mean global temperature (5-year average) is 14.5°C.
The Nobel Peace Prize is awarded to the IPCC and Al Gore jointly for work on climate change.
Over 10,000 people, including representatives from over 180 countries, participate in the United Nations Climate Change Conference, COP 13, in Bali, Indonesia, for 2 weeks in December.

2008   Atmospheric CO_2 is rising at three times the rate of a decade earlier, and even faster than the IPCC AR4 report had projected just a year earlier.
The COP 14 meets in Poznan, Poland, with the CMP that is working on the next phase and revision of the Kyoto Protocols.

2009   The COP 15 meetings in Copenhagen will conclude the work begun in Bali and lead to the revision plan for the Kyoto Protocols.

2012   The first period of the Kyoto Protocols expires, and the new revised standards will take effect in the second period of work on greenhouse gas reduction.

2014   The IPCC Fifth Assessment Report will be delivered and will emphasize ice sheet melting, ocean chemistry changes, and geo-engineering.

Source: Adapted from Weart, Spencer (Lead Author); American Institute of Physics (Content Partner); Cutler J. Cleveland (Topic Editor). 2008. "Climate Change Timeline." In: Encyclopedia of Earth. Eds. Cutler J. Cleveland (Washington, D.C.: Environmental Information Coalition, National Council for Science and the Environment). [First published in the Encyclopedia of Earth May 13, 2008; Last revised May 21, 2008; Retrieved August 11, 2009]. http://www .eoearth.org/article/Climate_Change_Timeline *Editor's note:* This article was originally published by the American Institute of Physics and Spencer Weart as Timeline of Milestones. The original version contains detailed references and links to additional information on the history of climate change science.

Climate Change:
Science and Solutions

Program from the NCSE 8th National Conference on
Science, Policy, and the Environment, January 2008

Topic	Speakers	Weblink
Climate Change: Science to Solutions—What Do We Know? How Do We Act in Time and in Appropriate Scale?	Amb. Richard Benedick, President, National Council for Science and the Environment Mohan Munasinghe, Vice Chair, Intergovernmental Panel on Climate Change (IPCC); Chairman, Munasinghe Institute for Development (MIND)	http://ncseonline.org/ http://www.ipcc.ch/
Summarizing Global Change Science and the Likely Implications of Climate Change	Mohan Munasinghe, Vice Chair, Intergovernmental Panel on Climate Change (IPCC); Chairman, Munasinghe Institute for Development (MIND) Michael MacCracken, Chief Scientist for Climate Change Programs, The Climate Institute Thomas Lovejoy, President, The H. John Heinz III Center for Science, Economics, and the Environment Sarah James, Alaskan Gwitch'in Steering Committee and Goldman Environmental Prize Awardee Sherri Goodman, General Counsel, The CNA Corporation	http://www.climate.org/ http://www.heinzctr.org/ http://www.gwichinsteering committee.org/ http://www.cna.org/
Tackling Global Change: Key Social and Ecological Issues for Mitigation and Adaptation	Arden Bement, Jr., Director, National Science Foundation Margaret Leinen, Chief Scientific Officer, Climos Abigail Kimbell, Chief, US Forest Service Thomas Dietz, Director, Environmental Science and Policy Program; Assistant Vice President for Environmental Research, Michigan State University	http://www.nsf.gov/ http://www.climos.com/ http://www.fs.fed.us/ www.wildlifetrust.org http://environment.msu.edu/

Topic	Speakers	Weblink
Tackling Global Change: Key Energy and Technology Issues for Climate Stabilization	Mark Myers, Director, US Geological Survey Leon Clarke, Senior Research Economist and Staff Scientist IV, Joint Global Change Research Institute (JGCRI), University of Maryland / Pacific Northwest National Laboratory Paul Epstein, Associate Director, Center for Health and the Global Environment, Harvard Medical School Frank Princiotta, Director, Air Pollution Prevention and Control Division, Office of Research and Development, US Environmental Protection Agency Lewis Milford, President, Clean Energy Group David Rodgers, Deputy Assistant Secretary, Energy Efficiency and Renewable Energy, US Department of Energy	http://www.usgs.gov/ http://globalchange.umd.edu/ http://chge.med.harvard.edu/ http://www.epa.gov/ http://www.cleanegroup.org/ http://www.energy.gov/
Perspectives of the Next Generation of Climate Change Leaders	Douglas Cohen, US Partnership for Education for Sustainable Development, National Youth Initiatives Eban Goodstein, Project Director, Focus the Nation The Envirolution: Alex Gamboo, Timothy Polmateer, Antuan Cannon DoRight Enterprises: Scott Beall, Madeleine Skaller, James Smith Jessy Tolkan, Energy Action Coalition	http://www.uspartnership.org http://www.nationalteachin.org/ http://www.focusthenation.org/ http://www.envirolution.org/ http://www.scottbeall.com/dorightsummary.htm http://energyactioncoalition.org/
Climate Change: Science to Solutions- The Case for Business Leadership	Karim Ahmed, Secretary / Treasurer, National Council for Science and the Environment James E. Rogers, Chairman of the Board, President and Chief Executive Officer, Duke Energy Corporation	http://www.duke-energy.com/
Solutions: Engaging Communities Large and Small	Peter Senge, Founding Chairperson, Society for Organizational Learning Rev. Richard Cizik, Vice-President, National Association of Evangelicals Michael Crow, President, Arizona State University Bill McKibben, Author, Scholar-in-residence in Environmental Studies at Middlebury College Dan Seligman, Director, Apollo Alliance, Washington Office	http://www.solonline.org/ http://www.nae.net/ http://www.asu.edu/ http://www.billmckibben.com/ http://apolloalliance.org/
Solutions: Science and Policy on a Global Scale	Reid Detchon, Executive Director, Energy Future Coalition Amb. Richard Benedick, President, National Council for Science and the Environment Stephen Schneider, Melvin and Joan Lane Professor for Interdisciplinary Environmental Studies, Stanford University Robert Corell, Director, Global Change Program, The H. John Heinz III Center for Science, Economics, and the Environment Jonathan Pershing, Director, Climate, Energy and Pollution Program, World Resources Institute	http://www.energyfuturecoalition.org/ http://stephenschneider.stanford.edu/ http://www.wri.org/climate

Topic	Speakers	Weblink
NCSE Lifetime Achievement Awards Ceremony	Margaret Leinen, NCSE Board of Directors Robert W. Corell, Director, Global Change Program, The H. John Heinz III Center for Science, Economics, and the Environment	
8th John H. Chafee Memorial Lecture on Science and the Environment	Amb. Richard Benedick, President, National Council for Science and the Environment John P. Holdren, President and Director, Woods Hole Research Center; Teresa and John Heinz Professor of Enviornmental Policy, Harvard University	http://ncseonline.org/2008 conference/cms.cfm?id=2121 http://www.whrc.org/
American Perspectives on Climate Change	Jon Krosnick, Professor of Communication, Political Science, and Psychology, Stanford University	
Developing Political Solutions to Climate Change	Ray Suarez, Senior Correspondent, The News Hour Ross C. "Rocky" Anderson, Mayor, Salt Lake City, Utah (2000–2006) Lynn Scarlett, Deputy Secretary, US Department of the Interior Representative Jay Inslee, US House of Representatives, 1st District, Washington State	http://www.pbs.org/newshour/ http://www.doi.gov/ http://www.house.gov/inslee/ issues/environment/global warmingindex.html
Presidential Candidates Forum: What Will the Next President do to Manage Climate Change?	Vijay Vaitheeswaran, Global Correspondent, The Economist Hillary Clinton Campaign: Todd Stern, Partner, Wilmer Cutler Pickering Hale and Dorr. John Edwards Campaign: Elgie Holstein, Serco, Inc. Dennis Kucinich Campaign: Representative Dennis Kucinich, U.S. House of Representatives, 10th District Ohio State Barack Obama Campaign: Dan Esty, Hillhouse Professor of Environmental Law and Policy, jointly with Yale Law School and School of Forestry and Environmental Studies	http://ncseonline.org/2008 conference/Presentations/ Friday/Presidential%20 Forum%20Transcript2.doc http://www.economist.com/

CONCURRENT SYMPOSIA

Topic	Speakers	Weblink
Beyond Kyoto: Elements of a 2020 International Agreement	Amb. Richard Benedick, President, National Council for Science and the Environment Dilip R. Ahuja, Professor, Indian National Institute of Advanced Studies Scott Barrett, Professor and Director, International Policy Program, Johns Hopkins University Malachy Hargadon, Environment Counselor, Delegation of the European Commission Richard Moss, Vice President and Managing Director for Climate Change for the United States, World Wildlife Fund Jonathan Pershing, Director, Climate, Energy and Pollution Program, World Resources Institute	http://ncseonline.org/2008 conference/cms.cfm?id=1863 (no details) http://www.nias.res.in/ http://www.sais-jhu.edu/index .html http://www.worldwildlife.org/

Topic	Speakers	Weblink
Climate Change and International Development	Mohan Munasinghe, Vice Chair, Intergovernmental Panel on Climate Change (IPCC); Chairman, Munasinghe Institute for Development (MIND) Ralph Cicerone, President, National Academy of Sciences Thomas Schelling, Nobel Laureate (Economics); Distinguished University Professor, University of Maryland Adrian Vazquez, Executive Director, Commiszsion for Environmental Cooperation	http://ncseonline.org/2008 conference/cms.cfm?id=1864 http://www.nationalacademies .org/
Role of Philanthropic Foundations: Promoting Strategic Initiatives on Climate Change	Sharon Alpert, Program Officer, Environment Program, Surdna Foundation Andrew Bowman, Director, Climate Change Initiative, Doris Duke Charitable Foundation Kathleen Welch, Deputy Director, Environment Program, The Pew Charitable Trusts Eric Heitz, President, The Energy Foundation Elizabeth Chadri, Program Officer for Conservation and Sustainable Development Program on Global Security and Sustainability, The John D. and Catherine T. MacArthur Foundation	http://ncseonline.org/2008 conference/cms.cfm?id=1865 http://www.ddcf.org/page.asp ?pageId=675 http://www.ef.org/programs .cfm?program=climate http://www.macfound.org/ site/c.lkLXJ8MQKrH/b.101 3733/k.9901/International_ Grantmaking__Conservation _and_Sustainable_ Development.htm
Business and Finance: Opportunities and Challenges from Climate Change	H. Jeffrey Leonard, President, and Chief Executive Officer, Global Environment Fund Bruce Schlein, Vice President, Environmental Affairs, Citibank Mindy Lubber, President, CERES Ben Lashkari, Director, Environmental and Commodity Markets, Swiss Re Mark Tercek, Director, Goldman Sachs' Environmental Markets Initiative	http://ncseonline.org/2008 onference/cms.cfm?id=1866 http://www.citigroup.com/citi/ environment/ http://www.ceres.org/page.aspx ?pid=705 http://www2.goldmansachs.com/ citizenship/environment/center -for-environmental-markets/ index.html
Forging Alliances between Business and Society: US Climate Action Program	Kevin Bryan, Meridian Institute Helen Howes, Vice President of Environment, Health and Safety, Exelon Corporation Tim Juliani, Markets and Business Strategy, Pew Center on Global Climate Change Eric Haxthausen, Senior Policy Advisor, Climate Change, The Nature Conservancy Michael Parr, Senior Manager, Government Affairs, DuPont	http://ncseonline.org/2008 conference/cms.cfm?id=1867 http://www.us-cap.org/

Topic	Speakers	Weblink
Legislative Agenda for Addressing the Carbon Problem	L. Jeremy Richardson, 2007–2008 AAAS Roger Revelle Fellow in Global Stewardship	http://ncseonline.org/2008 conference/cms.cfm?id=1868
	Margaret Turnbull, Astrobiologist, Space Telescope Science Institute	http://www.climatestrategies.us/
	Kenneth Colburn, CCS Senior Consultant, Co-Director and Facilitator of Projects, Center for Climate Strategies	
	Lexi Shultz, Representative for Climate Policy, Union of Concerned Scientists	
	Holmes Hummel, AAAS Congressional Fellow, Office of Representative Inslee	
	Alex Barron, ACS Congressional Fellow, Office of Senator Lieberman	
Engaging State and Local Government: Developing and Implementing Climate Action Plans	Dan Kammen, Professor, Energy Resources Group, University of California–Berkeley	http://ncseonline.org/2008 conference/cms.cfm?id=1869
	Ross C. "Rocky" Anderson, Mayor, Salt Lake City, Utah (2000–2008)	
	Gary Radloff, Director of Policy and Strategic Communications, Wisconsin Department of Agriculture, Trade and Consumer Protection	
Climate Scientists and Decisionmakers: The Communication Interface	Rebecca J. Romsdahl, Assistant Professor, Earth System Science & Policy program, University of North Dakota	http://ncseonline.org/2008 conference/cms.cfm?id=1870
	Stacy Rosenberg, Assistant Professor, Department of Politics & Environmental Studies, SUNY Potsdam	
	Deborah Cowman, Assistant Research Scientist, Institute for Science, Technology and Public Policy, Texas A&M University	
	Chris Pyke, Constructive Technologies Group, Inc. (CTG)	
	Kit Batten, Director of Environmental Policy, Center for American Progress	
	David Bookbinder, Senior Attorney, Sierra Club	
	Roger Pulwarty, Research Associate, National Drought Information System, NOAA	
Communicating Climate Science to the Public Through the Media	Deborah Potter, Executive Director, NewsLab	http://ncseonline.org/2008 conference/cms.cfm?id=1903
	David Malakoff, Editor/Correspondent, NPR Science Desk	http://www.neefusa.org/
	Stephen Schneider, Melvin and Joan Lane Professor for Interdisciplinary Environmental Studies, Stanford University	
	Joe Witte, Meteorologist, WJLA-TV	
	Doyle Rice, Weather Editor, USA Today	
	Sara Espinoza, Program Manager, Earth Gauge, National Environmental Education Foundation	

Topic	Speakers	Weblink
Science for Carbon Management	Eric Sundquist, Research Geologist, US Geological Survey	http://ncseonline.org/2008 conference/cms.cfm?id=1871
	Richard A. Birdsey, Program Manager, Global Change Research, U.S. Forest Service	
	Sandra Brown, Senior Scientist, Ecosystems Services Unit, Winrock International	
	Stephen Faulkner, Wetland Research Ecologist, National Wetlands Research Center, U.S. Geological Survey	
	Bryan Hannegan, Vice President, Environment, Electric Power Research Institute	
	Brian McPherson, Associate Professor of Civil and Environmental Engineering, University of Utah	

Acknowledgments

The National Council for Science and the Environment (NCSE) thanks our sponsors for supporting the Eighth National Conference on Science, Policy and the Environment: *Climate Change: Science and Solutions*, which served as the basis for this book.

NCSE thanks the scores of scientists and engineers, educators, decision makers, and society leaders who, without financial compensation, shared their time and knowledge by serving as organizers, speakers, discussants, moderators, and chairs of the plenary sessions, symposia, breakout sessions, and workshops. We especially thank the members of the conference planning committee who met many times via conference call and e-mail. This book would not be possible without the help of students and other volunteers who provided detailed and valuable notes on the breakout sessions and symposia and ensured that the sessions ran smoothly.

The success of the Eighth National Conference would not have been possible without the participation of over 1,350 individuals from across the United States and around the world. NCSE thanks each of the conference participants for exhibiting creativity, patience, and endurance in developing an impressive set of targeted recommendations for reconnecting the environment and human health in research, understanding, and decision making.

The Eighth National Conference would not have been a success without the dedication and imagination of the NCSE staff, especially David Blockstein (Conference Chair), Kelly McManus (Conference Program Coordinator), Nicole Buell (Program Assistant), Lindsey Ehrler (Breakout Session Coordinator), Christopher Prince (Meeting Manager), Shelley Kossak (Development Director), and especially Peter Saundry (Executive Director).

Finally, NCSE thanks Island Press, especially editor Todd Baldwin for taking on this project to turn the conference into a book; copy editor Lou Doucette; Jessica Heise; Emily Davis; and others in the production team. Most importantly, NCSE thanks Leo Wiegman for his incredible dedication, ability to synthesize science for a lay audience, wonderful sense of wit, attention to detail, and perseverance in a project that turned out to be many times more work than we ever imagined or budgeted.

Sponsors

USDA Forest Service
US Geological Survey

Partners

Centers for Disease Control
Environmental Protection Agency
NASA
National Oceanic and Atmospheric Agency

Patrons

Duke Energy
Environ
National Institutes of Standards and Technology
The Ocean Foundation
Toyota

Supporters

Center for Environmental Education
Disney Animal Kingdom
Lighting Science
Mary Kay
Robert and Patricia Switzer Foundation

Media Sponsor

Imaging Notes

Collaborating Organizations

American Council for an Energy-Efficient Economy • Au Sable Institute of Environmental Studies • Biogeosciences.org • Chesapeake Climate Action Network • The Climate Conservancy • Environment for the Americas • EPA Emerging Leaders Network • Focus the Nation • Gary Braasch Photography • The Geological Society of America • The H. John Heinz III Center for Science, Economics, and the Environment • The Humane Society of the United States • National Spiritual Assembly of the Baha'is of the United States • The Population-Health-Environment Policy and Practice Coalition • SoL Sustainability Consortium • US Partnership for Education for Sustainable Development

About the Authors

DAVID E. BLOCKSTEIN is an ecologist and conservation biologist who over the past 20 years has shifted his focus from birds to the political environment. As Senior Scientist with the National Council for Science and the Environment, he chairs NCSE's annual National Conference on Science, Policy and the Environment. The 2008 National Conference: Climate Change: Science and Solutions forms the basis for this book. David organizes the academic environmental and energy communities as Executive Secretary of the Council of Environmental Deans and Directors (CEDD) and the Council of Energy Research and Education Leaders (CEREL). With these groups, he is developing a Climate Adaptation and Mitigation E-Learning (CAMEL) community to help educate all college students about climate science and solutions.

As the 1987–88 Congressional Science Fellow of the American Institute of Biological Sciences and American Society of Zoologists, he worked with the House of Representatives Science Committee to prepare the legislation to study and conserve biological diversity. In 2008, David received the AIBS Lifetime Achievement Award.

David has a B.S. in wildlife ecology from the University of Wisconsin and a M.S. and Ph.D. in ecology from the University of Minnesota. He has conducted research on evolution, ecology and conservation of pigeons and doves, including the extinct Passenger Pigeon. David and his family live in the people's republic of Takoma Park, Maryland, outside Washington, DC, but within the beltway.

LEO WIEGMAN, a Hudson Valley resident, founded E to the fourth in 2008 (etothefourth.com) and serves as instigator-in-chief for this environmental communications service. In addition to this book, Leo is a contributor to energy related periodicals and writes frequently on sustainability topics. Previously, Leo worked in book publishing for many years, most recently as an editor at W. W. Norton & Company publishing non-fiction books in basic sciences, photography, the environment and current affairs for general readers and for college courses. In 2009, Leo was elected Mayor of the Village of Croton-on-Hudson where he also served as Trustee from 2001 to 2007. Leo is a member of the Board of Trustees of Teatown Lake Reservation in Ossining and serves on a steering committee for local watershed conservation and an intermunicipal energy action council. Born in the Netherlands, Leo grew up in New Hampshire, and is a graduate of Tufts University, where he majored in eclecticism.

Index

Italicized page numbers refer to tables and insight boxes.

United States
  international comparisons of, 107, 128, 137, *138*,
    139, 141–42, 148, 212, *213*
  and Kyoto Protocol, 13–15, 217, 220, 222–25, *224*,
    225, 227–28
  largest nuclear power user, 123–25
  in life cycle assessments (LCAs), 123–25, 128–29,
    129
  waste-to-energy facilities, *19*
United States Climate Action Partnership (USCAP),
    165–67, *167*
Uranium, 123–25
Ürge-Vorsatz, Diana, 131
USA Today, 207–8
US Global Change Research Program (USGCRP),
    262–63, 274, 276
Utah Population and Environment Coalition, 189
Utah Vital Signs Project 2007, 189
Utilities, 42, 95, 105, 114–15, 146, *149*, 164–65,
    180–81, 186, 188, 194

Vandenberg, Michael, 95–97
Vaughn, Mace, 72
Vazquez, Adrian, 226
Vector-borne diseases, 81, *82*, 83, 226
Viederman, Stephen, 178
Volcanic activity, 33, 35, 38, 40, *40*, 41

Wackernagel, Mathis, 94
Walking, 114, 131, 186, *187*
Warner, John, 194, 224
Waste disposal, 18, 94, 104
  ballast water discharge, 74, 83, 225
  in developing nations, 232
  in life cycle assessments (LCAs), 122, *122*
  local initiatives for, 186, *187*, 188, 192, *194*
  mitigation of emissions in process, 105, *106*, *113*,
    116
  of radioactive waste, 108, *109*, 114, 123–24
  recycling, *19*, *121*, 131, 163, 186, *187*
  "waste equals food," 122
  waste-to-energy facilities, *19*, *20*, 141
Wave energy, *113*, 129
Wedges-within-the-stabilization-triangle game, 105,
    110–16, *111*, *112*
  and business, 162
  carbon capture and storage (CCS) systems,
    145–46

and Kyoto Protocol, 225, *225*
  in life cycle assessments (LCAs), 124–25, 130–33,
    131, *132*
  no-regrets climate wedges, 112–16, *113*, 130–33
Welch, Kathleen, 180
Western Climate Initiative (WCI), 160, *161*, 168, 192
West Nile virus, 81, *82*
Whales, 52, 75–76
Whistler (British Columbia), 189
White roofs, 93, 114, 245
Wilcoxen, Peter, 222
Wildlife. *See* Biodiversity
Wind power, 141, 148, *148*, *225*
  and carbon neutrality, 198
  immediate actions, 251–52
  increase in use of, *213* (*see also* Energy
    consumption)
  in life cycle assessments (LCAs), *125*, 131–33,
    132
  local initiatives for, *187*, 193
  mitigation of emissions through use of, 108–10,
    109, *110*, *112*, 113, *113*, *115*, 116, *117*
Wirth, Timothy, 222
Wisconsin Department of Agriculture, Trade, and
    Consumer Protection, 193
Witte, Joe, 207
Woods Hole Oceanographic Institution, 52–53, *54*,
    55
World Bank, 216, 227, 261
World Climate Conference (1979), 9
World Economic Forum (Davos, Switz.), 162
World Health Organization (WHO), 23, 114, 261
World Meteorological Organization (WMO), 9–10,
    12, 15
World Resources Institute, 140–41, 224, 227
Worldwatch Institute, 132, 176
World Wildlife Fund (WWF), 18, 223–24
Worm, Boris, 86

Yohe, Gary, 77
Yucca Mountain (Nev.), 124

Zero-emission buildings, *130*, 142, 193
Zoning regulations, 192, 205
Zooplankton, 52, *57*, 58, *59*
Zoning regulations, 192, 205
Zooplankton, 52, *57*, 58, *59*